W9-BCC-566

621.385
B411ud

BETHANY
COLLEGE
LIBRARY

DIGITAL TELEPHONY

DIGITAL TELEPHONY

JOHN BELLAMY

*Manager of Switching Systems Development
Building Automation Inc.
United Technologies*

*Adjunct Professor of Electrical Engineering
and Computer Science
Southern Methodist University*

1807 175 YEARS OF PUBLISHING 1982

A Wiley-Interscience Publication
JOHN WILEY & SONS
New York   Chichester   Brisbane   Toronto   Singapore

Copyright © 1982 by John Wiley & Sons, Inc.

All rights reserved. Published simultaneously in Canada.

Reproduction or translation of any part of this work
beyond that permitted by Sections 107 or 108 of the
1976 United States Copyright Act without the permission
of the copyright owner is unlawful. Requests for
permission or further information should be addressed to
the Permissions Department, John Wiley & Sons, Inc.

Library of Congress Cataloging in Publication Data:

Bellamy, John, 1941-
   Digital telephony.

   Bibliography: p.
   Includes index.
   1. Digital communications. 2. Digital modula-
tion. 3. Telephone switching systems, Electronic.
I. Title.
TK5103.7.B44   621.385   81-11633
ISBN 0-471-08089-6   AACR2

Printed in the United States of America

10 9 8 7 6 5

As a reader of technical books,
I have sometimes wondered why
so many of these books are dedicated
to members of the author's family
and not to members of the technical community.
My wife and daughters know why.

To Judy, Michelle, Joy, Kelly, and Cary

621.385
B414d

PREFACE

In a strict sense, the term "digital telephony" refers to the use of digital technology in the message path of voice communications networks. In this case the term "digital" refers to a method of encoding the signal, that is a form of modulation. Hence digital telephony implies voice transmission and switching applications, not data communications. However, one of the most attractive features of voice digitization is that the encoded signals (pulses) are divorced from the analog waveform of the source. Thus the digital transmission and switching equipment of a voice network is inherently capable of servicing any traffic of a digital nature, particularly data traffic.

As noted by several reviewers, the title *Digital Telephony* is somewhat restrictive for the material covered in this book. A more representative title perhaps could be: "Digital Networks for Voice and Data Communications." At the risk of narrowing the market, I retained the original title in anticipation of a broadened meaning for the term "telephony." Dramatic changes are occurring in the telephone networks of the world—in terms of the technology being used, and in terms of the services being provided. In both cases the changes have been spawned by the data processing industry. The development of low-cost digital electronics has enabled economical use of a new and, in many cases, superior technology for conventional voice telephone systems. At the same time, the rapid growth of data communications has created expanded use of the "telephone" network. It is clear that a single (integrated) digital network is evolving that will supply both voice and data services. I suspect that this network will always be referred to as a telephone network—even if some day it carries more traffic of an inherently digital nature than of a voice nature.

A second motivation for including the term "telephony" in the title is to call attention to an established, but new and exciting industry. The newness arises because in many countries around the world, the United States in particular, the market for telephone equipment and services has been opened up to new companies in competition with the established suppliers. The industry is exciting because of the revolution in technology and service offerings taking place.

This book provides an introduction to all aspects of digital communications with an emphasis on voice applications: voice digitization, digital transmission, digital switching, network synchronization, network control, and network analysis. For the most part, the book is an outgrowth of the author's work, study, and teaching in the areas listed. Because telephony in the United States has been the province of the common carriers who have traditionally trained their own engineers, publicly available textbooks covering much of the material presented here have been unavailable. In preparing this book the author relied heavily on articles in the Bell System Technical Journal, the IEEE Transactions on Communications, and the proceedings of technical conferences.

It should be emphasized that this is not a technical book in the traditional, analytical sense of communications theory. Since numerous books covering communications theory are already available (see the bibliography), this book stresses the application and operational aspects of communications system design. Some basic theory is presented in both qualitative and, when appropriate, quantitive terms. The purpose of this volume, however, is to introduce terminology and explain how the various systems work and how applications influence implementations. In most cases the concepts are supported by citing example implementations in the telephone network. The examples are mostly taken from the U.S. network, although some examples pertaining to international (CCITT) standards are also included.

This book is primarily intended for graduate electrical engineers as either a textbook or a reference. The electrical engineering student is most capable of appreciating occasional references to theory and its influence on the practice. However, because analytical rigor is waived in favor of operational descriptions, less technically oriented readers should have no difficulty understanding the principles.

Some areas of the book delve more deeply into analytical and implementational details than do others. No doubt, some of the imbalance occurs because the author has had more experience in particular areas than in others. On the other hand, a conscious effort has been made to emphasize subjects not widely covered elsewhere. In particular, switching system design and traffic analysis are two subjects that have not received widespread public exposure (at least not in the United States). Some aspects of digital modulation and radio system design (in Chapter 6) also contain extra detail. These subjects can be easily skipped without losing continuity.

Most of the material in this book has been used for graduate courses in electrical engineering at Southern Methodist University covering telecommunications, data communications, and specifically, digital telephony. There is too much material for students with no previous exposure to the terminology, practice, and systems concepts of telecommunications. Chapter 6 (covering digital radio and modulation) and

Chapter 9 (covering traffic analysis) can be omitted and still provide a coherent, full semester of study in the other areas. A complete list of topics is as follows:

Chapter 1: Overview of public analog telephone network in the United States followed by a brief review of the introduction of digital technology into the network.

Chapter 2: A detailed discussion of the basic advantages and disadvantages of digital implementations for voice communications.

Chapter 3: Descriptions and comparisons of the most common voice digitization algorithms.

Chapter 4: Fundamentals of digital wire-line transmission and multiplexing.

Chapter 5: Basic concepts and operations of digital switching machines.

Chapter 6: Digital modulation and radio fundamentals.

Chapter 7: Network synchronization, control, and management requirements and implementations.

Chapter 8: High-level descriptions of several digital networks and a discussion of the future of digital telephony.

Chapter 9: Fundamentals of traffic analysis for designing networks.

The appendixes cover the derivation of equations, PCM voice coding relationships, fundamentals of digital communications theory, and traffic tables.

JOHN BELLAMY

Dallas, Texas
October 1981

ACKNOWLEDGMENTS

To Arthur A. Collins I owe the greatest measure of gratitude for introducing me to most of the subjects in this book. He provided the inspiration, resources and research environment that made this book possible. To whatever degree I have managed to pull together a comprehensive overview of digital telephony, the credit goes to Mr. Collins and colleagues at Arthur A. Collins, Inc.: Robert D. Pedersen, Richard L. Christensen, and Jon W. Bayless. In addition, I would like to thank coworkers, Paul Hartmann, Charles Hogge, Drew Crosset, and Brian Bynum of Rockwell International for teaching me something about digital radio systems, and Robert Aaron at Bell Telephone Laboratories for reviewing my manuscript.

Most importantly, I acknowledge Uncle Bleak, whose bequeath provided the opportunity to work full time at the outset of this project.

I must also acknowledge the forbearance of my students at Southern Methodist University in putting up with initial drafts of the manuscript and offering many helpful suggestions. Foremost among those students who gave the extra effort are Herb Frizzell, Marvin Lucas, Tom Ekberg, Tommy Fox, Larry Hannay, Jo Hambrick, Rick Longley, and Jacob Hsu.

In assembling much of the background material in Chapter 1, I relied heavily on the following books prepared by members of the technical staff at Bell Telephone Laboratories: *Engineering and Operations in the Bell System* and *Transmission Systems for Communications*.

J. B.

CONTENTS

CHAPTER SIX   DIGITAL RADIO

CHAPTER SEVEN   NETWORK SYNCHRONIZATION, CONTROL, AND MANAGEMENT

CHAPTER ONE

BACKGROUND
AND TERMINOLOGY

Beginning in the 1960s, telecommunications in the United States began undergoing radical changes in several different areas. First, the conventional analog telephone network was being called upon to provide many new and different services, most of which emanated from the data processing industry. Second, the marketplace and the regulatory agencies stimulated competition in both old and new areas of traditionally monopolistic services. Third, digital technology emerged to implement many of the fundamental transmission and switching functions within the U.S. telephone network and other networks around the world. The main purpose of this book is to describe the design, application, and operational aspects of this new digital equipment. As background the present, predominantly analog, telephone network is reviewed to provide a framework for the new digital equipment being introduced.

For the most part the following discussions describe individual systems and technical reasons for transitions from conventional analog equipment to seemingly less natural digital counterparts. Thus one purpose of this book is to describe how digital technology improves and expands the capabilities of various subsystems within voice telephone networks. Another purpose of the book is to describe the ultimate benefits to be derived when an entire network is implemented with digital techniques. A great degree of synergism exists when individual systems are designed into one cohesive network utilizing digital implementations throughout. The synergistic effect benefits conventional voice services and other existing or yet to arise needs for the public network.

Most of the equipment descriptions and design examples presented in this book come from material authored by engineers at Bell Telephone Laboratories. The basic principles, however, are by no means unique to the Bell system, or to public telephone networks in general. The concepts and implementation examples are applicable to any communications network: public or private, voice or data. An inherent attribute of a digital network is that it can be designed independently of its application.

1

It should be emphasized that the use of digital technology for voice services is motivated by desires to improve the quality, add new features, and reduce the costs of standard voice services. Digitization of the network is not arising from the needs of the data processing industry for better data transmission services. Indeed, most of the digital technology being introduced to the network at the present is inaccessible to data traffic, except through analog channels. Some digital transmission facilities in the United States are available through the Bell Systems Digital Dataphone Service network, but these facilities require special interfaces (data ports). Of course, a digital network is a natural environment for data communications services. As the present network evolves with more digital implementations, or as new digital networks are developed, the performance and economics of data transmission will improve significantly.

If we consider the extensive amount of analog equipment now in place, a long time will pass before the public network in the United States evolves to an all-digital implementation. However, as new private networks are developed, all-digital implementations are not only feasible but also economically desirable, particularly if significant amounts of nonvoice service are accommodated. The major impediment to private digital networks at this time (1981) is the absence of suitable long-distance transmission facilities. With new digital microwave systems in existence already and with the anticipated emergence of digital fiber optics and satellite services, however, the prospects for private digital networks for voice and data are close at hand.

1.1 TERMINALS, TRANSMISSION, AND SWITCHING

The three basic elements of a communications network are: terminals, transmission systems, and switches. The first part of this chapter provides an overview of these elements as implemented in analog telephone networks. Then, the last part of this chapter provides a brief overview of digital implementations *within* the analog network. Following a detailed discussion of the motivation for digital implementations in Chapter 2, the next four chapters describe the operation and design of the basic elements of a digital voice network. Chapter 3 discusses digital voice terminals and the most common algorithms used to convert analog voice signals into digital bit streams. Chapter 4 presents the basics of digital transmission systems. Fundamentals of digital switching follow in Chapter 5. Basic digital modulation techniques and their application to point-to-point digital microwave systems are described in Chapter 6. A discussion of various synchronization and control considerations for digital networks is provided in Chapter 7.

The main emphasis of Chapter 7, and the book as a whole, involves circuit switching as traditionally implemented for voice telephone networks. A circuit-switched network is one that assigns a complete end-to-end connection in response to each request for service. Each connection, with its associated network facilities is held for the duration of the call. Chapter 8 describes a different type of a network, commonly referred to as a packet-switched network, that is particularly suited to servicing data traffic. Chapter 8 also discusses basic considerations to be made when designing a network to accommodate efficiently both voice and data traffic. The last chapter presents the basics of traffic theory: the fundamental mathematics for analyzing and predicting telecommunications network performance.

Telecommunications Standards Organizations

Successful operation of large national or international telecommunications networks requires a set of standards to ensure proper interworking of equipment. In North America standards are essentially set by the Bell System since over 80% of U.S. telephones are serviced by it. Other U.S. organizations that help establish standards are the United States Independent Telephone Association (USITA) and the Rural Electrification Authority (REA). Most of the rest of the world relies on two international committees established under the auspices of the International Telecommunication Union (ITU). These committees are: the International Telegraph and Telephone Consultative Committee (CCITT), and the International Radio Consultative Committee (CCIR). CCITT establishes recommendations for telephone, telegraph, and data-transmission circuits and equipment. CCIR is concerned with coordinating the use of the radio spectrum and radio equipment interfaces to wire-line facilities. In the United States use of the radio spectrum is controlled by the Federal Communications Commission (FCC).

Bell System standards and CCITT standards are sometimes but not always incompatible. North American standards are often incorporated into CCITT recommendations as a subset. Furthermore, equipment manufacturers for the North American network often follow CCITT recommendations when they do not conflict with Bell System requirements.

1.2 THE ANALOG TELEPHONE NETWORK

Existing analog telephone networks in the United States and around the world represent remarkable engineering achievements. Because the present networks have evolved over a long time, a great deal of diversity in equipment types and implementations exists. The use of digital techniques for basic voice communications represents a transition to a new

TABLE 1.1   UNITED STATES TELEPHONE INDUSTRY STATISTICS FOR 1980 [1, 2]

The number of telephones in the United States increased 2.8% to 181 million phones (81% of the phones are serviced by the Bell System).

Telephone company plant investment increased 8.6% to $169 billion.

Telephone industry revenues totaled $61 billion.

Total telephone industry construction expenditures were $22.3 billion.

More than 1 million people in the United States were directly employed by the industry.

The Bell System alone has enough wire to go to the sun and back three times.

There are 1600 independent telephone companies the largest of which is General Telephone and Electronics with 8.6% of the market and 15 million phones.

technology which reduces costs and improves performance for traditional services. The digital transition also represents an opportunity to evolve into more unified networks capable of accommodating new kinds of traffic. However, the transition is significantly complicated by the need for new equipment, both analog and digital, to be compatible with the existing plant.

Although the following discussion is basically technical in nature, a perspective of the transition to digital technology in the United States requires some mention of the nature and size of this country's telecommunications industry. Table 1.1 has been included for two reasons: first, to impress upon the reader the enormity of the industry and the size of the existing network; second, to provide an indication of the size of the potential market, and therefore, the reasons for the intense interest in telecommunications. Many companies that were formerly only users of telecommunications services are now supplying both services and equipment.

The speed of any transition is dependent on both the motivation behind it and the magnitude of the change involved. The statistics of Table 1.1 demonstrate that the conversion of existing implementations, manufacturing facilities, and manpower training is a formidable obstacle to widespread or rapid change in the industry. The statistics also indicate, however, that new equipment and small conversion percentages represent large markets to most corporations.

1.2.1   Network Hierarchy

Alexander Graham Bell invented the first practical telephone in 1876. It soon became apparent, however, that the telephone was of little use without some means of changing connections on an "as needed" basis. Thus the first switching office was established in New Haven, Connecticut only two years later. This switching office, and others following, was located at a central point in a service area and provided switched connections for all subscribers in the area. Because of their locations in the service areas, these switching offices are often referred to as central offices.

As telephone usage grew and subscribers desired longer distance connections, it became necessary to interconnect the individual service areas with trunks between the central offices. Again, switches were needed to interconnect these offices, and a second level of switching evolved. Continued demand for even longer distance connections, along with improved long-distance transmission facilities, stimulated even more levels of switching. In this manner the public telephone network in the United States has evolved to a total of five levels. These levels are listed in Table 1.2.

At the lowest level of the network are class 5 switching offices, also called central offices (CO) or end offices (EO). The next level of the network is composed of class 4 toll offices. The toll network contains three more levels of switching: primary centers, sectional centers, and regional centers. There are ten regional centers in the United States and two in Canada.

In general terminology, a switched communications network is composed of switching nodes and transmission links between the nodes. In a symbolic representation of the public telephone network, the switching nodes represent the various switching offices, and the transmission links represent interoffice trunks. Figure 1.1 depicts a hierarchical switching network, like the backbone public network, except that only three levels of switching are shown. In general terms, the structure shown in Figure 1.1 is referred to as hierarchical tree network.

Figure 1.1 depicts a single switch at the top of the hierarchy. Actually, the North American network has 12 switching offices (regional centers) at the top of its hierarchy. These highest level switching centers (shown in Figure 1.2) are directly interconnected (there is no higher level switching node). The highest level interconnections, with $(12) \cdot (11)/2 =$

TABLE 1.2 SWITCHING HIERARCHY OF NORTH AMERICA

Class	Functional Designation	Total Number in United States [3]
1	Regional center	10 (plus 2 in Canada)
2	Sectional center	67
3	Primary center	230
4	Toll center	1,300
5	End office	19,000

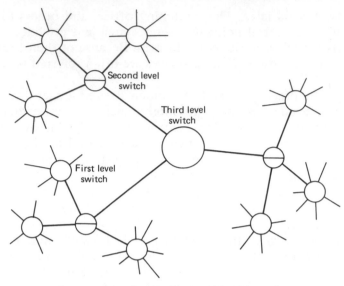

Figure 1.1   Three level switching hierarchy.

66 routes directly between every pair of regional centers, are referred to as a "mesh."

A significant deficiency of the hierarchical structure shown in Figure 1.1 is that only one path exists between any two switching offices. Thus interconnected tree networks are highly vulnerable to failures in either the switching offices or the transmission links. To circumvent this vulnerability, and to provide more economical interconnections

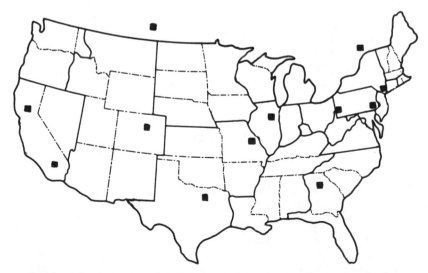

Figure 1.2   Location of regional switching centers in North America.

between pairs of switching offices with heavy traffic volumes, the back-bone network is augmented with many high-usage trunks. High-usage trunks are used for direct connections between switching offices with high volumes of interoffice traffic. Normally, traffic between two such offices is routed through the direct trunks. If the direct trunks are busy (which may happen frequently if they are highly utilized), the backbone hierarchical network is still available for alternate routing.

Traffic is always routed through the lowest available level of the network. This procedure not only uses fewer network facilities but also implies better circuit quality because of shorter paths and fewer switching points. Figure 1.3 shows the basic order of selection for alternate routes. The direct interoffice trunks are depicted as dashed lines, while the backbone, hierarchical network is shown with solid lines.

In addition to the high-usage trunks, the backbone network is also augmented with additional switching facilities called tandem switches. These switches are employed at the lowest levels of the network and provide switching between end offices. Tandem switches are not part of the toll network, as indicated in Figure 1.4, but are part of what is referred to as an exchange area. Generally speaking, an exchange area is an area within which all calls are considered to be local calls (i.e., toll free).

In general terms, any switching machine in a path between two end offices provides a tandem switching function. Thus toll switches also

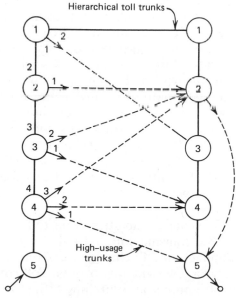

Figure 1.3 Alternate routing in North American network.

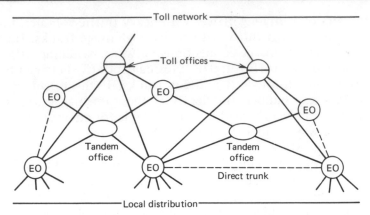

Figure 1.4 Exchange area network.

provide tandem switching functions. Within the telephone industry, however, the term "tandem" refers specifically to intermediate switching within the exchange area.

The basic function of a tandem office is to interconnect those central offices within an exchange area having insufficient interoffice traffic volumes to justify direct trunks. Tandem offices also provide alternate routes for exchange area calls that get blocked on direct routes between end offices. Although Figure 1.4 depicts tandem switching offices as being distinct from end offices and toll offices, tandem switching offices are usually colocated with end offices or class 4 toll offices. Operationally, however, exchange area switching and toll switching are separate. Hence, toll connecting trunks are not used for exchange area calls and tandem switches are not used for toll connections. (Exchange area equipment and toll network equipment may use common facilities but are functionally separate.)

The separation of the toll and exchange area switching functions has evolved primarily as a matter of convenience for billing and maintenance purposes. Because newer switching machines use programmable computers to process connect requests, this convenience is becoming less necessary, and the two switching functions can be combined. Furthermore, as usage sensitive pricing, also called measured service or message unit accounting, is introduced into the exchange area for local calls, the distinction between the toll network and the exchange area will become less defined.

The separation of exchange facilities from toll facilities has had an important effect on the transmission and switching equipment utilized in the respective applications. Exchange area connections are usually short and only involve a few switching offices. Toll connections, on the other hand, may involve numerous switching offices with relatively long transmission links between them. Thus, for comparable end-to-end quality, individual exchange area equipment need not provide as much quality as do toll network counterparts.

In passing it should be mentioned that the five-level network just described, pertains only to the public switched network. By including private switching equipment, such as private branch exchanges (PBXs), another level of switching can be identified below the end-office level. Furthermore, user-controlled key sets (line selectable telephones) represent yet another level of switching.

1.2.2 Switching Systems

In general terms the equipment associated with any particular switching machine can be categorized as providing one of the following functions:

1 Signaling.
2 Control.
3 Switching.

The basic function of the signaling equipment is to monitor the activity of the incoming lines and forward appropriate status or control information to the control element of the switch. Signaling equipment is also used to place control signals onto outgoing lines under direction of the switch control element.

The control element processes incoming signaling information and sets up connections accordingly. The switching function itself is provided by a switching matrix: an array of selectable crosspoints used to complete connections between input lines and output lines. These basic constituents of a switching machine are shown in Figure 1.5.

Electromechanical Switching

A majority of switching offices in North America, and around the world, are equipped with one of two basic types of electromechanical

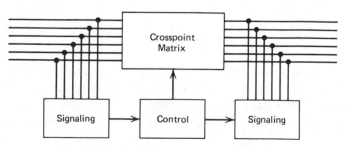

Figure 1.5 Switching system components.

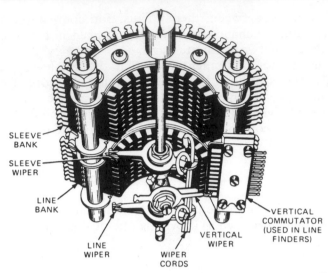

Figure 1.6 Step-by-step switching element. (Copyright 1977, Bell Telephone Laboratories. Reprinted by permission.)

switches: step-by-step* and crossbar. As shown in Figure 1.6, cross-points of a step-by-step switch are wiper contacts that move in direct response to dial pulses. As the pulses of the first digit enter the switch they immediately "step" the vertical wiper to a horizontal row corresponding to the first digit. After the proper row is selected, the wiper is rotated across another set of contacts until an idle line to the next stage of switching is located. The next set of dial pulses, representing the second digit, then "steps" the second stage in like manner. The process continues through however many stages are needed for a particular switch size.

As the name implies, a step-by-step switch uses *direct progressive control:* successive segments of a path through the switch are established as each digit is dialed. With progressive control, the control elements of the switch are integrated into the switching matrix. This feature is very useful for implementing a variety of switch sizes and allowing relatively easy expansion. A progressive control switch, however, has a number of significant limitations:

1 A call may be blocked even though an appropriate path through the switch exits but is not attempted because an unfortunate path gets selected in an early stage.

2 Alternate routing for outgoing trunks is not possible. That is, the

*A step-by-step switch is also referred to as a "Strowger switch" in honor of its inventor Almon B. Strowger.

outgoing line is directly selected by incoming dial pulses and can not be substituted.

3 Signaling schemes other than dial pulses (e.g., push-button tone signaling) are not directly usable.

4 Number translation is impossible.

In contrast to a step-by-step switch, a crossbar switch is one that uses centralized, common control for switch path selection. As digits are dialed, the control element of the switch receives the entire address before processing it. When an appropriate path through the switch is determined (which may involve number translation or alternate routing), the control element transfers the necessary information in the form of control signals to the switching matrix to establish the connection. The fundamental feature, and advantage, of a common control switch is that the control function implementation is separate from the switch implementation. Common control crossbar systems introduced the ability to assign logical addresses (telephone numbers) independently of physical line numbers.

The crosspoints of a crossbar switch (Figure 1.7) are mechanical contacts with magnets to set up and hold a connection. The term "crossbar" arises from the use of crossing horizontal and vertical bars to initially select the contacts. Once established, the switching contacts are held by electromagnets energized with direct current passing through the established circuit. When the circuit is opened, the loss of current causes the crosspoints to be released automatically.

Because of the operational limitations of progressive control, step-by-step switches have been used primarily in smaller class 5 switching offices. Crossbar switches, on the other hand, have been used predominantly in metropolitan areas and within the toll network. In some cases step-by-step switches have been augmented with common control by receiving the digits into special control equipment. After processing the request, the control equipment generates pulses that set up a connection as if the switch was receiving dial pulses directly. All modern switches now use common control no matter what type of a switching matrix is used.

Stored Program Control

Initially the control functions of a switching system were provided by operators. Now, however, routine control functions of the public telephone network are implemented with automatic control almost exclusively. Foremost among the automatic control switches are various versions of step-by-step and crossbar. As indicated previously, these switching systems use electromechanical components for both the switching matrix and the control elements. In some cases the electromechanical control elements in these switches represent rudimentary

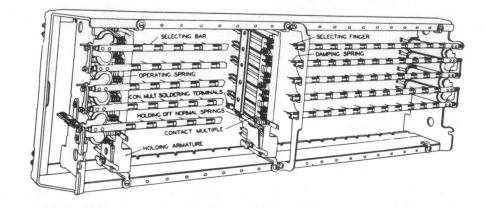

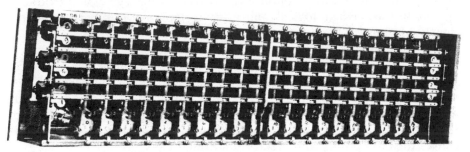

Figure 1.7 Crossbar switching element. (Copyright 1977, Bell Telephone Laboratories. Reprinted by permission.)

forms of special purpose digital computers. The hardwired electromechanical logic, however, has limited capabilities and is virtually impossible to modify.

A major milestone for telephony was established in 1965 when the Bell System installed its first computer controlled switching system: the No. 1 ESS.* This switching system uses a stored program digital computer for its control functions. The stored program control (SPC) feature of the No. 1 ESS allows the offering of many new services and the possibility of even more services in the future. If your telephone company can provide you with customized features such as abbreviated dialing, call forwarding, call waiting, or three-way calling, it is because your local end office is equipped with an ESS machine from Western Electric (Bell System manufacturing) or a similar SPC machine from other manufacturers such as the Automatic Electric Division of

---

*Computer controlled PBXs were available before 1965. The No. 1 ESS represents the first instance of computer control in the public network hierarchy.

General Telephone and Electronics (GTE). At the present time the SPC switching machines are being installed in the United States at a rate of better than one a day—primarily in metropolitan areas. However, the Bell System alone has almost 10,000 local switching offices. Thus, it will be a while before everyone can avail themselves of the benefits of stored program control.

Stored program control not only provides significant advantages to the users, but also simplifies many administrative and maintenance tasks for the operating companies. A large part of line administration that formerly required many manual modifications (mainframe cross-connects) can now be accomplished with changes in computer data tables of an SPC switch. Furthermore, physical line numbers are easily independent of the logical (directory) line numbers, thus making number changes easy. Some of the other benefits of SPC to an operating company are: automated record keeping, lower blocking probabilities, generation of traffic statistics, automated call tracing, and message unit accounting (per call charges as opposed to flat rate billing for unlimited local calling).

The switching matrix of the No. 1 ESS and other presently available class 5 switches in the Bell System (such as the No. 3 ESS) are implemented with electromechanical reed relays. Thus the term ESS, which stands for electronic switching system, refers in general to computer controlled switching and not to the nature of the switching matrix itself. Similarly, the EAX terminology of GTE stands for electronic automated exchange and does not necessarily imply anything about the nature of the matrix. However, the No. 4 ESS, which was first installed in 1976, is a high-capacity toll switch using computer control and digital electronics for its switching matrix. Thus the No. 4 ESS is "electronic" in its control and its switching matrix. Furthermore, the recently announced No. 5 EAX of GTE and the No. 5 ESS of the Bell System also utilize digital logic circuits for the crosspoint matrix.

Private Branch Exchanges

In the United States the term "private branch exchange" (PBX) refers generically to any switching system owned or leased by a business or organization to provide both internal switching functions and access to the public network. Thus a PBX in the United States may use either manual or automatic control. The term "PABX" is also used in the United States, and particularly in other countries, to refer specifically to automatically controlled PBXs.

The historical development of PBX systems has followed closely that of switches in the public network. PBXs with computerized control became available in 1963 (before the No. 1 ESS) when the Bell System's No. 101 ESS was first installed. Since that time a large number of independent manufacturers have developed computer controlled PBXs [or

computerized branch exchanges (CBXs) as they are sometimes called].
In fact, the PBX market is currently one of the most competitive and
innovative businesses in all telecommunications.

The use of computer control for PBXs has proved to be even more
rewarding for users than computer control in the public network. Not
only are customized calling features (abbreviated dialing, etc.) usually
provided, but numerous facilities for cost management are also available. Some of the more useful features in an advanced "smart" PBX
are the following:

1 Accounting summaries by individual employee or department.
2 Multiple classes of service with priorities and access restrictions to
   area codes, WATS lines, and so on.
3 Least-cost routing to automatically select tie lines, foreign exchange circuits, WATS, DDD, and so forth.
4 Automatic callback when circuits are available.
5 Traffic monitoring and analysis to determine the utilization of
   existing circuits or to ascertain blocking probabilities and network cost effectiveness.

A vast majority of new PBX offerings now use electronic time
division switching in lieu of older electromechanical space division
techniques. The Customer Switching System (CSS) 201 (trademark
DIMENSION PBX) of the Bell System uses analog time division switching while most other new switches use digital time division switching.*
Chapter 5 describes the basic operation of both analog and digital time
division switching.

1.2.3 Transmission Systems

Functionally, the communications channels between switching systems
are referred to as trunks. These channels are implemented with a variety
of facilities including: pairs of wires, coaxial cable, point-to-point
microwave radio links; and most recently optical fibers. Most of the
network investment in transmission facilities, however, lies in local distribution from end offices to telephone customer premises. These
facilities are almost exclusively implemented with dedicated wire pairs.
These wire pairs are commonly referred to as either station loops, subscriber loops, local loops, or customer loops.

Open Wire

A classical picture of the telephone network in the past consisted of
telephone poles with crossarms and glass insulators used to support

---

*Engineers at Bell Telephone Laboratories are currently developing a digital PBX
code named the ANTELOPE System.

uninsulated open-wire pairs. Except in rural environments, the open wire has been replaced with multipair cable systems. The main advantage of an open-wire pair is its relatively low attenuation (a few hundredths of a decibel per mile at voice frequencies). Hence, open wire is particularly useful for long rural customer loops. The main disadvantages are having to separate the wires with crossarms to prevent shorting, and the need for large amounts of copper. (A single open-wire strand has a diameter that is five times the diameter of a typical strand in a multipair cable. Thus open wire uses roughly 25 times as much copper as does cable.) As a result of maintenance problems and a continuing decrease in electronics costs, open wire in rural environments is being replaced with cable systems using amplifiers to offset attenuation on long loops.

Paired Cable

In response to overcrowded crossarms and high maintenance costs, multipair cable systems were introduced as far back as 1883. Today a single cable may contain anywhere from 6 to 2700 wire pairs. Figure 1.8 shows the structure of a typical cable. When telephone poles are used, a single cable can provide all the circuits required on the route— thereby eliminating the need for crossarms. More recently the preferred means of cable distribution is to bury it directly in the ground (buried cable) or use underground conduit (underground cable).

Table 1.3 lists the most common wire sizes to be found within paired cable systems. The lower-gauge (higher-diameter) systems are used for longer distances where signal attenuation and DC resistance can become limiting factors. Figure 1.9 shows attenuation curves [5] for the common gauges of paired cable as a function of frequency. An important

TABLE 1.3 WIRE GAUGE AND RESISTANCE OF COMMON PAIRED CABLE.[a]

Gauge	Diameter (in.)	DC Resistance (ohms/1000 ft)
26	0.016	40.81
24	0.020	25.67
22	0.025	16.14
19	0.036	8.051

[a] Data in this table were obtained from Reference 4 for annealed copper at 20°C. Note that the loop resistance of a pair is twice the resistance of a single wire given in the table.

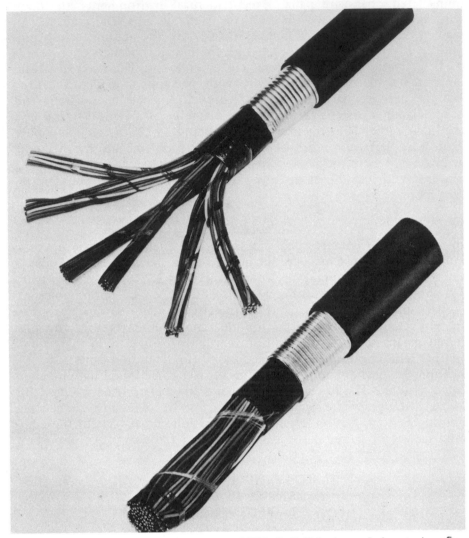

Figure 1.8 Multipair cable. (Copyright 1977, Bell Telephone Laboratories. Reprinted by permission.)

point to notice in Figure 1.9 is that the cable pairs are capable of transmitting much higher frequencies than required by a voice signal (approximately 3 kHz).

The exchange areas of the telephone network use paired cable as a transmission medium almost exclusively. The only exceptions are wideband media for longer, high-capacity interoffice trunks. In the past, interoffice exchange area transmission utilized individual wire pairs for each voice channel. However, greater traffic volumes, increasing copper

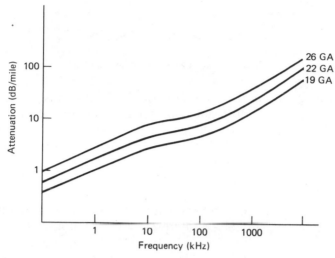

Figure 1.9 Attenuation versus frequency of common gauges of paired cable.

costs, and decreasing electronics costs have all stimulated the use of interoffice carrier systems that utilize cable pair bandwidth more fully by multiplexing: providing more than one voice channel per cable pair. Foremost among these systems are digital T-carrier systems described later in this chapter.

Two-Wire versus Four-Wire

All wire-line transmission in the telephone network is based on transmission through pairs of wires. As shown in Figure 1.10, transmission through a single wire (with a ground return) is possible and has been used in the past. However, the resulting circuit is too noisy for customer acceptance. Instead, balanced pairs of wires as shown in Figure 1.11 are used with signals propagating as a voltage difference between the two wires. The electrical currents produced by the difference signal flows through the wires in opposite directions called a "metallic current." In contrast, induced noise or interference is coupled equally into both wires of a pair and propagates along the pair in one direction.

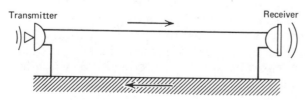

Figure 1.10 Single-wire transmission with ground return.

Figure 1.11　Two-wire transmission.

Current propagating in the same direction in both wires is referred to as common mode or longitudinal current. Longitudinal currents are not coupled into a circuit output unless there is an imbalance in the wires that converts some of the longitudinal signal (noise or interference) into a difference signal. Thus the use of a pair of wires for each circuit provides much better circuit quality than does single-wire transmission. Some switching systems use single-wire (unbalanced) transmission to minimize the number of contacts. Unbalanced circuits are only feasible in small switches where noise and crosstalk can be controlled.

Virtually all subscriber loops in the telephone network are implemented with a single pair of wires. The single pair provides for both directions of transmission. If users on both ends of a connection talk simultaneously, their conversations are superimposed on the wire pair and can be heard, though not usually understood, at the opposite ends. In contrast, wire-line transmission over longer distances, as between switching offices, usually involves two pairs of wires: one pair for each direction. Longer distance transmission requires amplification and often involves multiplexing. These operations are implemented most easily if the two directions of transmission are isolated from each other.* Thus interoffice trunks, particularly in the toll network, typically use two pairs of wires and are referred to as four-wire systems. The use of two-wire pairs does not necessarily imply the use of twice as much copper as a two-wire circuit. Four-wire systems are frequently used with some form of multiplexing to provide multiple channels in one direction on one wire pair. Thus, a net savings in copper results.

Sometimes the bandwidth of a single pair of wires is separated into two subbands that are used for the two directions of travel. These systems are referred to as "derived four-wire systems." Hence, the term "four-wire" has evolved to imply separate channels for each direction of transmission—even when wires may not be involved. For example, radio systems that necessarily use separate channels for each direction are also referred to as "four-wire systems."

The use of four-wire transmission has a direct impact on the switching systems of the toll network. Since most toll network circuits are four-wire, the switches are designed to separately connect both directions of transmission. Hence, two paths through the switch are needed for each connection. A two-wire switch, as used in local switching, requires only one path through the switch for each connection.

*Modest amounts of bidirectional amplification can be realized with "negative resistance" amplifiers.

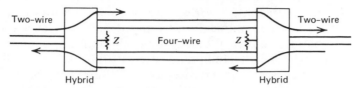

Figure 1.12   Interconnection of two-wire and four-wire circuits.

Two-Wire to Four-Wire Conversion

At some point in a long distance connection it is always necessary to convert from two-wire transmission of local loops to four-wire transmission on long-distance trunks. Usually the conversion occurs at the trunk interface of the end office switch. Sometimes, however, two-wire transmission is used for interoffice trunks so that the conversion occurs at the next switching office.

The interconnection of two-wire and four-wire facilities for a typical long distance connection is shown in Figure 1.12. The basic conversion function is provided by hybrid circuits that couple the two directions of transmission as shown. Hybrid circuits have been traditionally implemented with specially interconnected transformers [3]. More recently, however, electronic hybrids have been developed [6]. Ideally a hybrid should couple all energy on the incoming branch of the four-wire circuit into the two-wire circuit, and none of the incoming four-wire signal should be transferred to the outgoing four-wire branch.

When the impedance matching network Z exactly matches the impedance of the two-wire circuit, near perfect isolation of the two four-wire branches can be realized. However, the two-wire circuit is normally a switched connection. Thus the matching network can only approximate the typical impedance of the two-wire facilities. The effect of an impedance mismatch is to allow some energy on the incoming branch of a four-wire circuit to be coupled to the outgoing branch and returned to the source as an echo. The effect and control of echos in the network is discussed in later sections of this chapter.

Half-Duplex versus Full-Duplex

The data communications terms "half-duplex" and "full-duplex" are closely related to, but not synonomous with, the telephony terms "two-wire" and "four-wire." A half-duplex circuit is one that provides transmission in two directions, but only in one direction at a time.* A full-duplex circuit is one that provides transmission in both directions simultaneously. Obviously, an end-to-end four-wire circuit pro-

---

*The CCITT definition of half-duplex differs from North American usage in that a half-duplex operation in CCITT context is determined by the mode of operation of the terminal and not the transmission link itself. A CCITT half-duplex circuit can provide simultaneous transmission in both directions—but only one direction is used at a time.

vides full-duplex capabilities. However, two-wire circuits can also be used for full-duplex communication by partitioning the available bandwidth into separate frequency bands for each direction of transmission (derived four-wire). This technique is often utilized when a full-duplex data communications circuit is desired over dial-up (two-wire) facilities.

On the other hand, the existence of four-wire circuits does not necessarily imply that full-duplex transmission can be achieved through simultaneous use of both pairs. As discussed later, very long-distance circuits require echo suppressors that effectively disable one pair of a four-wire circuit while the other pair is in use. Thus, only one pair can be used at a time. Long-distance (four-wire) leased lines for data communications typically have the echo suppressors removed so that both pairs can be used simultaneously.

Loading Coils

The attenuation curves shown in Figure 1.9 indicate that the higher frequencies of the voice spectrum (up to 3 kHz) experience more attenuation than the lower frequencies. This frequency dependent attenuation distorts the voice signal and is referred to as amplitude distortion. Amplitude distortion becomes most significant on long cable pairs where the attenuation difference is greatest.

The usual method of combating amplitude distortion on intermediate length (3–15 miles) wire pairs is to insert artificial inductance into the lines. The extra inductance comes from loading coils that are inserted at 3000, 4500, or 6000 ft intervals. Figure 1.13 shows the effect of loading coils on a 24 gauge loop. Notice that the voiceband response up to 3 kHz is greatly improved, but the effect on higher frequencies is devastating.

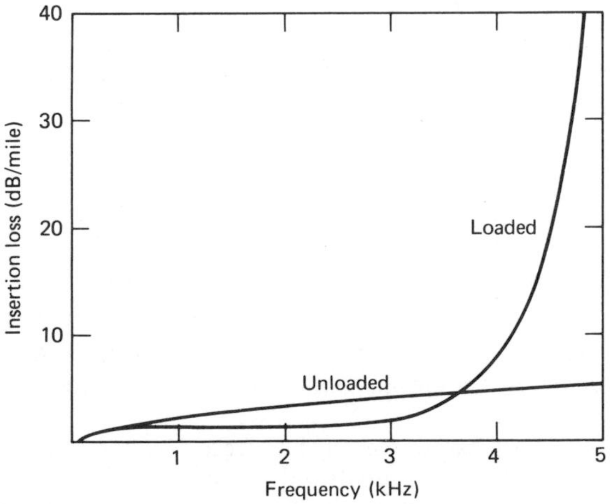

Figure 1.13   Effect of loading on 24-gauge cable pair.

Loading coils are used extensively on interoffice exchange trunks since almost all of these trunks are greater than 3 miles in length and most are less than 15 miles long. Loading coils are also used on the longer, typically rural, subscriber loops.

1.2.4 Pair-Gain Systems

The vast majority of customer loops in the public telephone network are implemented with a single pair of wires dedicated to each station (not including extensions). The average length of these loops is about two miles. On some of the longer loops, wire costs are reduced by sharing the wire pairs. One common method that allows users to directly share a wire pair is to use party lines. However, owing to access contention and often noticeable degradations of voice quality, party lines are becoming less popular. An alternate means of sharing wire pairs is to use a pair-gain system. In contrast to party lines, pair-gain systems provide loop sharing in a manner that is transparent to users. Two basic types of pair-gain systems are shown in Figure 1.14: concentration and multiplexing.

Concentration

The first form of a pair-gain system in Figure 1.14 depicts a basic line concentration system. When viewed from the station set end of the system, a pair-gain system provides concentration by switching some number of active stations to a smaller number of shared output lines. At the

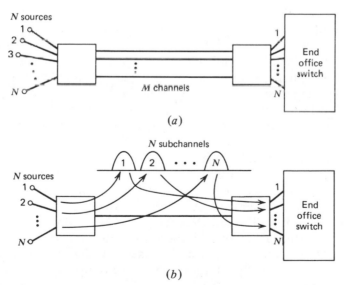

Figure 1.14 Pair gain systems: concentration and multiplexing. (*a*) Concentration ($N > M$). (*b*) Multiplexing.

other end of the system, deconcentration (expansion) occurs by switching from the shared lines to individual inputs of the switching office corresponding to the active stations. Expanding the traffic back to the original number of stations ensures that the system is operationally transparent to both the switch and the user. Notice that a definition of which end provides concentration and which end provides expansion is dependent on the point of view.

Since a concentrator is incapable of simultaneously connecting all stations it services, a certain amount of blocking is necessarily introduced by concentration. When the activity of individual stations is low enough, significant amounts of concentration can be achieved with acceptable blocking probabilities. For example, 40 stations that are each active only 7.5% of the time can be concentrated onto 10 lines with a blocking probability of .001.* This is an acceptable degradation in service since an equally active called station is busy 75 times as often.

Notice that a concentration system requires the transfer of control information between the concentrator/expander switch terminals. When one end of the system establishes a new connection to one of the shared lines, the other end must be informed to set up the appropriate reverse connection.

Multiplexing

As shown in Figure 1.9 the inherent bandwidth of a typical wire pair is considerably greater than that needed for a single voice signal. Thus, multiplexing can be used to carry multiple voice channels on a single pair of wires. The increase in attenuation implied by the higher frequencies is offset by amplifiers in the multiplex equipment and at periodic points in the transmission lines. The particular multiplexing technique shown in Figure 1.14*b* is a frequency division multiplex system. Another form of multiplexing, time division multiplexing of digital voice signals, is also used in pair-gain systems and is discussed later in this chapter.

As shown in Figure 1.14*b*, there is a one-to-one relationship between the customer lines and the subchannels of the multiplexer. Thus, unlike the concentration system, there is no possibility of blocking in a multiplexing type of pair-gain system. Also, there is no need to transfer switching information since the same one-to-one relationship defines the correspondence between customer lines at one end and switching office lines at the other end. A major drawback of multiplexing pair-gain systems is that the subchannels are highly underutilized if the sources are relatively inactive. In these situations a combination of concentration and multiplexing can be used. A digital subscriber loop multiplexer described in a later section is an example of such a system.

---

*A discussion of traffic analysis is provided in Chapter 9—from which this result can be obtained.

1.2.5 Modulation/Multiplexing

The introduction of cable systems into the transmission plant to increase the circuit packing density of open wire is one instance of multiplexing in the telephone network. This form of multiplexing, referred
to as "space division multiplexing," involves nothing more than bun-
dling more than one pair of wires into a single cable. The telephone
network uses two other forms of multiplexing, both of which use
electronics to pack more than one voice circuit into the bandwidth
of a single transmission medium. By far the most prevalent form of
electronic multiplexing in use today is frequency division multiplexing
(FDM) of analog signals. The second form of electronic multiplexing,
time division multiplexing (TDM), is used primarily for digital signals
and is discussed in a later section.

Frequency Division Multiplexing

The form of multiplexing alluded to in Section 1.2.4 and in Figure
1.14*b* for a pair-gain system is frequency division multiplexing. Actually, FDM in cable systems is not as common as digital time division
multiplexing. However, in most wideband transmission media described
later, analog frequency division multiplexing predominates.

As indicated in Figure 1.14*b*, an FDM system divides the available
bandwidth of the transmission medium into a number of narrower
bands or subchannels. Individual voice signals are inserted into the subchannels by amplitude modulating appropriately selected carrier frequencies. As a compromise between realizing the largest number of
voice channels in a multiplex system and maintaining acceptable voice
fidelity, the telephone companies have established 4 kHz as the standard bandwidth of a voice circuit.* If both sidebands produced by
amplitude modulation are used (as in N1 or N2 carrier systems on
paired cable), the subchannel bandwidth is 8 kHz, and the corresponding carrier frequencies lie in the middle of each subchannel. Since
double-sideband modulation is wasteful of bandwidth, single-sideband
(SSB) modulation is used whenever the extra terminal costs are justified. The carrier frequencies for single-sideband systems lie at either
the upper or lower edge of the corresponding subchannel, depending
on whether the lower or upper sideband is selected. The A5 channel
bank multiplexer of the Bell System uses lower-sideband modulation.

FDM Hierarchy

In order to standardize the equipment in the various broadband transmission systems of the network, the Bell System established an FDM
hierarchy as provided in Table 1.4. Each level of the hierarchy is im-

---

*Actually, the usable bandwidth of a voice channel is closer to 3 kHz owing to
guard bands needed by FDM separation filters.

TABLE 1.4   FDM HIERARCHY OF THE BELL NETWORK

Multiplex Level	Number of Voice Circuits	Formation	Frequency Band (kHz)
Voice channel	1		0–4
Group	12	12 Voice circuits	60–108
Supergroup	60	5 Groups	312–552
Mastergroup	600	10 Supergroups	564–3,084
Mastergroup Mux	1,200–3,600	Various	312,564–17,548
Jumbogroup	3,600	6 Mastergroups	564–17,548
Jumbogroup Mux	10,800	3 Jumbogroups	3,000–60,000

plemented using a set of standard FDM modules. The multiplex equipment is independent of particular broadband transmission media.

All multiplex equipment in the FDM hierarchy use single-sideband modulation. Thus, every voice circuit requires approximately 4 kHz of bandwidth. The lowest-level building block in the hierarchy is a channel group consisting of 12 voice channels. A channel group multiplex uses a total bandwidth of 48 kHz. Figure 1.15 shows a block diagram of an A5 channel group multiplexer, the most common A-type channel bank used for first-level multiplexing. Twelve modulators using 12 separate carriers generate 12 double-sideband signals as indicated. Each channel is then bandpass filtered to select only the lower sideband of each double-sideband signal. The composite multiplex signal is produced by superposing the filter outputs. Demultiplex equipment in a receiving terminal uses the same basic processing in reverse order.

Notice that a sideband separation filter not only removes the upper sideband, but also restricts the bandwidth of the retained signal: the

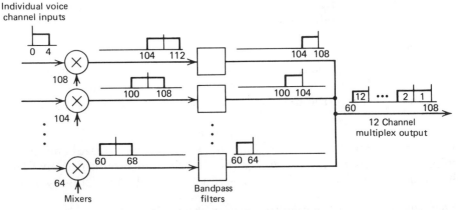

Figure 1.15   A5 channel bank multiplexer.

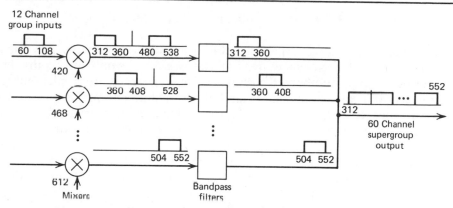

Figure 1.16 LMX group bank multiplexer.

lower sideband. These filters therefore represent a basic point in the telephone network that defines the bandwidth of a voice circuit. Since frequency division multiplexing is used on all long-haul circuits, long-distance connections provide somewhat less than 4 kHz of bandwidth. (The loading coils discussed previously also produce similar bandwidth limitations into a voice circuit.)

As indicated in Table 1.4, the second level of the FDM hierarchy is a 60 channel multiplex referred to as a supergroup. Figure 1.16 shows the basic implementation of an LMX group bank which multiplexes five first-level channel groups. The resulting 60 channel multiplex output is identical to that obtained when the channels are individually translated into 4 kHz bands from 312 to 552 kHz. Direct translation requires 60 separate single-sideband systems with 60 distinct carriers. The LMX group bank, however, uses only five single-sideband systems plus five lower-level modules. Thus two-stage multiplexing, as implied by the LMX group bank, requires more total equipment but achieves economy through the use of common building blocks.*

Because a second-level multiplexer packs individual first-level signals together without guard bands, the carrier frequencies and bandpass filters in the LMX group bank must be maintained with high accuracy. Higher-level multiplexers do not pack the lower-level signals as close together. Notice that a mastergroup, for example, does not provide one voice channel for every 4 kHz of bandwidth. It is not practical to maintain the tight spacing between the wider bandwidth signals at higher frequencies. Furthermore, higher-level multiplex signals include pilot tones to monitor transmission link quality and aid in carrier recovery.

*Bell engineers have developed a new multiplex unit to generate a supergroup directly [3]. Development of this system is indicative of the increasing number of high-traffic routes justifying the development of additional equipment optimized to the application.

1.2.6 Wideband Transmission Media

As discussed previously, paired cable transmission systems are used extensively in the exchange areas of the network. Applications involving either a large number of circuits or long distances are implemented more economically with wider bandwidth media. The maximum number of voice circuits multiplexed into a single cable pair is 24 for N3 analog carrier or 96 for T2 digital carrier. A single tube of an L5E coaxial cable system, on the other hand, can carry up to 13,200 voice signals. In addition to coaxial cable systems, other wideband transmission media are: terrestial microwave radio, satellites, waveguides, and the rapidly developing technology of fiber optics. At the present time available light sources for fiber optics (lasers and light-emitting diodes) function best in a pulsed mode of operation. Thus fiber optic systems are being developed primarily for digital modes of transmission. Analog fiber optic systems are being developed for cable TV where terminal costs must be minimized.

Coaxial Cable

Coaxial cable systems have been used predominantly to satisfy long-haul requirements of the toll network. The first commercial system was installed in 1941 for transmission of 480 voice circuits over a 200 mile stretch between Minneapolis, Minnesota and Stevens Points, Wisconsin [3]. To combat attenuation, repeater amplifiers were installed at 5.5 mile intervals. Considering the maximum capacity of 12 voice circuits on open-wire or cable at the time, the introduction of "coax" was a significant development. Since that time coaxial cable capacity has been steadily increased by (1) using larger diameter cables (0.375 inch) to reduce attenuation; (2) decreasing the distance between repeaters; and (3) improving the noise figure, linearity, and bandwidth of the repeater amplifiers.

A summary of the analog coaxial cable systems used in the Bell System is provided in Table 1.5. Notice that each system reserves one

TABLE 1.5 COAXIAL CABLE SYSTEMS IN THE BELL NETWORK

System Designation	Pairs per System[a]	Signal Designation	Repeater Spacing (miles)	Capacity per Pair	Total Capacity
L1	3/6	Mastergroup	8	600	1,800
L3	5/6	Mastergroup Mux	4	1,860	9,300
L4	9/10	Jumbogroup	2	3,600	32,400
L5	10/11	Jumbogroup Mux	1	10,800	108,000

[a]The number of pairs are shown as working/total.

pair of tubes as spares in the event of failure—a particularly important consideration since each tube carries a high volume of traffic.

Microwave Radio

Much of the impetus for terrestial microwave radio systems came from the need to distribute television signals nationwide. However, as the volume of long-distance traffic increased steadily, radio systems also proved to be the most economical means of distributing voice circuits in the long-distance network. Beginning in 1948, when the first system was installed between New York and Boston, the number of microwave radio systems has grown to supply 60% of the voice circuit miles in the U.S. toll network [3]. These radio systems require line-of-sight transmission with repeater spacings typically 26 miles apart. The major advantage of radio systems is that a continuous right-of-way is not required—only small plots of land spaced 20–30 miles apart for towers and equipment shelters. A major cost of guided transmission, for example, wire pairs, coax, or waveguide, is the right-of-way costs. In many metropolitan areas, microwave routes are heavily congested and can not be expanded with presently allocated frequency bands. Thus coaxial cable systems, fiber optics, and possibly some waveguide systems are, sometimes, the only alternatives for high-capacity transmission.

The frequency bands allocated by the FCC for common carrier use in the United States are listed in Table 1.6. Of these bands, 4 and 6 GHz have been the most popular. The 2 GHz band has not been used extensively because the relatively narrow allocated channel bandwidths do not permit implementation of economical numbers of voice circuits. The basic drawback of the 11 GHz band is its vulnerability to rain attenuation. However, 11 GHz radios are used in some short-haul applications.

The microwave radio systems of the Bell network are listed in Table 1.7. Notice that each radio system is designed to carry one of the multiplex hierarchies described previously. All of these radios except the AR-6A use low-index frequency modulation (FM) of the signal gener-

TABLE 1.6 MICROWAVE FREQUENCIES ALLOCATED FOR COMMON CARRIER USE IN THE UNITED STATES

Band (MHz)	Total Bandwidth (MHz)	Channel Bandwidths (MHz)
2,110–2,130	20	3.5
2,160–2,180	20	3.5
3,700–4,200	500	20
5,925–6,425	500	30
10,700–11,700	1000	40, 20

TABLE 1.7 BELL SYSTEM ANALOG MICROWAVE RADIOS

System	Band (GHz)	Voice Circuits	Application
TD-2	4	600–1500	Long haul
TD-3	4	1200	Long haul
TH-1	6	1800	Short/long haul
TH-3	6	2100	Short/long haul
TM-1	6	600–900	Short/long haul
TJ	11	600	Short haul
TL-1	11	240	Short haul
TL-2	11	600–900	Short haul
AR-6A	6	6000	Long haul (SSB)

ated by the FDM multiplexer equipment. Thus, the FM radios transmit the SSB-FDM signal as a baseband signal with a bandwidth as indicated in Table 1.4. FM modulation was chosen to permit the use of nonlinear power amplifiers in the transmitters and to take advantage of FM signal-to-noise ratio performance.

Examination of Tables 1.6 and 1.7 indicates that the bandwidth utilized per voice circuit in TD-2 radios is 13.3 kHz and 14.3 kHz in TH-3 radios. The indicated channel densities have been achieved through continual refinement of the FM equipment. Although small increases in circuit capacity may be possible, the state of the art is approaching the theoretical maximum efficiency for FM. Hence FM will always require a significant increase in the 4 KHz bandwidth of the individual single-sideband voice circuits. In contrast, Bell System engineers are currently testing a single-sideband radio, the AR-6A, that greatly improves the bandwidth utilization. This radio provides as many as 6000 voice circuits in the 30 MHz channels at 6 GHz.

Since many voice circuits are carried by each radio channel, microwave systems usually include extra equipment and extra channels to maintain service despite outages that may result from any of the following:

1 Atmospheric induced multipath fading.
2 Equipment failures.
3 Maintenance.
4 Rain attenuation.

On some routes, the most frequent source of outage in a microwave radio system arises from multipath fading. Figure 1.17 depicts a simple model of a multipath environment arising as a result of atmospheric refraction. As indicated, the model involves two rays: a primary ray and

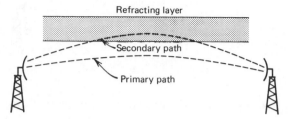

Figure 1.17   Two-ray model of multipath propagation.

a delayed secondary ray. If the secondary ray arrives out of phase with respect to the primary ray, the composite signal is effectively attenuated. The amount of attenuation is dependent on both the magnitude and the phase of the secondary ray. Quite often only nominal amounts of fading occur and can be accommodated by excess signal power in the transmitter, called a fade margin. In some instances, however, the received signal is effectively reduced to zero, which implies that the channel is temporarily out of service.

Frequency Diversity

Fortunately, exceptionally deep fades normally affect only one channel (carrier frequency) at a time. Thus, a backup channel including a spare transmitter and a spare receiver can be used to carry the traffic of a faded primary channel. Selection of and switching to the spare channel is performed automatically without a loss of service. This remedy for multipath fading is referred to as frequency diversity. Notice that frequency diversity also provides hardware backup for equipment failures.

A fully loaded TD-3 radio system uses 12 channels: 10 main channels and two backup channels for protection. This is referred to variously as 2 for 10, 10 by 2, or 10 X 2 protection switching. Some short-haul systems use one-for-one protection switching because it is simpler to implement. However, since only half of the required bandwidth is actually carrying traffic new systems with one-for-one protection are only allowed in uncongested environments.

Except in maintenance situations, protection switching must be automatic in order to maintain service continuity. A typical objective is to restore service within 30 ms to minimize noticeable effects in the message traffic. A more critical requirement is to restore service before the loss of signal is interpreted by some signaling schemes as a circuit disconnect. Inadvertent disconnects occur if the outage lasts for more than 1 to 2 seconds.

Space Diversity

Since deep fades only occur when a secondary ray arrives exactly out of phase with respect to a primary ray, it is unlikely that two paths of

Figure 1.18   Space diversity.

different lengths experience fading simultaneously. Figure 1.18 depicts a technique, called space diversity, using different path lengths to provide protection against multipath fading. As indicated, a single transmitter irradiates two receive antennas separated by some distance on the tower. Although the path length difference may be less than a meter, this difference is adequate at microwave frequencies, which have wavelengths on the order of tenths of meters.

Rain is another atmospherically based source of microwave fading. As already mentioned, rain attenuation is a concern mostly in higher-frequency radios (11 GHz and above). Unfortunately neither frequency diversity (at the high frequencies) nor space diversity provides any protection against rain fades. Sometimes, a lower-frequency channel can be used as a protection channel for a higher-frequency channel.

Satellites

Following the launch of the first international communications satellite in 1965, the use of satellites for international telephone traffic has grown phenomenally. Domestic use of satellites in the United States is beginning to show comparable growth—mostly for private line service and television program distribution.

In the interest of promoting competition in the use of satellites, the FCC established Docket 16495 in 1972. This docket allowed Western Union, RCA, and American Satellite Corporation to offer leased line satellite services without an immediate competitive threat from AT&T and has come to be known as the "open skies policy." More recently, IBM [7] and Xerox [8] have announced intentions to establish satellite based corporate communications services. As these systems are developed and AT&T is allowed to offer more satellite services,* coast-to-coast transmission costs will drop and new types of offerings will occur.

In one sense a satellite system is a microwave radio system with only one repeater: the transponder in outer space. In fact, some satellite systems use the same 4 and 6 GHz frequency bands used by terrestial microwave radios. In another sense, however, the broadcast nature of the down link offers additional opportunities for new services not available from point-to-point terrestial systems. Distribution of network television programming is one application particularly suited to the broadcast nature of satellites. Direct broadcast satellites (DBS)

*The prohibition against AT&T offering satellite services expired in July 1979.

to home receivers is now technically and economically viable. If regulatory issues can be resolved a vast array of new nationwide television services will develop.

One drawback to satellite communications is the inherent propagation delay of the long transmission path. For a stationary satellite, this delay (not including ground links) is 250 ms up and down. A complete circuit with satellite links for both directions of travel therefore implies greater than a one-half second round-trip propagation time. Delays of this magnitude are noticeable in a voice conversation but not prohibitive. The effects of the propagation delays can be alleviated somewhat by pairing each satellite circuit with a ground based circuit in the opposite direction. Thus, the round trip delay involves only one satellite link. In this manner, existing terrestial circuits can be used advantageously while demands for more capacity are economically satisfied by new satellite systems.

Long transmission delays can have significant, almost debilitating, effects on many data communications circuits. The most popular form of error control for data communications involves retransmission of errored data blocks. As commonly implemented, only one block at a time is transmitted; the source waits for a response from the opposite end of the circuit before sending the next block. Thus, transmission delays much greater than a block transmission time lead to low throughputs and inefficient channel utilizations.

1.2.7 Transmission Impairments

One of the most difficult aspects of designing an analog telephone network is determining how to allocate transmission impairments to individual subsystems within the network. Using subjective evaluations by listeners, certain objectives for end-to-end transmission quality can be established in a relatively straightforward manner [9]. After tempering the goals with economic feasibility, the end-to-end objectives can be established. However, considering the myriad of equipment types and connection combinations in a network, designing the individual network elements to meet these objectives in all cases is a complex problem. A great deal of credit is due the common carriers in the United States for having developed a nationwide network with the level of consistent performance it has.

The major factors to be considered in establishing transmission objectives for the analog network are: signal attenuation, noise, interference, crosstalk, distortion, echos, singing, and various modulation and carrier related imperfections.

Signal Attenuation

Subjective listening tests have shown that the preferred acoustic-to-acoustic loss [10] in a telephone connection should be in the neighborhood of 8 dB. A study of local telephone connections [11] demonstrates that the typical local call has only 0.6 dB more loss than ideal. Surveys of the toll network [12] indicate that the average toll connection has an additional 6.7 dB of loss. This same survey also showed that the standard deviation of loss in toll connections is 4 dB (most of which is attributable to the local loops). Thus 84% of all toll connections have less than 19.3 dB loss which is about 11 dB worse than ideal.*

Since trunks within the toll network use amplifiers to offset transmission losses it would be straightforward to design these trunks with zero decibel nominal insertion loss. However, as discussed later, echo and singing considerations dictate a need for certain minimum levels of net loss in most analog trunk circuits.

Interference

Noise and interference are both characterized as unwanted electrical energy fluctuating in an unpredictable manner. Interference is usually more structured than noise since it arises as unwanted coupling from just a few signals in the network. If the interference is intelligible, or nearly so, it is referred to as crosstalk.[†] Some of the major sources of crosstalk are: coupling between wire pairs in a cable, the effects of nonlinear components on FDM signals, inadequate filtering or carrier offsets in FDM equipment, and intersymbol interference in time division multiplex equipment.

Crosstalk, particularly if intelligible, is one of the most disturbing and undesirable imperfections that can occur in a communications network. Crosstalk in analog systems is particularly difficult to control since voice signal power levels vary considerably (i.e., across a dynamic range of 40 dB). The absolute level of crosstalk energy from a high-level signal must be small compared to a desired low-level signal. In fact, crosstalk is most noticeable during speech pauses when the power level of the desired signal is zero.

Two of the most common forms of crosstalk of concern to communications engineers are "near-end crosstalk" (NEXT) and "far-end crosstalk" (FEXT). Near-end crosstalk refers to coupling from a transmitter into a receiver at a common location. Often this form of crosstalk is most troublesome because of a large difference in power levels between the transmitted and received signals. Far-end crosstalk refers to unwanted coupling into a received signal from a transmitter at a distant location. Both forms of crosstalk are illustrated in Figure 1.19.

*The loss distributions are not exactly normal. A standard deviation is just a convenient means of parameterizing the variability.

[†]Crosstalk is also used to characterize signal-like interferences in nonvoice networks. For example, crosstalk in a data circuit would refer to an interfering signal being coupled in from another similar data circuit.

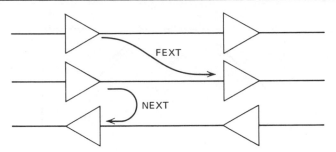

Figure 1.19   Near-end and far-end crosstalk.

Noise

The most common form of noise analyzed in communications systems is white noise with a Gaussian (normal) distribution of amplitude values. This type of noise is both easy to analyze and easy to find since it arises as thermal noise in all electrical components. Battery systems used to power customer loops are also a source of this type of noise. White noise is truly random in the sense that a sample at any instant in time is completely uncorrelated to a sample taken at any other instant in time. The other most common forms of noise in the telephone network are impulse noise and quantization noise in digital voice terminals (Chapter 3). Most impulse noise occurs as a result of switching transients in electromechanical switching offices. Step-by-step and the less common panel switches are the most frequent culprits. More modern switches (e.g., the No. 1 ESS) which use glass encapsulated reed relays for cross-points produce much less noise. Whereas white noise is usually quantified in terms of average power, impulse noise is usually measured in terms of so many impulses per second. Impulse noise is usually of less concern to voice quality than background (white) noise. However, impulse noise tends to be the greatest concern in a data communications circuit.

The power level of any disturbing signal, noise or interference, is easily measured with a root mean square voltmeter. However, disturbances at some frequencies within the passband of a voice signal are subjectively more annoying than others. Thus, more useful measurements of noise or interference power in a speech network take into account the subjective effects of the noise as well as the power level. The two most common such measurements in telephony use: a C-message weighting curve or a psophometric weighting curve as shown in Figure 1.20. These curves essentially represent filters that weight the frequency spectrum of noise according to its annoyance effect to a listener. C-message weighting represents the response of the 500 type telephone set. As far as perceived voice quality is concerned, only the noise that gets passed by the telephone set is important. Notice that disturbances between 1 and 2 kHz are most perceptible.

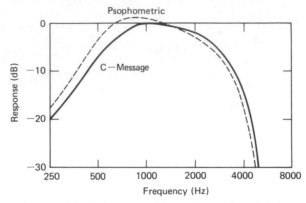

Figure 1.20 C-message and psophometic weighting.

C-message weighting is used in North America while psophometric weighting is the European (CCITT) standard.

A standard noise reference used by telephone engineers is one picowatt, which is 10^{-12} watts or -90 dBm (dBm is power in decibels relative to a milliwatt). Noise measured relative to this reference is expressed as so many decibels above the reference (dBrn). Thus, a noise level of 30 dBrn corresponds to -60 dBm or 10^{-9} W of power. If the readings are made using C-message weighting, the power level is expressed by the abbreviation dBrnc. Similarly, psophometrically weighted picowatts are expressed by the abbreviation pWp. The relationships between various noise power measurements are given in Table 1.8.

The quality of an analog voice circuit is usually not specified in terms of the classical signal-to-noise ratio. The reason is that relatively low levels of noise or interference are noticeable during pauses in speech when there is no signal. On the other hand, high levels of noise can occur during speech and be unnoticeable. Thus, absolute levels of noise are more relevant than signal-to-noise ratios for specifying voice quality.* The objectives for maximum noise levels in the AT&T network are 28 dBrnc for connections up to 60 miles in length and 34 dBrnc for 1000 mile circuits.†

Distortion
In a previous section signal attenuations were considered with the tacit assumption that the received waveforms were identical in shape to the source waveforms but merely scaled down in amplitude. Actually, the

---

*It is common practice in the industry to specify the quality of a voice circuit in terms of a test tone to noise ratio. However, the test tone must be at a specific power level so the ratio, in fact, specifies absolute noise power.

†These noise power values are related to a particular point in a circuit called a "zero-transmission-level point," discussed later.

TABLE 1.8   RELATIONSHIPS BETWEEN VARIOUS NOISE MEASUREMENTS

To Convert			
From		To	
dBm		dBrn	Add 90 dB
dBm	3 kHz flat	dBrnc	Add 88 dB
dBm	3 kHz flat	dBp	Add 87.5 dB
dBrn	3 kHz flat	dBrnc	Subtract 2 dB
dBc		dBp	Subtract 0.5 dB
pW	3 kHz flat	pWp	Multiply by 0.562

received waveforms generally contain certain distortions not attributable to external disturbances such as noise and interference but which can be attributed to internal characteristics of the channel itself. In contrast to noise and interference, distortion is deterministic; it is repeated every time the same signal is sent through the same path in the network. Thus distortions can be controlled or compensated for once the nature of the distortion is understood.

There are many different types and sources of distortion within the telephone network. The telephone companies have minimized those types of distortion that most affect the subjective quality of speech. More recently they have also become concerned with distortion effects on data transmission. Some distortions arise from nonlinearities in the network such as: carbon microphones, saturating voice-frequency amplifiers, and unmatched compandors (Chapter 3). Other distortions are linear in nature and are usually characterized in the frequency domain as either amplitude distortion or phase distortion.

Amplitude distortion refers to attenuating some frequencies in the voice spectrum more than others. The loading coils discussed earlier represent one means of eliminating amplitude distortion on long voice-frequency wire pairs. Amplitude distortion is also introduced by spectrum limiting filters in FDM equipment. Ideally these filters should uniformly pass all voiceband frequencies up to 4 kHz and reject all others. Practical designs, however, imply the need for gradual attenuation "roll-offs" beginning at about 3 kHz. Figure 1.21 shows the amplitude response of a typical toll connection.

Phase distortion is related to the delay characteristics of the transmission medium. Ideally a transmission system should delay all frequency components in a signal uniformly so the proper phase relationships exist at the receiving terminal. If individual frequency components experience differing delays, the time domain representation at the output becomes distorted because superposition of the frequency terms is altered at the output. For reasons not discussed here the delay of an

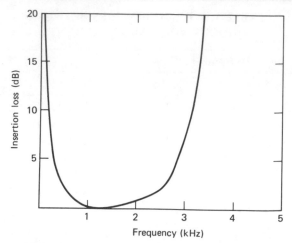

Figure 1.21   Amplitude response of typical toll connection.

individual frequency components is usually referred to as its envelope delay. For a good explanation of envelope delay see Reference 13.

Uniform envelope delay relates to a phase response that is directly proportional to frequency. Thus systems with uniform envelope delay are also referred to as linear phase systems. Any deviation from a linear phase characteristic is referred to as phase distortion. The perceptual effects of phase distortion to a voice signal, however, are small. Thus only minimal attention need be given to the phase response of a voice network. The phase response and corresponding envelope delay of a typical toll connection is shown in Figure 1.22.

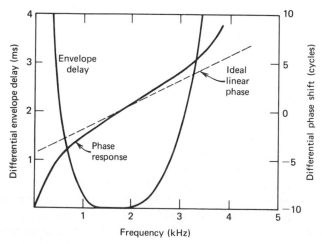

Figure 1-22   Envelope delay and phase response of typical toll connection.

In addition to the distortions just mentioned, carrier systems sometimes introduce other frequency related distortions such as frequency offsets, jitter, phase hits and signal dropouts. The effects of these imperfections are adequately controlled for voice traffic but sometimes present difficulties for high-rate data traffic.

Echos and Singing

Echos and singing both occur as a result of transmitted signals being coupled into a return path and fed back to the respective sources. The most common cause of the coupling is an impedance mismatch at a four-wire to two-wire hybrid. As shown in Figure 1.23, mismatches cause signals in the incoming branch of a four-wire circuit to get coupled into the outgoing branch and return to the source. It is impractical to provide good impedance matches at this point for all possible connections since the switched network involves many different local loops and trunks—each with its own characteristic impedance.

If only one reflection occurs, the situation is referred to as "talker echo." If a second reflection occurs, "listener echo" results. When the returning signal is repeatedly coupled back into the forward path to produce oscillations singing occurs. Basically, singing results if the loop gain at some frequency is greater than unity. If the loop gain is only slightly less than unity, a near-sing condition causes damped oscillations. Singing and near-singing conditions have a disturbing effect on both the talker and the listener. Talker echo is usually the most noticeable and troublesome.

The degree of echo annoyance experienced by a talker is dependent on both the magnitude of the returning signal and the amount of delay involved [14, 15]. On short connections the delay is small enough that the echo merely appears to the talker as natural coupling into his ear. In fact, a telephone is purposely designed to couple some speech energy (called sidetone) into the earpiece. Otherwise, the telephone seems dead to a talker. Near-instantaneous echos merely add to the sidetone and go unnoticed. As the round-trip delay increases, however, it becomes necessary to increasingly attenuate the echos to eliminate the annoyance to a talker. Hence, long-distance circuits require significant attenuation to minimize echo annoyance. Fortunately, an echo experiences twice as much attenuation as does the signal since it traverses twice

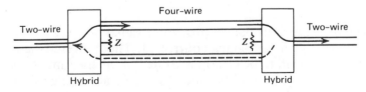

Figure 1.23 Generation of echos at two-wire to four-wire interface.

the distance. Intermediate-length connections are typically designed with 2 to 6 dB of path attenuation depending on the delay.

Connections longer than 1800 miles (or 45 ms round-trip delay) require more attenuation for echo control than can be tolerated for the forward speech signal. In these situations echo suppressors are used to dynamically switch in high levels of attenuation (35 dB typically) as a function of voice activity. Echo suppressors operate on four-wire circuits by comparing the speech volume in each path and inserting the attenuation in the path with the lowest power level. Thus, returning echos are greatly attenuated while the forward path can be designed with little or no net loss.

One drawback of echo suppressors for voice circuits is that they may clip beginning portions of speech segments. If a party at one end of a connection begins talking at the tail end of the other party's speech, the echo suppressor does not pass the new talkspurt until it has had time to reverse directions. Newer echo suppressors are able to reverse transmission directions rapidly (2 to 5 ms [14]) and therefore minimize clipping. Even with the newer echo suppressors, a party can not "break in" to another party's speech unless he talks louder than the other party.

Recent developments in electronics technology have paved the way for a new form of echo control to be used in voice communications circuits. The new devices, called echo cancellors, are tuned to eliminate only reflected signals and not another party's speech. An echo cancellor operates by storing transmitted speech for a period of time equal to the round-trip delay of the circuit. Then, the stored signal is properly attenuated and subtracted from the incoming return signal. Thus echo cancellation requires knowledge of circuit length, the echo return loss, and continuous storage of the transmitted signal. The complex electronics required to perform these operations have only recently become economically feasible. Bell System engineers have developed an integrated circuit echo cancellor for use on satellite circuits [16]. For a good tutorial on all types of echo control see Reference 17.

In general, the procedures used to control echos also control singing. On some fairly short connections, however, no echo control is necessary, and singing may become a problem. Singing can also occur in idle circuits and still be a problem if it overloads amplifiers or causes intermodulation in frequency division multiplex equipment.

1.2.8 Power Levels

As indicated in previous paragraphs, voice signal power in long-distance connection needs to be rigidly controlled. The delivered signal power must be high enough to be clearly audible but, at the same time, not be so strong that circuit instabilities such as echo and singing result.

To maintain rigid control on the end-to-end power level of a circuit involving a variety of transmission systems, telephone companies necessarily control the net attenuation and amplification of each transmission system. These systems are designed for a certain amount of net loss depending somewhat on a transmission link's position in the network hierarchy but mostly on its length. Generally speaking, longer links have greater amounts of net loss up to the point that echo suppressors are used. When echo suppressors are used, net loss is unnecessary. However, because toll-connecting trunks are designed on a net loss basis, all toll connections have some net loss, even if echo suppressors are used in the longest link of the circuit.

In order to administer the net loss of transmission links, the transmission levels of various points in a transmission system are specified in terms of a reference point. CCITT recommendations call this point the zero-relative-level point and the North American term is a zero-transmission-level point (0-TLP). The reference point may not exist as an accessible point but has long been considered to be at the sending end terminal of a two-wire switch. In North America the sending end of a four-wire switch is defined to be a −2 dB TLP. Hence, a 0 dB TLP is only a hypothetical point on a four-wire circuit. Nevertheless, it is useful in relating the signal level at one point in the circuit to the signal level at another point in the circuit.

If a 0 dBm (1 mW) test tone is applied at a 0-TLP, the power level at any other point in the circuit is determined directly (in decibels referred to 1 mW) as the TLP value at that point. It should be emphasized, however, that TLP values do not specify power levels—only the gain or loss at a point relative to the reference point.

Signal power levels or noise power levels are not normally expressed in terms of local measured values. Instead, powers are expressed in terms of their value at the 0-TLP. For example, if an absolute noise power of 100 pW (20 dBrn or −70 dBm) is measured at a −6 dB TLP, it is expressed as 26 dBrn0. (The "0" indicates that the specification is relative to the 0-TLP). If noise power is measured with C-message weighting, it is designated as so many dBrnc0. Similarly, psophometric weighted noise is commonly expressed in units of picowatts psophometrically weighted (dBm0p or pWp0).

In a survey of voice signals in the Bell System the average power level during active speech has been determined as −16 dBm0 [18]. In comparison, the noise power objectives of the AT&T network given previously is 34 dBrnc0 (−56 dBmc0) for a 1000 mile circuit.

1.2.9 Signaling

The signaling functions of a telephone network refer to the means for transferring network related control information between the various terminals, switching nodes, and users of the network. There are two basic aspects of any signaling system: specially encoded electrical wave-forms (signals) and how these waveforms should be interpreted. The most common control signals are dial tone, ringback, and busy tone. These signals and their meaning are well established and may never change. The signaling procedures used internally to the network are not constrained by user convention and are often changed to suit particular characteristics of transmission and switching systems. As a result the public network uses a wide variety of old and new signaling schemes to transfer control information between switching offices.

Signaling Functions

Signaling functions can be broadly categorized as belonging to one of two types: supervisory or information bearing. Supervisory signals con-vey status or control of network elements. The most obvious examples are: request for service (off-hook), ready to receive address (dial tone), call alerting (ringing), call termination (on-hook), request for an opera-tor (hook flash), called party ringing (ringback), and network or called party busy tones. Information bearing signals include: called party address, calling party address, and toll charges. In addition to call related signaling functions, switching nodes communicate between themselves and network control centers to provide certain functions related to network management. Network related signals may convey status such as maintenance test signals, all trunks busy, equipment failures, or they may contain information related to routing and flow control. Chapter 7 discusses some of the basic considerations of net-work management for routing and flow control.

In-Channel Signaling

Signals are transmitted with one of two basic techniques: in-channel signaling or common channel signaling. In-channel signaling (sometimes referred to as "per trunk signaling") uses the same transmission facili-ties or channel for signaling as for voice. Common channel signaling, as discussed in the next section, uses one channel for all signaling functions of a group of voice channels. In the past most signaling systems in the telephone network were the in-channel variety. Recently, however, the trend is toward common channel signaling.

In-channel signaling systems can be further subdivided into in-band and out-of-band techniques. In-band systems transmit the signaling information in the same band of frequencies used by the voice signal. The main advantage of in-band signaling is that it can be used on any transmission medium. The main disadvantage arises from a need to eliminate mutual interference between the signaling waveforms and a user's speech. The most prevalent example of in-band signaling is single frequency (SF) signaling, which uses a 2600 Hz tone as an on-hook signal for interoffice trunks. Although normal speech rarely produces a pure 2600 Hz signal, inadvertent disconnects have occurred as a result of user-generated signals.

Two other common examples of in-band signaling are addressing by dual tone multifrequency (DTMF)* signals from push-button telephones or multifrequency (MF) signaling between switching offices. These signaling techniques are never used during times when speech is present so that there is little chance of misinterpretation between signaling and speech.

In-channel but out-of-band signaling uses the same facilities as the voice channel but a different portion of the frequency band. Thus out-of-band signaling represents a form of frequency division multiplexing within a single voice circuit. The most common instance of out-of-band signaling is direct current signaling as used on most customer loops. With this form of signaling, the central office recognizes the off hook condition by the flow of direct current in the line. Other commonly used loop signals are dial pulses generated by a rotary dial at a rate of 10 pulses per second and a 20 Hz ringing voltage from the central office. All of these signals use lower frequencies than those generated in speech. Thus there is no possibility of one being mistaken for the other. The major disadvantage of out-of-band signaling is its dependence on the transmission system. For example, single sideband carrier systems filter out the very low frequencies associated with each voice channel. Thus, the on-hook/off-hook signal must be converted to something like SF signaling for FDM transmission. Out-of-band signaling is also implemented with frequencies above the cut-off frequency of voice separation filters but below the 4 kHz limit of a channel. CCITT recommends the use of 3825 Hz for this purpose.

Common Channel Interoffice Signaling (CCIS)

Common channel signaling uses a single dedicated control channel for all signaling functions of an associated group of channels. The Bell System installed their first common channel signaling facilities in 1976 [19]. These facilities are referred to as Common Channel Interoffice Signaling (CCIS). Current plans are to equip more and more of the network with CCIS facilities, mostly in conjunction with stored program control switching offices. Eventually 15,000 offices in the United States will be tied together with a CCIS signaling network [20]. CCITT has also established common channel signaling recommendations which are referred to as system No. 6 [21].

The following are the main advantages of common channel signaling:

1 Only one set of signaling facilities is needed for each associated trunk group instead of separate facilities for each individual circuit.

---

*Within the Bell System DTMF signaling is referred to as Touch-Tone signaling, a registered service mark.

2   A single dedicated control channel allows transfer of information such as address digits directly between the control elements (computers) of switching offices. In-channel systems, on the other hand, must have the control information switched from the common control equipment of the originating office onto the outgoing channel, and then the receiving office must switch the incoming control information from the voice channel into its common control equipment. The simpler procedure for transferring information directly between switch processors is one of the main motivations for CCIS.

3   Since separate channels are used for voice and control there is no chance of mutual interference. With CCIS the control channels will initially use dedicated voiceband circuits with data modems.

4   Since the control channel of a common channel system is inaccessible to users, a major means for fraudulent use of the network is eliminated.

5   Connections involving multiple switching offices can be set up more rapidly since forwarding of control information from one office can overlap a circuit set up through the node. With in-channel systems the associated circuit must first be established before the control information can be transferred across it.

6   The channel used for common channel signaling does not have to be associated with any particular trunk group. In fact, the control information can be routed to a centralized control facility where requests are processed, and from which switching offices receive their connection control information. Figure 1.24 depicts a common channel signaling network that is disassociated from the message network structure. One advantage of centralized control is its ability to process requests with knowledge of networkwide

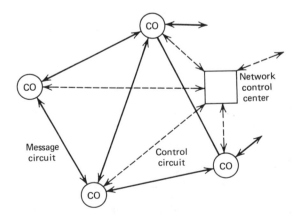

Figure 1.24 Disassociated common channel signaling network.

traffic conditions. Centralized control is also attractive for managing multiple switching offices that are too small to warrant call processing facilities of their own. The transition from in-channel signaling to disassociated common channel signaling at the network level is analogous to the lower level transition from direct progressive control switches (step-by-step) to common control switches.

The major disadvantages of common channel signaling are as follows:

1 Control information pertaining to an established circuit, such as a disconnect, must be relayed from one node to the next in a store-and-forward fashion. An in-channel disconnect signal, on the other hand, automatically propagates through the network enabling all nodes involved in the connection to simultaneously process the disconnect and release the associated facilities.

2 If one node in a common channel system fails to relay the disconnect information properly, facilities downstream from the disconnect will not be released. Thus, a high degree of reliability is required for the common channel both in terms of physical facilities (duplication) and in terms of error control for the data link.

3 Since the control information traverses a separate path from the voice signal, there is no automatic test of the voice circuit as when the voice channel is used to transfer control information. CCIS systems include special provisions for testing a voice circuit when it is set up.

As a final note, it should be pointed out that some signaling functions originating or terminating with an end user inherently require in-channel implementations. For example, dial tone, ringback, and busy tones must be in-channel signals to get to the user. Furthermore, a user sometimes needs the capability to access certain control elements in the network that are associated with an established connection. The most practical means of accomplishing this is to use in-channel signals. For example, a request for operator assistance on an existing connection typically uses an in-channel hook flash to alert the operator. In addition, data terminals wishing to disable echo suppressors in dial-up connections need to send special signaling tones that get recognized by echo suppressors in the circuit.

1.2.10 Interfaces

The design, implementation, and maintenance of any large and complex system requires a partitioning of the system into manageable subsystems or modules. Associated with each module is an interface that defines the requirements of the various inputs and outputs from an external point of view. A complete interface specification includes mechanical, electrical, and operational details of the inputs and outputs. Ideally, a well defined interface allows interconnection to a device with no knowledge of the device's internal operation. Well established interfaces are a fundamental requirement to maintain compatibility between old and new equipment in the network. The sheer complexity of the telephone network implies the existence of a large number of interfaces. Because of the diversity of applications and environments, special situations sometimes exist that require special considerations on both sides of an otherwise standard interface. For example, extra long customer loops cannot be driven by a standard switch interface because they require the installation of range extenders at the central office.

One of the main sources of complexity and variation in interfaces within the network can be traced to the various signaling procedures in use. Oftentimes, the incompatibility of signaling schemes requires an interface adapter. One commonly used adapter converts push-button (DTMF) dialing into dc dial pulses for step-by-step offices. Push-button to dial-pulse conversion is a common enough requirement that standard modules are available to match the interfaces.

Subscriber Loop Interface

The most common interface in the network, other than station set connections, is the subscriber loop interface at the central office. Owing to features of conventional station sets on one side and electromechanical switches on the other, this interface contains numerous characteristics particularly burdensome to electronic crosspoint switches. The basic functional requirements of this interface in the existing analog network are as follows:

Battery
Application of dc power to the loop (48 V typical) to make possible dc signaling and provide bias current for carbon microphones.

Overvoltage Protection
Equipment and personnel protection from lightning strikes and power line induction or shorts.

Ringing
Application of a 20 Hz signal at a level of 86 V rms for ringer excitation. The signal is typically applied for 2 seconds and then removed for 4 seconds.

Supervision
Detection of off-hook by the flow of current in the line.

Test
Access for outward test of the loop or inward test of the switch.

New switches designed to operate in existing environments with existing maintenance procedures must provide these functions in a conventional manner. The result is that the termination costs of the customer loops can represent more than 50% of the total cost of a digital switch [22]. Semiconductor manufacturers are working diligently to provide an economical implementation of this interface. The problem is inherently difficult for solid state electronics (analog or digital) because of the high voltages and currents involved.

1.2.11 Special Services

The main concern of telephone companies, voice service, is sometimes referred to as POTS: plain old telephone service. In addition to POTS the telephone companies also provide a number of special services such as distribution of radio and TV network programs, telephoto, teletype, facsimile, and data transmission. Some of these services, television in particular, have required special considerations in the design of the network. Television program distribution was a major stimulus for the nationwide microwave radio network in the United States. Most other special services, data in particular, evolved after the applicable facilities were designed specifically for voice. Thus, those services using conventional voice circuits are forced to present "voicelike" signals to the network. The rest of this section discusses the basic considerations involved in using the telephone network for data communications.

Data Transmission

Foremost among the implications of using the public (analog) telephone network for data traffic is the need for a modulator/demodulator (modem) to provide some form of carrier transmission. Although the metallic circuit provided in customer loops is capable of transmitting frequencies down to and including direct current, the rest of the network does not generally pass frequencies below 300 Hz. Transformer couplings, as in two-wire to four-wire hybrids, and FDM separation filters are two of the main sources of low-frequency rejection. The modulation formats most often used in modems are frequency shift keying for the least expensive, lower data rate modems and various forms of phase shift keying, amplitude and phase shift keying, and single-sideband amplitude modulation for higher rate modems.

A second consideration when using the telephone network for data is the bandwidth restriction of approximately 3 KHz imposed by FDM separation filters, loading coils, and band-limiting filters in digital voice terminals (Chapter 3). The main implication of a restricted bandwidth is a limitation on the signaling rate or baud rate, which in turn directly relates to the data rate. A common signaling rate is 2400 symbols per second using carriers between 1700 and 1800 Hz. Symbol rates of 4800 symbols per second are also used in lower-sideband modems with a carrier at 2850 Hz.

As mentioned, user acceptance of voice quality does not require stringent control of the phase response (envelope delay) of the channel. High-speed data transmission, however, requires comparatively tight tolerance on the phase response to prevent intersymbol interference. Thus voice channels used for high-speed data transmission require special treatment. The special treatment is available at additional monthly rates from the telephone companies in the form of C-type conditioning for leased lines. C-type conditioning is available in several different grades that provide various amounts of control of both phase and amplitude distortion. Comparable functions are also available in some modems in the form of equalizers.

Most medium rate (synchronous) modems include fixed equilization circuitry designed to compensate for the phase distortion in a typical connection. For higher data rates, automatically adjustable equalization is needed. An automatic equalizer first undergoes a training sequence in which the characteristics of the transmission channel are determined by measuring the response to known test signals. Then, equalizing circuitry in the receiver of each modem is adjusted to provide compensation for the amplitude and phase distortions in the channel. High-speed transmission over dial-up lines requires automatic equalization in the modems since the channel characteristics change with each connection.

Another form of conditioning, referred to as a D-type conditioning, has been offered recently by the telephone companies. This form of conditioning provides levels of noise and harmonic distortion lower than normally provided in leased lines, even with C-type conditioning. D-type conditioning usually does not involve special treatment of any particular line. Instead, the telephone company tests a number of different circuits until one with suitable quality is found. Sometimes this means avoiding a cable that includes pairs from a noisy switching office. Unlike voice, which is reasonably tolerant of impulse noise, a data circuit is more susceptible to impulse noise than to the normal background (white) noise. As more and more older equipment (e.g. step-by-step switches) is phased out, the quality of network transmission improves.

Even when the best leased lines available use both types of conditioning, some residual amounts of noise and distortion always exist. The remaining imperfections generally preclude any data rate higher than 9600 bps. One common way of obtaining 9600 bps is to signal at 2400 symbols per second (baud rate) and encode 4 bits into one of 16 symbols during each signal interval. Dial-up connections are generally limited to 4800 bps because: they necessarily pass through switching equipment, present different characteristics to equalize each time they are set up, and typically contain more noise than leased lines.

Another important consideration for data transmission over long-distance circuits is the effect of echo suppressors. As mentioned in Section 1.2.7, an echo suppressor blocks the signal in a return path when the corresponding forward path of a four-wire circuit is active. Thus, operative echo suppressors effectively preclude a full-duplex operation. Even in a half-duplex mode of operation the echo suppressors must be given 100 ms of deactivation time to reverse the direction of propagation. For these reasons the common carriers have provided a means of disabling the echo suppressors by using an in-channel control signal from the data terminals. Echo suppressors are disabled by transmitting a pure tone between 2010 and 2240 Hz for 400 ms. The echo suppressors remain disabled as long as there is continuous energy in the channel. Thus, the modems can switch to signaling frequencies and begin full-duplex data transmission after the suppressors are first disabled. If the energy in the channel is removed for 100 ms., the echo suppressors are reactivated. Hence, rapid line turnaround is required for half-duplex operations.

1.3 THE INTRODUCTION OF DIGITS

Voice digitization and transmission first became feasible in the late 1950s when the economic, operational, and reliability features of solid state electronics became available. In 1962 Bell System personnel established the first commercial use of digital transmission when they began operating a T1 carrier system for use as a trunk group in a Chicago area exchange [23]. Since that time a whole family of T-carrier systems (T1, T1C, T1D, T2, T4) has been developed—all of which involve time division multiplexing of digitized voice signals. Following the development of T-carrier systems for interoffice transmission, Western Electric and numerous independent manufacturers developed digital pair-gain systems for long customer loop applications. However, the pair-gain systems have not been used nearly as extensively as T-carrier systems. Projections for T-carrier usage indicate that 50% of the exchange area trunks will be digital by 1985 [24].

The world's first commercially designed digital microwave radio system was established in Japan by Nippon Electric Company (NEC) in 1968 [25]. In the early 1970s digital microwave systems began to appear in the United States for specialized data transmission services. Foremost among these systems was a digital network developed by Data Transmission Company (Datran).* The first digital microwave link in the U.S. public telephone network was supplied by NEC of Japan for

*The Datran facilities were taken over in 1976 by another specialized common carrier: Southern Pacific Communications.

a New York Telephone link between Brooklyn and North Staten Island in 1972 [25]. Digital microwave systems have since been developed and installed by several U.S. manufacturers for use in intermediate length toll and exchange area circuits.

In addition to transmission systems, digital technology has proven to be equally useful for implementing switching functions. The first country to use digital switching in the public telephone network was France in 1970 [26]. The first application of digital switching in the public network of the United States occurred in early 1976 when the Bell System began operating its No. 4ESS [27] in a class 3 toll office in Chicago. Two months after that Continental Telephone Company began operation in Ridgecrest, California of an IMA2 digital toll switch supplied by TRW-Vidar [28]. The first digital end office switch in the United States became operational in July of 1977 in the small town of Richmond Hill, Georgia [29]. This switch was supplied by Stromberg-Carlson for Coastal Utilities, Inc., a small independent telephone company.

1.3.1 Voice Digitization

The basic voice coding algorithm used in T-carrier systems, and most other digital voice equipment in telephone networks around the world, is shown in Figure 1.25. The first step in the digitization process is to periodically sample the waveform. As discussed at length in Chapter 3, all of the information needed to reconstruct the original waveform is contained in the samples—if the samples occur at an 8 kHz rate. The second step in the digitization process involves quantization: identifying which amplitude interval of a group of adjacent intervals a sample value falls into. In essence the quantization process replaces each continuously variable amplitude sample with a discrete value located at the middle of the appropriate quantization interval. Since the quantized samples have discrete levels, they represent a multiple level digital signal.

For transmission purposes the discrete amplitude samples are converted to a binary code word. (For illustrative purposes only, Figure 1.26 shows 4 bit code words.) The binary codes are then transmitted as binary pulses. At the receiving end of a digital transmission line the binary data stream is recovered, and the discrete sample values are reconstructed. Then a low-pass filter is used to "interpolate" between sample values and recreate the original waveform. If no transmission errors have occurred, the output waveform is identical to the input waveform except for a small amount of quantization distortion: the difference between a sample value and its discrete representation. By having a large number of quantization intervals (and hence enough bits in a code word to encode them), the quantization intervals can be small enough to effectively eliminate quantization effects.

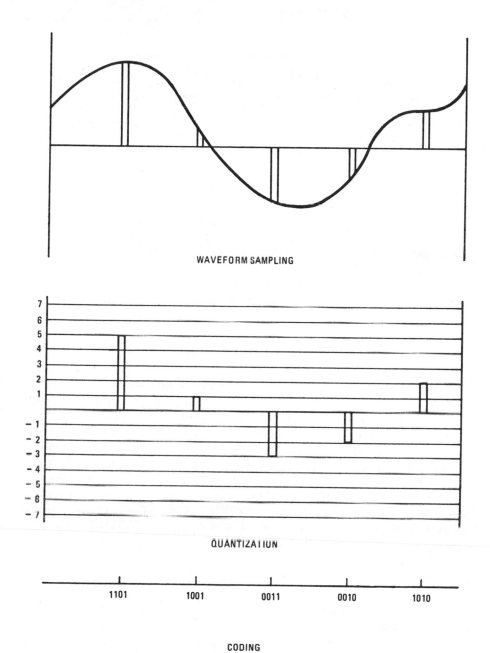

WAVEFORM SAMPLING

QUANTIZATION

1101   1001   0011   0010   1010

CODING

Figure 1.25   Voice digitization process.

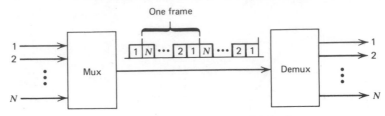

Figure 1.26   Time division multiplexing.

It is worth noting that the bandwidth requirements of the digital signal increase as a result of the binary encoding process. If the discrete, multiple-amplitude samples are transmitted directly, the bandwidth requirements are theoretically identical to the bandwidth of the original signal. When each discrete sample is represented by a number of individual binary pulses, the signal bandwidth increases accordingly. The two-level pulses, however, are much less vulnerable to transmission impairments than are the multiple-amplitude pulses (or the underlying analog signal).

1.3.2   Time Division Multiplexing

Basically, time division multiplexing (TDM) involves nothing more than sharing a transmission medium by establishing a sequence of time slots during which individual sources can transmit signals. Thus the entire bandwidth of the facility is periodically available to each user for a restricted time interval. In contrast, FDM systems assign a restricted bandwidth to each user for all time. Normally, all time slots of a TDM system are of equal length. Also, each subchannel is usually assigned a time slot with a common repetition period called a frame interval. This form of time division multiplexing (as shown in Figure 1.26) is sometimes referred to as synchronous time division multiplexing to specifically imply that each subchannel is assigned a certain amount of transmission capacity determined by the time slot duration and the repetition rate. In contrast, another form of TDM (referred to as "statistical or asynchronous time division multiplexing") is described in Chapter 8. With this second form of multiplexing, subchannel rates are allowed to vary according to the individual needs of the sources. All of the backbone digital links of the public telephone network (T-carrier and digital microwave) use a synchronous variety of TDM.

Time division multiplexing is normally associated only with digital transmission links. Although TDM can be used conceivably for analog signals by interleaving samples from each signal, the individual samples are usually too sensitive to all varieties of transmission impairments. In contrast, time division switching of analog signals is more feasible than analog TDM transmission because noise and distortion of the switching equipment is more controllable. As discussed in Chapter 5, analog TDM techniques are used in some PBXs.

T-Carrier Systems

The volume of interoffice telephone traffic in the United States has been growing more rapidly than local traffic. This rapid growth puts severe strain on interoffice transmission facilities that are designed for lower traffic volumes. Telephone companies are then faced with the necessary task of expanding the number of interoffice circuits. T-carrier systems evolved primarily as a cost effective means for interoffice transmission: both for initial installations and for relief of overloaded interoffice trunk groups.

Despite the need to convert the voice signals to a digital format at one end of a T1 line and back to analog at the other, the combined conversion and multiplexing cost of a digital TDM terminal is lower than the cost of a comparable analog FDM terminal. Thus T1 lines are most economical for intermediate length transmission. For very short distances wire-line transmission with individual pairs for each circuit are most economical. Over very long distances, however, terminal costs are relatively unimportant so that bandwidth-efficient (analog) FDM techniques are justified. The first T-carrier systems were designed specifically for exchange area trunks at distances between 10 and 50 miles.

A T-carrier system consists of terminal equipment at each end of a line and a number of regenerative repeaters at intermediate points in the line. The function of each regenerative repeater is to restore the digital bit stream to its original form before transmission impairments obliterate the identity of the digital pulses. The line itself, including the regenerative repeaters, is referred to as a span line. The terminal equipment is referred to as D-type (digital) channel banks, which come in numerous versions. The transmission lines are wire pairs using 16 to 26 gauge cable. A block diagram of a T-carrier system is shown in Figure 1.27.

The first T1 systems used D1A channel banks for interfacing, converting, and multiplexing 24 analog circuits. A channel bank at each end of a span line provides interfacing for both directions of transmission. Incoming analog signals are time division multiplexed and digi-

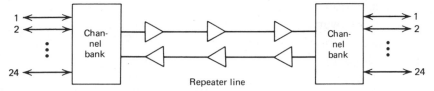

Figure 1.27   T1-Carrier system.

tized for transmission. When received at the other end of the line, the incoming bit stream is decoded into analog samples, demultiplexed, and filtered to reconstruct the original signals. Each individual TDM channel is assigned 8 bits per time slot. Thus, there are $(24)(8) = 192$ bits of information in a frame. One additional bit is added to each frame to help identify the frame boundaries, thereby producing a total of 193 bits in a frame. Since the frame interval is 125 μs, the basic T1 line rate becomes 1.544 Mbps. This line rate has been established as the fundamental standard for digital transmission in North America and Japan. The standard is referred to as a DS-1 signal (for digital signal –1).

A similar standard of 2.048 Mbps has been established by CCITT for most of the rest of the world. This standard evolved from a T1-like system that provides 32 channels at the same rate as the North American channels. Only 30 of the channels in the CCITT standard, however, are used for voice. The other two are used for frame synchronization and signaling. Signaling and control information for T1 systems are inserted into each voice channel (or transmitted separately by CCIS facilities). Digital signaling and control techniques for both systems are discussed in Chapter 7.

As indicated, all popular wire sizes can be used for T1 transmission systems. In general, however, not all wire pairs within a cable can be used for T1 transmission because of crosstalk considerations. Near-end crosstalk is usually the limiting consideration since transmitted signal powers are typically 30 dB above received signal powers. When both directions of transmission are included in one cable pair, care in selecting the pairs is needed to ensure that go and return signal paths are not next to each other. If the two directions of transmission are isolated from each other by using separate cables or by using cables with internal shielding, all pairs in a cable can be used for T-carrier transmission.

The greatly increased attenuation of a wire pair at the frequencies of a DS-1 signal (772 kHz center frequency) mandate the use of amplification at intermediate points of T1 span line. In contrast to an analog signal, however, a digital signal can not only be amplified but it can also be detected and regenerated. That is, as long as a pulse can be detected it can be restored to its original form and relayed to the next line segment. For this reason T1 repeaters are referred to as regenerative repeaters. The basic functions of these repeaters are:

1   Equalization.
2   Clock recovery.
3   Pulse detection.
4   Transmission.

Equalization is required because the wire pairs introduce certain amounts of both phase and amplitude distortion that cause intersymbol interference if uncompensated. Clock recovery is required for two basic purposes: first, to establish a timing signal to sample the incoming pulses; second, to transmit outgoing pulses at the same rate as at the input to the line.

Regenerative repeaters are normally spaced every 6000 ft in a T1 span line. This distance was chosen as a matter of convenience for converting existing voice frequency cables to T-carrier lines. Interoffice voice frequency cables typically use loading coils that are spaced at 6000 ft intervals. Since these coils are located at convenient access points (manholes) and must be removed for high-frequency transmission, it was only natural that the 6000 ft interval be chosen. One general exception is that the first regenerative repeater is typically spaced 3000 ft from a central office. The shorter spacing of this line segment is needed to maintain a relatively strong signal in the presence of impulse noise generated by some switching machines.

The operating experience of T1 systems has been so favorable that they have been continually upgraded and expanded. One of the initial improvements produced T1C systems which provide higher transmission rates over 22 gauge cable. A T1C line operates at 3.152 Mbps for 48 voice channels: twice as many as a T1 system.

Another level of digital transmission became available in 1972 when the T2 system was introduced. This system is primarily designed for switched toll network connections. In contrast, T1 systems were originally designed only for exchange area transmission. The T2 system provides for 96 voice channels at distances up to 500 miles. The line rate is 6.312 Mbps which is referred to as a DS-2 standard. The transmission media is special low capacitance 22 gauge cable. By using separate cables for each direction of transmission and the specially developed cables, T2 systems can use repeater spacings up to 14,800 ft in low-noise environments. The freedom to select repeater spacings also reflects the application to toll transmission where conversion of existing wire line facilities was not a consideration.

TDM Hierarchy

In a manner analogous to the FDM hierarchy, the Bell System has established a digital TDM hierarchy that has become the standard for North America. Starting with a DS-1 signal as a fundamental building block, all other levels are implemented as a combination of some number of lower-level signals. The designation of the higher-level digital multiplexers reflects the respective input and output levels. For example, an M12 multiplexer combines four DS-1 signals to form a single DS-2 signal. Table 1.9 lists the various multiplex levels, their bit rates, and the transmission media used for each. Notice that the bit rate of a high-level multiplex signal is slightly higher than the combined rates of the lower-level inputs. The excess rate is included for certain control

TABLE 1.9 DIGITAL TDM SIGNALS OF NORTH AMERICA AND JAPAN

Digital Signal Number	Number of Voice Circuits	Multiplexer Designation	Bit Rate (Mbps)	Transmission Media
DS-1	24	D channel bank (24 analog inputs)	1.544	T1 paired cable 1A radio (DUV)
DS-1C	48	M1C (2 DS-1 inputs)	3.152	T1C paired cable
DS-2	96	M12 (4 DS-1 inputs)	6.312	T2 paired cable
DS-3	672	M13 (28 DS-1 inputs)	44.736	3A-RDS 11-GHz radio
DS-4	4032	M34 (6 DS-3 inputs)	274.176	T4M coax WT4 waveguide DR18 18 GHz radio

and synchronization functions discussed in Chapter 7. A similar digital hierarchy has also been established by CCITT as an international standard. As shown in Table 1.10, this hierarchy is similar to the North American standard but involves different numbers of voice circuits at all levels.

Digital Pair-Gain Systems

Following the successful introduction of T1 systems for interoffice trunks, most major manufacturers of telephone equipment developed digital TDM systems for local distribution. These systems are most applicable to long rural loops where the cost of the electronics is offset by the savings in wire pairs. In general, the prove-in distance for carrier systems is primarily dependent on the wire savings [30]. At times, however, extenuating circumstances can virtually dictate that a pair-gain system be used. For example, a castastrophic loss of cable facilities or

TABLE 1.10 DIGITAL TDM HIERARCHY OF CCITT

Level Number	Number of Voice Circuits	Multiplexer Designation	Bit Rate (Mbps)
1	30		2.048
2	120	M12	8.448
3	480	M23	34.368
4	1920	M34	139.264
5	7680	M45	565.148

unexpected growth can be most economically accommodated by adding electronics, instead of wire, to produce a pair-gain system. The possibility of trailer parks and apartment houses springing up almost overnight causes nightmares in the minds of cable plant forecasters. Pair-gain systems, however, provide networking alternative to dispel those nightmares.

Digital pair-gain systems are also useful as alternatives to switching offices in small communities. Small communities are often serviced by small automatic switching systems normally unattended and remotely controlled from a larger switching office nearby. These small community switches are referred to as community dial offices (CDOs). A CDO typically provides only limited service features to the customers and often requires considerable maintenance. Because digital pair-gain systems lower transmission costs for moderate-sized groups of subscribers, they are a viable alternative to a CDO: stations in the small community are serviced from the central office by way of pair-gain systems. A fundamental consideration in choosing between pair-gain systems and remote switching involves the traffic volumes and calling patterns within the small community. The basic techniques of analyzing traffic patterns and determining trunk group sizes are provided in Chapter 9.

The first two digital pair-gain systems used in the Bell System are the subscriber loop multiplex (SLM) system [31, 32] and, its successor: the subscriber loop carrier (SLC-40) system [31, 33]. Although these systems use a form of voice digitization (delta modulation) different from that used in T-carrier systems (pulse code modulation), they both use standard T1 repeaters for digital transmission at 1.544 Mbps. Both systems also convert the digitized voice signals back into individual analog interfaces at the end-office switch to achieve system transparency. Figures 1.28 and 1.29 show block diagrams of these systems. Notice that the SLM system provides both concentration and multiplexing (80 subscribers for 24 channels) while the SLC-40 is strictly a multiplexer (40 subscribers assigned in a one-to-one manner to 40 channels). The switching function at the end office of the SLM is provided by a 24 X 80 crossbar switch. Both systems are designed to operate with two pairs of 22 gauge wire. System lengths up to 50 miles are possible with

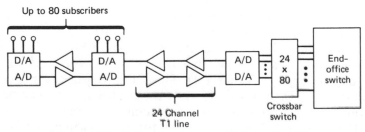

Figure 1.28 Subscriber loop multiplexer (SLM).

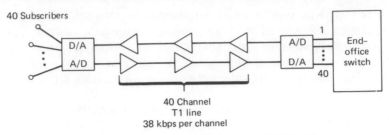

Figure 1.29 Subscriber loop carrier (SLC-40).

remote power. Notice also that the SLM system allows access to the digital line at more than one point. The SLC-40 system, on the other hand, requires all subscriber line interfaces at a common point.

A more recent pair-gain system to be used in the Bell System is the SLC-96 which began serving customers in 1979 [34]. Unlike the SLM and SLC-40 systems, the SLC-96 uses the same voice digitization technique (pulse code modulation) used by T-carrier systems. Thus the SLC-96 is compatible with other digital transmission and switching equipment being incorporated into the network. Not only is this system applicable to rural environments, but it is also economical for roughly 10% of new suburban routes [34]. When interfaced to anticipated digital end-office switches, the number of applicable suburban routes increases to 35%.

1.3.3   Data Under Voice

After the technology of T-carrier systems had been established, AT&T began offering leased digital transmission services for data communications. This service, known as Dataphone Digital Service (DDS) [35], uses T1 transmission links with special terminals (channel banks) that provide direct access to the digital line. A major drawback of DDS service arises because T-carrier systems are used only for exchange area and short toll network trunks. Without some form of long-distance digital transmission, the digital circuits in widely separated exchange areas can not be interconnected. AT&T's response to long-distance digital transmission needs was the development of a special radio terminal called the 1A Radio Digital Terminal (1A-RDT) [36, 37]. This terminal is specifically designed to encode one DS-1 signal (1.544 Mbps) into less than 500 kHz of bandwidth. As shown in Figure 1.30, a signal of this bandwidth can be inserted below the lowest frequency of a mastergroup multiplex (Table 1.4). Since this frequency band is normally unused in TD or TH radio systems, the DS-1 signal can be added to existing analog routes without displacing any voice channels. The use of frequencies below those used for voice signals leads to the designation: "data under voice" (DUV).

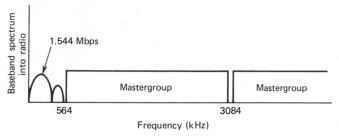

Figure 1.30   Data under voice.

It is important to point out that data under voice represents a special development specifically intended for data transmission and not for voice services. In fact, DUV is used only to provide long-distance digital transmission facilities for Digital Dataphone Service. Although the DDS network uses common technology (such as T1 lines) and common facilities (such as TD radios) of the voice network, the system is functionally separate from the voice network. That is, the voice network does not use DDS circuits, or does the DDS network use digital circuits within the voice network.

Data under voice technology is an example of how a common carrier can augment existing facilities to provide cost effectively a new service—sometimes to the consternation of competing companies. The Specialized Common Carriers (SCCs) and Value Added Networks (VANs) in the United States that compete with the common carriers and specialize in data transmission feel that AT&T should not be allowed to charge for services such as DDS based only on the cost of the facilities added to provide that service. The SCCs and VANs feel that this practice, called incremental pricing, is anticompetitive since they must develop stand-alone facilities for their offerings. The common carriers, on the other hand, counter that incremental pricing is justified since the public benefits from the common use of the facilities (the basic rationale for regulated monopolies). Thus data under voice represents just one technical development involving important regulatory considerations.

1.3.4   Digital Microwave Radio

In contrast to data under voice systems, which piggyback 1.544 Mbps onto an analog radio for data services, a common carrier digital microwave system uses digital modulation exclusively to transmit and receive higher-level digital multiplex signals for voice traffic. Digital radios use the same frequency bands allocated for analog radios as listed in Table 1.6. Thus a primary design requirement of a digital radio is that it must confine its radiated power to prevent excessive interference into adjacent, possibly analog, channels. Moreover, the FCC has stipulated that

TABLE 1.11   MINIMUM VOICE CIRCUIT REQUIREMENTS OF DIGITAL
RADIOS IN THE UNITED STATES

Frequency Band (MHz)	Minimum Number of Circuits	Equivalent Number of DS – 1 Signals	Resultant Bit Rate (Mbps)[a]	Channel BW (MHz)
2,110–2,130	96	4	6.144	3.5
2,160–2,180	96	4	6.144	3.5
3,700–4,200	1152	48	73.7	20
5,925–6,425	1152	48	73.7	30
10,700–11,700	1152	48	73.7	40

[a]The actual bit rate is usually slightly greater owing to extra framing and synchronization bits. Furthermore, most radio systems provide more voice circuits than the minimum.

the digital radios must be capable of providing roughly the same number of voice circuits as existing analog radios. Table 1.11 lists the minimum number of voice circuits and the resulting bit rate that must be provided by digital radios in each of the common carrier microwave bands [38].

Despite the design constraints imposed by compatibility requirements with analog radios, digital radios have proven to be more economical than analog transmission in several applications. Because of lower terminal costs (multiplexing and demultiplexing), digital radio systems are currently less expensive than analog radio systems for distances up to about 300 miles, and on longer routes requiring channel access (drop-and-insert) at intermediate points in the route [39]. Depending on the cirumstances and the total number of voice circuits required on a particular route, digital radio systems can even be more economical than T-carrier systems for distances as short as 8 to 10 miles [39]. The major impetus for digital radio in the United States has been the emergence of digital toll switches like the No. 4ESS. The interconnection of digital radio signals to a digital switch avoids costly lower level multiplex equipment.

1.3.5   Digital Switching

The original research into digital switching at Bell Laboratories was reported by Earle Vaughan in 1959 [40]. Laboratory models were developed to demonstrate the concept of integrating digital time division multiplex transmission systems with time division switching systems. Unfortunately the necessary solid state electronics had not matured sufficiently at that time so commercial development of digital switching was not pursued, and development of the No. 1ESS continued along the lines of space division electromechanical technology.

Almost 10 years later, however, Bell Labs began development of a digital toll switch, the No. 4ESS.

When placed in service in January of 1976, the No. 4ESS provided several new capabilities for the toll network. First, it was the first toll switch to be designed for stored program control at the outset.* Second, its capacity was three times that of the prevailing electromechanical switch at the time: the No. 4A cross bar. The larger capacity of the No. 4ESS meant that many metropolitan areas could consolidate toll traffic into one switch instead of several. Third, the digital time division design of the No. 4ESS allowed direct connection to digital T-carrier lines. This last feature illustrates the particular attraction of digital switching for toll and tandem offices of the Bell network. By 1976, when the first No. 4ESS was installed, it was clear that digital transmission was dominating the exchange area and shorter intertoll trunks. Thus significant economies and improvements in transmission quality resulted by eliminating channel banks at the interface between digital trunks and a switching system.

The early development of digital end-office switches in the United States was undertaken by independent equipment manufacturers with the first system being placed in service in 1977. These systems are primarily designed for the smaller switching offices of the independent telephone companies. Digital switches are particularly attractive to rural telephone companies because they can provide significant copper savings when directly connected to digital pair gain transmission systems. The first large switching systems to be introduced into the North American network were provided by Northern Telecom in the U.S. and Canada. Development of other large systems for metropolitan applications is underway in the form of GTE's No. 5EAX and AT&T's No. 5ESS. These systems are slated for introduction in 1981. Table 1.12 lists switching machines currently available for the North American network. Comparable equipment is also available worldwide from Japanese and European manufacturers.

The switching systems listed in Table 1.12 pertain only to equipment used in the public network. A vast array of digital switching equipment is also available for private exchanges. The first commercially available digital switch to be used in the United States was an Automatic Call Distributor† (ACD) supplied by Collins/Rockwell in 1974. Since that time numerous companies, including several newcomers to the tele-

---

*Stored program control was first implemented in the toll network beginning in 1969 by retrofitting crossbar switches [41].

†An automatic call distributor is a switching machine designed to connect incoming requests to any one of a number of available attendants. If all attendants are busy, calls are put on hold and automatically switched to attendants as they become available. ACDs are commonly used for operator assistance, airline reservations, service organizations, or any other application where requests can be serviced by any one of a number of attendants.

**TABLE 1.12 DIGITAL CENTRAL OFFICE SWITCHING SYSTEMS
OF NORTH AMERICA**

Manufacturer	Designation	Date of Introduction	Application	Maximum Terminations
AT&T	4ESS	1976	Toll	107,000
AT&T	5ESS	1982[a]	Local	100,000
GTE	3EAX	1978	Toll/tandem	60,000
GTE	5EAX	1981[a]	Local	150,000
NEC America	NEAX-61	1979	Local/toll	32,000
North Electric (ITT)	DSS-1	1978	Local	6,000
North Electric (ITT)	DSS-2		Local/toll	64,000
Northern Telecom	DMS-10	1977	Local	6,000
Northern Telecom	DMS-100	1979	Local	100,000
Northern Telecom	DMS-200	1978	Toll	100,000
Stromberg Carlson	DCO	1977	Local	16,000
Stromberg Carlson	DTM		Tandem	17,500
Vidar	IMA2[b]	1976	Toll	1,536
Vidar	ITS4	1977	Toll/tandem	4,800
Vidar	ITS4/5	1978	Toll/local	12,768

[a]Projected dates of availability.
[b]Has been superceded by the ITS4.

phone equipment business, have introduced digital PBX equipment. In fact, the majority of PBX equipment now being offered is digital [42]. A notable exception, however, is the DIMENSION PBX of the Bell System which uses analog time division switching. (Western Electric is currently working on a digital PBX code named the ANTELOPE System).

The functional essence of a digital time division switching matrix is illustrated in Figure 1.31. As indicated, all inputs are time division multiplex links. These links may represent digital pair-gain systems, T-carrier interoffice trunks, or the outputs of colocated channel banks used to interface analog lines to the digital switch. In any event the switching matrix itself is designed to service only TDM input links.

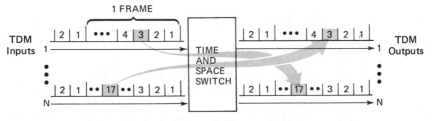

Figure 1.31 Digital time division switching operaton.

Basically, the switching matrix is required to transfer information arriving in a specified time slot (channel) on an incoming TDM link to a specified time slot on an outgoing TDM link. Since an arbitrary connection involves two different physical links and two different time slots, the switching process requires spatial translation (space switching) and time translation (time switching). Thus the basic operation is sometimes referred to as two-dimensional switching. Space switching is achieved with conventional digital logic selector (or multiplexer) circuits, and time switching is achieved by temporarily storing information in a digital memory or register circuit.

For convenience the inputs to the switching matrix of Figure 1.31 are all shown on the left, and the outputs are on the right. Since the links are inherently four-wire, each input time slot is paired with the same time slot on a corresponding output link. Hence, a full-duplex circuit also requires a "connection" in the reverse direction which is achieved by transferring information from time slot 17 of link N to time slot 4 of link 1 in the example shown. Operational and implementation details of a large range of digital time division switches are provided in Chapter 5.

REFERENCES

1 R. Smith and W. Sonneville, "Record '81 Telco Construction Set at $23.5 Billion, Up 5.6%," *Telephone Engineer and Management*, January 15, 1981, p 55.

2 N. R. Kleinfield, AT&T reprint from *New York Times*.

3 Technical Staff, Bell Laboratories, *Engineering and Operations in the Bell System*, Bell Telephone Laboratories Inc., 1977.

4 *Reference Data for Radio Engineers*, 5th ed., Howard Sams & Co., ITT, Indianapolis, 1972.

5 L. Jachimowicz, J. Olszewski, and I. Kolodny, "Transmission Properties of Filled Thermoplastic Insulated and Jacketed Telephone Cables at Voice and Carrier Frequencies," *IEEE Transactions on Communications*, March 1973, pp 203-209.

6 D. G. Messerschmitt, "An Electronic Hybrid with Adaptive Balancing for Telephony," *IEEE Transactions on Communications*, August 1980, pp 1399-1407.

7 George M. Dick, "SBS's Celestial Version of 'The System is the Solution'," *Datamation*, November 1977, pp 141-158.

8 W. M. Combs, C. P. Fort, and R. A. Zeitz, "Ending the Information Float," *Proceedings of Intelcom '79*, Horizon House International, 1979, pp 42-45.

9 A. J. Aikens and D. A. Lewinski, "Evaluation of Message Circuit Noise," *Bell System Technical Journal*, July 1970, pp 879-909.

10 J. L. Sullivan, "A Laboratory System for Measuring Loudness Loss of Tele-

phone Connections," *Bell System Technical Journal*, October 1971, pp 2663–2739.

11 W. K. MacAdam, "A Basis for Transmission Performance Objective in a Telephone Communications System," *AIEE Transactions on Communications and Electronics*, May 1978, pp 205–209.

12 F. P. Duffy and T. W. Thatcher, Jr., "Analog Transmission Performance on the Switched Telecommunications Network," *Bell System Technical Journal*, April 1971, pp 1311–1347.

13 W. R. Bennett and J. R. Davey, *Data Transmission*, McGraw-Hill, New York, 1965.

14 Members of Technical Staff, Bell Telephone Laboratories, *Transmission Systems for Communications*, 4th ed., Western Electric Company, 1971.

15 H. R. Huntly, "Transmission Design of Intertoll Telephone Trunks," *Bell System Technical Journal*, September 1953, pp 1019–1036.

16 D. L. Duttweiler and Y. S. Chen, "A Single-Chip VLSI Echo Cancellor," *Bell Systems Technical Journal*, February 1980, pp 149–160.

17 H. G. Suyderhoud, M. Onufry, and S. J. Campanella, "Echo Control in Telephone Communications," *National Telecommunications Conference Record*, 1976, pp 8.1-1–8.1-5.

18 F. T. Andrews, Jr. and R. W. Hatch, "National Telephone Network Transmission Planning in the American Telephone and Telegraph Company," *IEEE Transactions on Communications Technology*, June 1971, pp 302–314.

19 A. E. Ritchie and J. Z. Menard, "Common Channel Interoffice Signaling, An Overview," *Bell System Technical Journal*, February 1978, pp 221–236.

20 A. E. Joel, Jr., "Circuit Switching: Unique Architecture and Applications," *COMPUTER*, June 1979, pp 10–22.

21 J. J. Bernard, "CCITT International Signaling No. 6," *Telecommunications Journal*, Vol. 41, No. 2, 1974, p 69.

22 John C. McDonald, "Techniques for Digital Switching," *Communication Society Magazine of IEEE*, July 1978, pp 11–19.

23 K. E. Fultz and D. B. Penick, "The T1 Carrier System," *Bell System Technical Journal*, September 1965, pp 1405–1451.

24 M. R. Aaron, "Digital Communications—The Silent (R)evolution?" *IEEE Communications Magazine*, January 1979, pp 16–26.

25 "Supply Record on NEC PCM Microwave Equipment," Nippon Electric Company, Tokyo, May 1974.

26 J. B. Jacob, "The New Approaches and Solutions Offered by the Digital ESS.E 10 to Deal with Switching, Operation, and Maintenance Functions," *International Communications Conference Proceedings*, 1979, pp 46.6-1–46.6-5.

27 R. Bruce, P. Giloth, and E. Siegel Jr., "No. 4ESS—Evolution of a Digital Switching System," *IEEE Transactions on Communications*, July 1979, pp 1001–1011.

28 J. C. McDonald and J. R. Baichtal, "A New Integrated Digital Switching System," *National Telecommunications Conference Record*, 1976, p 3.2-1–3.2-5.

29 "First Class 5 Digital Central Office goes into Service in Georgia," *Telephony*, August 1, 1977, pp 24–25.

30 J. A. Stewart, "System Considerations for Short Haul Multi-Channel Subscriber Carrier," *International Conference on Communications Record*, 1974, p 24E-1.

31 F. T. Andrews, Jr., "Customer Lines Go Electronic," *Bell Labs Record*, February 1972, pp 59–64.

32 M. T. Manfred, G. A. Nelson, and C. H. Sharpless, "Digital Loop Carrier Systems," *Bell System Technical Journal*, April 1978, pp 1129–1142.

33 S. J. Brolin, G. E. Harrington, and G. R. Leopold, "The Subscriber Loop Carrier System Gets Even Better," *Bell Labs Record*, May 1977, pp 119–122.

34 S. Brolin, Y. S. Cho, W. P. Michaud, and D. H. Williamson, "Inside the New Digital Subscriber Loop System," *Bell Labs Record*, April 1980, pp 110–116.

35 L. A. Sibley, "Digital Data System—Status and Plans," *International Communications Conference Proceedings*, 1979, pp 54.1.1–3.

36 R. C. Prime and L. L. Sheets, "The 1A Radio Digital System Makes 'Data Under Voice' a Reality," *Bell Labs Record*, December 1973, pp 335–339.

37 R. R. Grady and J. W. Knapp, "1A Radio Digital Terminals Put 'Data Under Voice'," *Bell Labs Record*, May 1974, pp 161–166.

38 FCC *Memorandum Opinion and Order* released January 29, 1975 modifying FCC *Report and Order* of Docket 19311 released September 27, 1974.

39 W. S. Chaskin, "Digital Transmission Economics," *Digital Microwave Transmission Engineering Symposium*, Rockwell International, September 1978.

40 H. E. Vaughan, "Research Model for Time-Separation Integrated Communication," *Bell System Technical Journal*, July 1959, pp 909–932.

41 B. T. Fought and C. J. Funk, "Electronic Translator System for Toll Switching," *IEEE Transactions on Communications*, June 1970, pp 168.

42 C. H. Devine and B. L. Contryman, "Digital Technology Takes Over in Modern PABX Design," *Telephone Engineering and Management*, March 15, 1978, pp 80–83.

CHAPTER TWO

WHY DIGITAL?

The first chapter provides an overview of the existing analog telephone network and a brief introduction to the use of digital transmission and switching. This chapter discusses the basic technical advantages of digital implementations, the disadvantages, and some of the economic considerations involved in the transition to digital networks.

2.1 ADVANTAGES OF DIGITAL VOICE NETWORKS

A list of technical features of digital communications networks is provided in Table 2.1. These features are listed in the order that the author considers to be their relative importance for general telephony. In particular applications, however, certain considerations may be more or less significant. For instance, the last item, ease of encryption, is a dominant feature favoring digital networks for the military.

Most of the features of digital voice networks listed in Table 2.1 and discussed in the following paragraphs pertains to advantages of digital transmission or switching relative to analog counterparts. In some instances, however, the features pertain only to all-digital networks. Encryption, for example, is practical and generally useful only if the secure form of the message is established at the source and translated back into the clear only at the destination. Thus an end-to-end digital system that operates with no knowledge of the nature of the traffic (i.e. provides transparent transmission) is a requirement for digital encryption applications. For similar reasons end-to-end digital transmission is needed for data related applications. When a network consists of a mixture of analog and digital equipment, universal use of the network for services such as data transmission dictates conformance to the least common denominator of the network: the analog channel.

2.1.1 Ease of Multiplexing

As mentioned in Chapter 1, digital techniques were first applied to general telephony in interoffice T-carrier (time division multiplex) systems. In essence these systems trade electronics costs at the ends of a trans-

TABLE 2.1 TECHNICAL ADVANTAGES OF
DIGITAL COMMUNICATIONS NETWORKS

1 Ease of multiplexing

2 Ease of signaling

3 Use of modern technology

4 Integration of transmission and switching

5 Operability at low signal-to-noise/interference ratios

6 Signal regeneration

7 Accommodation of other services

8 Performance monitorability

9 Ease of encryption

mission path for the cost of multiple pairs of wires between them. (A trade that is more cost effective every year.) Although frequency division multiplexing of analog signals is also used to reduce cable costs, FDM equipment is typically more expensive than TDM equipment— even when the cost of digitization is included. After voice signals have been the digitized, TDM equipment costs are quite small by comparison. Since digitization occurs only at the first level of the TDM hierarchy, high-level digital TDM is even more economical than high-level FDM counterparts.

It should be pointed out that time division multiplexing of analog signals is also very simple and does not require digitization of the sample values. The drawback of analog TDM lies in the vulnerability of narrow analog pulses to noise, distortion, crosstalk, and intersymbol interference. These degradations can not be removed by regeneration as in a digital system. Hence, analog TDM is not feasible except for noiseless, distortion-free environments.* In essence the ability to regenerate a signal, even at the expense of greater bandwidth, is almost a requirement for TDM transmission.

2.1.2 Ease of Signaling

Control information (on hook/off hook, address digits, coin deposits, etc.) is inherently digital and, hence, readily incorporated into a digital transmission system. One means of incorporating control information into a digital transmission link involves time division multiplexing the

---

*Some analog TDM transmission systems have been developed and are in use for telephone applications. Farinon's Subscriber Ratio System, for example, uses time division multiplexed pulse width modulation [1]. As discussed in Chapter 5, analog TDM is also used in some switching systems.

control as a separate but easily identifiable control channel. Another approach involves inserting special control codes into the message channel and having digital logic in the receiving terminals decode that control information. In either case, as far as the transmission system is concerned, control information is indistinguishable from message traffic.

In contrast, analog transmission systems require special attention for control signaling. Many of the various analog transmission systems present unique and sometimes difficult environments for inserting control information. An unfortunate result is that many varieties of control signal formats and procedures have evolved. The control formats depend on the nature of both the transmission system and its terminal equipment. In some interfaces between network subsystems control information must be converted from one format to another. Signaling has traditionally represented a significant administrative and financial burden to the operating telephone companies.

The Bell system has developed a comprehensive signaling plan called Common Channel Interoffice Signaling (CCIS) [2] which overcomes many of the signaling problems in the present network. Although CCIS can be implemented with modems on analog circuits, the utmost benefits will be derived when higher-rate digital channels are available. CCIS and other signaling procedures are discussed more fully in Chapter 7.

In summary, digital systems allow control information to be inserted into and extracted from a message stream independently of the nature of transmission medium (cable, coax, microwave, satellite, etc.). Thus the signaling equipment can (and should) be designed separately from the transmission system. It then follows that control functions and formats can be modified independently of the transmission subsystem. Conversely, digital transmission systems can be upgraded without impacting control modules at either end of the link.

2.1.3   Use of Modern Technology

A multiplexer or switching matrix for time division digital signals is implemented with the same basic circuits used to build digital computers: logic gates and memory. The basic crosspoint of a digital switch is nothing more than an "AND" gate with one logic input assigned to the message signal and the other inputs used for control (crosspoint selection). Thus the dramatic developments of digital integrated circuit technology for computer logic circuits and memory are applicable directly to digital transmission and switching systems. In fact, many standard circuits developed for use in computers are usable directly in a switching matrix. The utmost advantages of modern technology are now becoming even more apparent as large scale integrated (LSI) circuits are developed specifically for telecommunications functions. With

LSI applied to all main functions in a network, the size, cost, and reliability improvement is dramatic [3, 4].

Even now, large digital switches implemented with small scale integration provide a cost effective replacement for large analog switches [5]. These cost comparisons include the penalty of converting from analog to digital and back to analog when the digital switch operates in an analog transmission environment. Obviously, significant savings in cost and complexity are obtained when digital transmission links are directly connected to a digital switch. The digital-to-analog conversion costs associated with the transmission link and the analog-to-digital conversion costs associated with the switch are eliminated.

The relative low cost and high performance of digital circuits allows digital implementations to be used in some applications that are prohibitively expensive when implemented with existing analog components. Completely nonblocking switches, for example, are not practical with conventional analog implementations, except in small sizes. In a modern digital switch the cost of the switching matrix itself can be relatively insignificant. Thus, for medium-size applications, the size of the switch matrix can be increased to provide nonblocking operations, if desired. The automatic call distributer developed by Collins-Rockwell [6] is an example of a digital switch operating in an analog environment. A digital implementation was chosen largely because it could economically provide a nonblocking operation.

The benefits of modern device technology are not confined to digital circuits alone. Analog integrated circuits have also progressed significanty—allowing traditional analog implementations to improve appreciably. One of the primary requirements of an analog component, however, is that it be linear. It appears, if only because of research and development emphasis, that fast digital components are easier to manufacture than linear analog counterparts. In addition, digital implementations appear to have an inherent functional advantage over analog implementations. This advantage is derived from the relative ease with which digital signals can be multiplexed. A major limitation with the full use of LSI components results from limited availability of external connections to the device. With time division multiplex techniques, a single physical pin can be used for multiple channel access into the device. Thus the same technique already used to reduce costs in short-haul transmission systems can also be used within a local module to minimize the interconnections and maximize the utilization of very large scale integration. In the end, a "switch on a chip" can only be possible if a great number of channels can be multiplexed onto a relatively small number of external connections.

Two promising transmission technologies being developed today are: satellites and fiber optics. Although most existing satellite systems use analog frequency division multiplexing, the advantages of a digital time

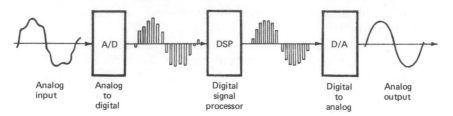

Figure 2.1   Digital signal processing of an analog signal.

division multiple access (TDMA) mode of operation indicate that future satellites will be digital [7]. Thus, even though land-based long-distance transmission may resist digitization for some time, the trend for space-borne counterparts is digital.

The present most exciting development in transmission systems in-volves the use of fiber optics for wide bandwidth local and interoffice transmission. The interface electronics to optical fibers (lasers, light emitting diodes, photodiodes) function primarily in an on-off (nonlinear) mode of operation. Although pulse width or amplitude modulation can be used by these devices to provide analog transmission, the develop-ment emphasis of fiber optics is clearly digital. Thus, in the case of fiber optics, the transmission link itself implies a digital mode of operation.

Digital Signal Processing

The preceding paragraphs emphasize the advantages of digital technol-ogy in implementing the transmission and switching systems of a net-work. Another promising application of digital technology is the area of signal processing. Basically, signal processing refers to an operation on a signal to enhance or transform its characteristics. Digital signal processing can be applied to either analog or digital waveforms. Ampli-fication, equalization, modulation, and filtering are common examples of signal processing functions.

Digital signal processing (DSP) refers to the use of digital logic and arithmetic circuits to implement signal processing functions on digitized signal waveforms. Sometimes analog signals are converted to digital representations for the express purpose of processing them digitally. Then the digital representations of the processed signals are converted back to analog. These operations are illustrated in Figure 2.1. Three instances of digital processing being used for functions formerly imple-mented with analog circuitry are DTMF tone detection, FDM demulti-plexing in transmultiplexers (discussed in Chapter 5), and data detection in high rate (14.4 kbps) voiceband modems. The main advantages of digitally processing signal waveforms are listed in Table 2.2.

Some of the signal processing operations of a telephone network that

TABLE 2.2 DIGITAL SIGNAL PROCESSING FEATURES

REPRODUCIBILITY

The immunity of digital circuits to small imperfections and parasitic elements implies that circuits can be produced with consistent operational characteristics without fine adjustments or aging tolerances.

PROGRAMMABILITY

A single basic structure can be used for a variety of signal types and applications by merely changing an algorithmic or parametric specification in a digital memory.

TIME SHARING

A single digital signal processing circuit can be used for multiple signals by storing temporary results of each process in random access memory and processing each signal in a cyclic (time divided) fashion.

AUTOMATIC TEST

Since the inputs and outputs of a digital signal processing circuit are digital data, tests can be performed routinely by comparing test responses to data patterns stored in memory.

VERSATILITY

Because of the decision making capabilities of digital logic, digital signal processing can perform many functions that are impossible or impractical with analog implementations.

---

can be implemented advantageously with digital signal processing are: tone generation/detection, echo control,* amplification/attenuation, equalization, filtering, companding,† and conversion between various digital voice encoding formats.

It is important to point out that digital signal processing in this context refers to the technology used to condition, manipulate, or otherwise transform a signal waveform (a digitized representation thereof). In another context signal processing refers to the interpretation of control signals in a network by the control processors of switching systems. In the latter case the logical interpretation of a control code is processed and not the underlying signal waveforms.

*Digital echo suppressors implemented in conjunction with Bell's No. 4 ESS toll switch are very economical in comparison to analog echo suppressors [8].

†Companding is a processing function used by digital voice coders discussed in Chapter 3.

2.1.4   Integration of Transmission and Switching

Traditionally the transmission and switching systems of telephone networks have been designed and administered by functionally independent organizations. In the operating telephone companies, these two
equipment classes are referred to as: outside plant and inside plant,
respectively. These equipments necessarily provide standardized interfaces, but, other than that, transmission equipment is functionally independent of switching equipment.

When time division multiplexing of digital voice signals was introduced into the exchange area and communications engineers began
considering digital switching, it become apparent that time division
multiplexing operations were very similar to time division switching
functions. In fact, as described in later chapters, the first stages of
digital switches generate first level TDM signals by nature—even when
interfaced to analog transmission links. Thus the multiplexing operations
of a transmission system can be easily integrated into the switching
equipment.

The basic advantage of integrating the two systems is shown in
Figure 2.2. The demultiplexing equipment (channel banks) at the
switching offices is unnecessary, and first stage switching equipment
is eliminated. If both ends of the digital TDM trunks are integrated into
a digital switch, the channel banks at both ends of the trunk are eliminated. In a totally integrated network voice signals are digitized at
or near the source and remain digitized until delivered to their destination. Furthermore, all interoffice trunks and internal links of a switching system carry TDM signals exclusively. Thus first level multiplexing
and demultiplexing is nonexistent except at the periphery of the
network.

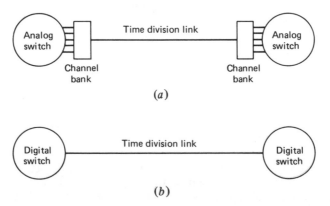

Figure 2.2 Integration of transmission and switching. (*a*) Nonintegrated transmission and switching. (*b*) Integrated time division switching and transmission.

Integration of transmission and switching functions not only eliminates much equipment, but it also greatly improves end-to-end voice quality. By eliminating multiple analog-to-digital and digital-to-analog conversions and by using low error rate transmission links, voice quality is determined only by the encoding process. In summary, the implementation benefits of a fully integrated digital network are:

1 Long-distance voice quality is identical to local voice quality in all aspects of noise, signal level, and distortion.
2 Since digital circuits are inherently four-wire, echos are eliminated, and true full-duplex, four-wire digital circuits are available.*
3 Cable entrance requirements and mainframe distribution of wire pairs is greatly reduced because all trunks are implemented as subchannels of a TDM signal.

2.1.5 Operability at Low Signal-to-Noise/Interference Ratios

Noise and interference in an analog voice network becomes most apparent during speech pauses when the signal amplitude is low. Relatively small amounts of noise occurring during a speech pause can be quite annoying to a listener. The same levels of noise or interference are virtually unnoticeable when speech is present. Hence it is the absolute noise level of an idle channel that determines analog speech quality. Subjective evaluations of voice quality [9, 10] have led to maximum noise level standards of 28 dBrncO (-62 dBm0) for short-haul systems and 34 dBrncO (-56 dBm0) for long-haul systems. For comparison, the power level of an active talker is typically -16 dBm0. Thus representative end-to-end, signal-to-noise ratios in analog networks are 46 and 40 dB for short- and long-haul systems, respectively. Signal-to-noise ratios on individual transmission systems are necessarily higher.

In a digital system speech pauses are encoded with a particular data pattern and transmitted at the same power level as active speech. Because signal regeneration virtually eliminates all noise arising in the transmission medium, idle channel noise is determined by the encoding process and not the transmission link. Thus speech pauses do not determine maximum noise levels as they do in an analog system. As discussed in Chapter 4, digital transmission links provide virtually error-free performance at signal-to-noise ratios of 15 to 25 dB, depending on the type of line coding or modulation used.

The ability of a digital transmission system to reject crosstalk is

---

*Echos are nonexistent if four-wire (analog or digital) circuits exist between station sets. If two-wire (analog) local loops are utilized in an otherwise all-digital network, echos still occur but can be controlled by matching the impedance of two-wire to four-wire hybrids to respective local loops.

sometimes more significant than its ability to operate in relatively high levels of random noise. One of the most troublesome considerations in the design and maintenance of an analog network is the need to eliminate crosstalk between conversations. The problem is most acute during pauses on one channel while an interfering channel is at maximum power. At these times relatively low-level crosstalk may be noticeable. The crosstalk is particularly undesirable if it is intelligible and therefore violates someone's privacy. Again, speech pauses do not produce low-amplitude signals on digital transmission links. The transmission links maintain a constant amplitude digital signal. Thus, low levels of crosstalk are eliminated by the regeneration process in a digital repeater or receiver. Even if the crosstalk is of sufficient amplitude to cause detection errors, the effects appear as random noise and, as such, are unintelligible.

This section emphasizes the ability of a digital transmission system to operate at relatively low signal-to-noise or interference ratios. Considering the typical need for greater bandwidths in digital systems we find that this ability is sometimes more of a requirement than an advantage. In some transmission media, such as wire pairs, greater bandwidths imply greater attenuation. Thus, all else being equal, the digital system must be capable of operating at lower received-signal levels. Still, the ability to reject crosstalk enables their use in congested transmission environments where the bandwith penalty might otherwise make a digital system uncompetitive with an analog system.

2.1.6 Signal Regeneration

As described more fully in the next chapter, the representation of voice (or any analog signal) in a digital format involves converting the continuous analog waveform into a sequence of discrete sample values. Each discrete sample value is represented by some number of binary digits of information. When transmitted, each binary digit is represented by only one of two possible signal values (e.g., a pulse versus no pulse or a positive pulse versus a negative pulse). The receiver's job is to decide which discrete values were transmitted and represent the message as a sequence of binary encoded discrete message samples. If only small amounts of noise, interference, or distortion are impressed upon the signal during transmission, the binary data in the receiver is identical to the binary sequence generated during the digitization or encoding process. As shown in Figure 2.3, the transmission process, despite the existence of certain imperfections, does not alter the essential nature of the information. Of course, if the imperfections cause sufficient changes in the signal, detection errors occur and the binary data in the receiver does not represent the original data exactly.

A fundamental attribute of a digital system is that the probability of

Figure 2.3 Signal regeneration in a digital repeatered line.

transmission errors can be made arbitrarily small by inserting regenerative repeaters at intermediate points in the transmission link. If spaced close enough together, these intermediate nodes detect and regenerate the digital signals before channel-induced degradations become large enough to cause decision errors. As demonstrated in Chapter 4, the end-to-end error rate can be made arbitrarily small by inserting a sufficient number of regeneration nodes in the transmission link.

The most direct benefit of the regeneration process is the ability to localize the effects of signal degradations. As long as the degradations on any particular regenerated segment of a transmission link do not cause errors, their effects are eliminated. In contrast, signal impairments in analog transmission accumulate from one segment to the next. Individual subsystems of a large analog network must be designed with tight controls on the transmission performance to provide acceptable end-to-end quality. An individual subsystem of a digital network, on the other hand, need only be designed to ensure a certain minimum error rate—usually a readily realizable goal.

When an all-digital network is designed with enough regeneration points to effectively eliminate channel errors, the overall transmission quality of the network is determined by the digitization process and not by the transmission systems. The analog-to-digital conversion process inherently introduces a loss of signal fidelity since the continuous analog source waveform can only be represented by discrete sample values. By establishing enough discrete levels, however, the analog waveform can be represented with as little conversion error as desired. The increased resolution requires more bits and consequently more bandwidth for transmission. Hence, a digital transmission system readily provides a trade-off between transmission quality and bandwidth. (A similar trade-off exists for frequency-modulated analog signals.)

2.1.7 Accommodation of Other Services

In a previous section it is pointed out that a digital transmission system readily accommodates control (signaling) information. This fact is representative of a fundamental aspect of digital transmission: any digitally encoded message (whether inherently digital or converted from analog) presents a common signal format to the transmission system. Thus the transmission system need provide no special attention to individual services and can, in fact, be totally indifferent to the nature of the traffic it carries.

In the present analog network the transmission standard is the 4 kHz voice circuit. All special services such as data or facsimile must be transformed to "look like voice." In particular, data signals must be converted to an analog format through the use of modems.

The standard analog channel has been optimized for voice quality. In so doing, certain transmission characteristics (such as the phase response and impulse noise) have received less attention than more noticeable voice quality impairments. Some less-emphasized considerations, phase distortion in particular, are critical for high-rate data services. Use of the analog network for nonvoice services may require special compensation for various analog transmission imparments. If the analog channel is too poor, it may be unusable for a particular application. In contrast, the only parameter of quality in a digital system is the error rate. Low error rate channels are readily obtainable. When desired, the effects of channel errors can be effectively eliminated with error control procedures implemented by the user.

An additional benefit of the common transmission format is that traffic from different types of sources can be intermixed in a single transmission medium without mutual interference. The use of a common transmission medium for analog signals is sometimes complicated because individual services require differing levels of quality. For example, television signals, which require greater transmission quality than voice signals, are not usually combined with FDM voice channels in a wideband analog transmission system [11].

Although telephone companies are primarily concerned with voice services (POTS), the rapid growth in data communications has stimulated increased attention to data transmission performance requirements. The inherent advantages of digital systems for data communications will help to stimulate even more growth in nonvoice services as digital channels become accessible.

2.1.8 Performance Monitorability

An additional benefit of the source independent signal structure in a digital transmission system is that the quality of the received signal can be ascertained with no knowledge of the nature of the traffic. The transmission link is designed to produce well defined pulses with discrete levels. Any deviation in the receive signal, other than nominal amounts planned for in the design, represents a degradation in transmission quality. In general, analog systems can not be monitored or tested for quality while in service since the transmitted signal structure is unknown. FDM multiplex signals typically include pilot signals to measure channel continuity and power levels. The power level of a pilot is an effective means of estimating the signal-to-noise ratio—only in a fixed noise environment. Hence, noise and distortion is sometimes determined by measuring the energy level in an unused message slot or at the edge of the signal passband. In neither case, however, is the quality of an in-service channel being measured directly.

One common method of measuring the quality of a digital transmission link is to add parity bits to the message stream. The redundancy introduced to the data stream by the parity bits enables digital logic circuits in a receiver to readily ascertain channel error rates. If the error rate exceeds some nominal value, the transmission link is degraded.

Another technique for measuring in-service transmission quality is used in T-carrier lines. This technique involves monitoring certain redundancies in the signal waveform itself. When the redundancy pattern at the receiver deviates from normal, decision errors have occurred A complete description of the line coding format used in T-carrier systems is provided in Chapter 4. Other methods of measuring transmission quality in digital systems are discussed in Chapters 4 and 6.

The response required when transmission degradation is detected depends mostly on how critical the particular transmission link is. At low levels of the network when only a few circuits are affected degradations produce routine maintenance reports. In more critical instances involving a larger number of circuits maintenance alarms are set, and traffic is routed around the failed equipment. At the highest levels of the network involving a very large number of circuits service interruptions are unacceptable, and the detection of degraded performance initiates automatic protection switching to a redundant system.

2.1.9 Ease of Encryption

Although most telephone users have little need for voice encryption, the ease with which a digital bit stream can be scrambled and unscrambled [12] means that a digital network provides an extra bonus for users with sensitive conversations. In contrast, analog voice is much more difficult to encrypt and is generally not nearly as secure as digitally encrypted voice. For a comparison of common analog voice encryption techniques see Reference 13. As mentioned previously, it is the encryptability of digital voice that has attracted military interest.

2.2 DISADVANTAGES OF DIGITAL VOICE NETWORKS

The first part of this chapter discussed the basic technical advantages of digital networks. To balance out the discussion this section reviews the basic technical disadvantages of digital implementations as listed in Table 2.3.

**TABLE 2.3   DISADVANTAGES OF
DIGITAL IMPLEMENTATIONS**

1   Increased bandwidth

2   A/D and D/A conversion

3   Need for time synchronization

4   Topologically restricted multiplexing

5   Incompatibility with existing analog equipment

2.2.1   Increased Bandwidth

In the brief introduction to voice digitization presented in Chapter 1, it is pointed out that transmission of samples of an analog waveform requires no more bandwidth than the underlying waveform (at least in theory). The bandwidth expansion comes when the samples are encoded into binary codes and transmitted with an individual pulse for each bit in the code. Thus a T1 system requires approximately eight times as much bandwidth as do 24 analog voice channels since each sample is represented by an 8 bit code word. In actual practice, a somewhat larger bandwidth is needed.

In some portions of the present network, such as the local loops, the bandwidth increase does not represent much of a penalty since the inherent bandwidth is underutilized. In long-haul systems, however, bandwidth is at a premium, and digital systems may be significantly inefficient in terms of the number of voice channels provided. As mentioned earlier, however, the ability of a digital system to overcome higher levels of noise and interference sometimes provides compensation for the bandwidth requirements—particularly in congested transmission environments where mutual interference can become a limiting consideration [11]. In exceptionally clean transmission systems, such as coaxial cable, noise and external interference is minimal, and analog systems are most applicable. Thus, when considered separately, some long-distance transmission systems might always favor analog implementations. In these cases digital implementations are justified only when the synergy of an all-digital network dictates the elimination of every analog subsystem.

The bandwidth increase imposed by voice digitization is directly dependent on the form of transmission coding or modulation used. With greater sophistication in the modulation/demodulation equipment, greater efficiency in terms of the bit rate in a given bandwidth is achievable. Basically, greater transmission efficiency is achieved by increasing the number of levels in the line code. With limited transmit power, however, the distances between discrete signal levels in the receiver are reduced dramatically. Thus, the transmitted signal is no longer as immune to noise and other imperfections as it is with lower information densities. Chapter 6 provides an overview of digital modulation formats and their corresponding bandwidth and signal power requirements.

2.2.2 Analog-to-Digital Conversion

Although the conversion costs of digital systems have always represented a significant portion of digital systems electronics cost, it is important to keep in mind that the digital systems in use today are installed, for the most part, in analog environments. Thus the conversion costs have been more than absorbed by the savings in other equipment such as multiplexers and switches. As more of the network is converted to digital, fewer voice conversions will be required, thereby reducing total conversion costs. Furthermore, electronics costs are continually dropping while other costs, such as wire and maintenance, become more significant. Thus the conversion costs as a percentage of network costs can be expected to decrease continually.

2.2.3 Need for Time Synchronization

Whenever digital information is transmitted from one place to another a timing reference or "clock" is needed to control the transfer. The clock specifies when to sample the incoming signal to decide which data value was transmitted. The optimum sample times usually correspond to the middle of the transmitted pulses. Thus, for optimum detection, the sample clock must be synchronized to the pulse arrival times. In general, the generation of a local timing reference for detecting the digital signal is not difficult. Chapter 4 discusses some of the design considerations needed to establish proper sample clocking in the receiver of a digital transmission link.

More subtle problems arise, however, when a number of digital transmission links and switches are interconnected to form a network. Not only must the individual elements of the network maintain internal synchronization, but also certain networkwide synchronization procedures must be established before the individual subsystems can interoperate properly. Chapter 7 discusses these basic network synchronization requirements and implementations.

The need for some form of synchronization is not unique to digital networks. Single sideband FDM transmission systems present similar requirements for carrier synchronization in analog networks. In analog systems, however, the need for synchronization is not an inherent requirement of the signal format but more a function of the transmission system and the type of modulation chosen.

2.2.4 Topologically Restricted Multiplexing

To the general public, the most apparent use of multiplexing is broadcast services for radio and television. In these systems the airspace is shared by using frequency division multiplexing of individual broadcast channels. With this system there are no operational restrictions to the geographic location of transmitters and receivers. As long as the transmitters confine their emissions to their assigned bandwidth and each receiver uses a sufficiently selective filter to pass only the desired channel, the network operates without mutual interference. Time division multiplexing, on the other hand, is not nearly as amenable to applications involving distributed sources and destinations. Since the time-of-arrival of data in a time slot is dependent on the distance of travel, distributed TDM systems require a guard time between time slots. FDM systems also require guardbands between the channels to achieve adequate channel separation. The width of the guardbands, however, is not dependent on the geographic location of the transmitters. In a TDM system the guard times must be increased as the geographic separation between transmitters increases. Furthermore, each time division source must duplicate the synchronization and time slot recognition logic needed to operate a TDM system. For these reasons, TDM has been used primarily in applications where all of the information sources are centrally located and a single multiplexer controls the occurrence and assignment of time slots.

Time division multiple access satellites are examples of an evolving application of TDM for distributed sources. These systems use sophisticated synchronization techniques so that each ground station times its transmission to arrive at the satellite at precisely defined times, allowing the use of small guard times between time slots. Notice that this application involves only one destination, the satellite. If an application involves multiple distributed sources and destinations with transmission in more than one direction, larger guard times are unavoidable. Figure 2.4 shows such an application but uses frequency division

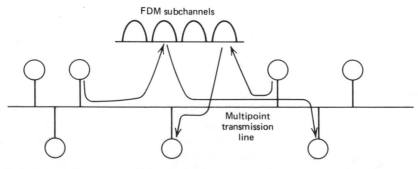

Figure 2.4 Frequency division multiplexing on distributed multipoint line.

multiplexing instead of time division multiplexing. Time division multiplex loops (using a series of unidirectional transmission links) are a means of interconnecting distributed nodes with a TDM system using no guard times. Time division multiplex loops are described in Chapter 4.

2.2.5 Incompatibility with Existing Analog Facilities

The digital equipment used at present in private and public telephone networks necessarily provides standard analog interfaces to the rest of the network. Sometimes these interfaces represent a major cost of the digital subsystem. The foremost example of this situation arises in digital end offices. The standard subscriber loop interface described in Chapter 1 is particularly incompatible with electronic switching matrices (analog or digital). The impact of this interface on local digital switching is discussed in Chapter 5.

As discussed previously, maximum advantages for voice quality and nonvoice services are not obtained in digital telephone systems unless the networks are all-digital. Hence it is highly desirable to implement the analog-to-digital and digital-to-analog conversions in the station sets themselves. Station set conversion offers numerous advantages including: readily available multiplexing for pair-gain systems; elimination of the conventional subscriber loop interface; end-to-end high-quality digital connections; and enhanced station set control capabilities from the central offices. (This feature is particularly useful for pay telephones, which present unique and cumbersome interface requirements in the present network.)

Although station set conversion is reasonable and desirable for networks being built from scratch, the overwhelming investment in the present loop plant for analog telephones precludes a widespread transition to digitization of local distribution facilities. The cable plant itself will support the higher bandwidth digital signals, but certain implementation and maintenance practices must be undone or modified for digital signals. Most notable of the present practices that complicate a transition to digital loops are: loading coils, build-out networks,* bridged taps,† high-resistance or intermittant splices, and overvoltage protection. Furthermore, crosstalk considerations at the higher frequencies may require separate cables for each direction of transmission or newer cable types with internal shields to isolate "go" and "return"

---

*A build-out network is a circuit added to a relatively short transmission line for the purpose of establishing overall transmission characteristics similar to a longer, standard-length line.

†A bridged tap is an unused pair of wires connected at some point of an in-use pair as an off-premises extension or for possible future reassignment of a cable pair.

directions of transmission.* For a discussion of a conversion strategy to provide a smooth transition to digital local loops in Great Britain, see Reference 14.

Owing to their insertion of artificial delay in the message path, digital switches complicate via net loss designs for echo control. This subject is covered more fully at the end of chapter 5.

2.3 ECONOMIC CONSIDERATIONS

The vast amounts of analog equipment already in use in the public telephone network represents the main impediment to widespread use of digital technology. The primary responsibility of the telephone companies is plain old telephone service (POTS). Hence as long as an operating telephone company is providing adequate voice service there is little incentive to change to a new technology. The impetus for change must be economic. Either the new equipment must provide proven savings in maintenance costs, or it must provide new services with a proven market to compensate the write-off of undepreciated equipment.

One result of the rapidly advancing technology and the increasing level of competition in the telephone industry is a trend toward shorter depreciation schedules for capitalized expenditures. If the regulatory authorities allow operating companies to depreciate their investments at a rate more in keeping with the competition (and the technology), telephone companies will replace technically obsolete equipment more rapidly.

One of the major headaches for manufacturing and operating telephone companies, alike, is maintaining support for the myriad of equipment types that have evolved over the years and remain in use. Although the evolution to an all-digital network provides the opportunity to eliminate much of the obsolete equipment, the short-term situation demands support for even more equipment types. Furthermore, digital equipment represents a radically different technology presenting new demands on personnel training, test equipment, and repair.

When evaluating digital switching in the end offices, a number of indirect economic benefits can be derived from the use of time division multiplexing of subscriber circuits. First, the number of digital subscriber carrier (pair gain) systems would increase significantly because the economic prove-in distance is reduced if the multiplexed signals can

---

*Local loop digitization can be implemented on two-wire lines by alternately transmitting and receiving bursts of data in a "ping-pong" fashion. As discussed in Chapter 5, this technique avoids near-end crosstalk problems because transmission and reception are separated in time.

be switched directly in a digital format [15]. If the telephones them-selves provide the analog-to-digital conversion, the prove-in distance for subscriber TDM goes to zero. In either case, moving the point of digit-ization closer to the periphery of the network and using subscriber multiplexing not only reduces wire costs but also reduces the number of wire pair terminations in the central office. The ubiquitous use of digital subscriber multiplexing radically simplifies the main distribution frame (MDF) and relieves floor space requirements in central offices. These considerations are particularly important when an operating com-pany experiences significant growth in its service area and must con-sider expansion of an existing building or a move to a new one. With digital switching and multiplexing, floor space and cable entry require-ments are greatly reduced—eliminating the need for building expansion.

Second, and probably more important, as the point of digitization moves toward the periphery of the network, a greater proportion of network costs are not incurred until the users are ready to begin paying for the services. In the present network only 20% of the operating com-pany expenditures are associated with the terminal equipment* (tele-phones, modems, PBXs, etc.) [16]. The rest of the costs must be in-curred long before revenues can be derived from their installation. Hence, long-range planning for anticipated demand is a critical aspect of telephone company operations. In an all-digital network, wire costs, switching hardware, and floor space can all be reduced by an order of magnitude [3, 4]. A large part of the cost of an all-digital network shifts to the telephone set itself—reducing the lag between expenditures and income.

REFERENCES

1 "SR-1536 Subscriber Radio, Technical Description," Farinon SR Systems, Quebec, Canada, 1977.

2 A. E. Ritchie and J. Z. Menard, "Common Channel Interoffice Signaling, An Overview," *Bell System Technical Journal*, February 1978, pp 221–236.

3 Arthur A. Collins, "Are We Committed to Change in Telecommunication Sys-tems," *SIGNAL*, August 1974, pp 62–65.

4 Robert D. Pedersen, "Digital Switched Networks," *New Technology Seminar— National Electronic Conference*, October 1974.

5 A. E. Ritchie and L. S. Tuomenoksa, "No. 4 ESS: System Objectives and Or-ganization," *Bell System Technical Journal*, September 1977, pp 1017–1029.

6 Hirvela, R. J., "The Application of Computer Controlled PCM Switching to Automatic Call Distribution," *IEEE Communications Systems and Technology Conference*, Dallas, Texas, May 1974.

*The 20% figure includes installation costs which are capitalized. Station equip-ment itself represents approximately 10% of the network costs.

7   James Spilker, *Digital Communications by Satellite*, Prentice-Hall, Englewood Cliffs, New Jersey, 1977.

8   R. C. Drechsler, "Echo Suppressor Terminal for No. 4 ESS," *1976 International Communications Conferences Record*, 1976, pp 36.9–36.12.

9   I. Nasell, "The 1962 Survey of Noise and Loss on Toll Connections," *Bell System Technical Journal*, March 1964, pp 697–718.

10 Bell Telephone Laboratories Members of Technical Staff, *Transmission Systems for Communications*, Western Electric Company, February 1970.

11 M. R. Aaron, "Digital Communications—The Silent (R)evolution?" *IEEE Communications Society Magazine*, Vol. 17, January 1979, pp. 16–26.

12 H. J. Hindin, "LSI-Based Data Encryption Discourages the Data Thief," *Electronics,* June 21, 1979, pp. 107–120.

13 N. S. Jayant, B. J. McDermott, S. W. Christensen, and A. M. Quinn, "A Comparison of Four Methods for Analog Speech Encryption," *International Communications Conference Record*, 1980, pp 16.6.1–16.6.5.

14 D. Sibbald, H. W. Silcock, and E. S. Usher, "Digital Transmission Over Existing Subscriber Cable," *International Symposium on Subscriber Loops and Services*, Atlanta, Georgia, 1978.

15 J. C. McDonald, "An Integrated Subscriber Carrier and Local Digital Switching System," *International Symposium on Subscriber Loops and Services*, 1978, pp 172–176.

16 Members of Technical Staff, Bell Telephone Laboratories, *Engineering and Operations in the Bell System*, Western Electric, 1977.

CHAPTER THREE

VOICE DIGITIZATION

Owing to the interesting nature of the subject and its usefulness in a variety of applications, the field of voice digitization has received intense study in the last 20 years. These studies have produced many different types of voice digitizers with numerous variations of each type. The choice of a particular type is primarily dependent on the application and the level of voice quality desired. Generally speaking, applications may be categorized as: (1) transmission, (2) switching, (3) storage, or combinations of these. Voice coders for transmission may be further classified according to wideband or narrowband applications. Wideband transmission applications involve T-carrierlike systems where standard telephone quality is desired. At present, economical telephone-quality voice digitizers require transmission rates between 32 and 64 kbps. Even greater data rates are required for digital transmission of program audio where higher quality is desired. For example, 384 kbps (including parity bits) has been proposed as a CCIR standard for sound broadcast signals [1].

Narrowband transmission applications generally arise when existing analog facilities are used such as the public telephone network or high frequency (HF) radio channels for secure voice applications. These applications typically restrict the transmission rates to a range between 2 kbps and 16 kbps. Except at the high end of this data rate range, these voice digitizers are much more sophisticated and use fundamentally different algorithms from those used in conventional telephony.

Since the internal bandwidth of a *switching* system is generally less constrained than the bandwidth of a *transmission* system, voice digitizers for switching applications typically use simple implementations with somewhat higher data rates. For example, the first digital PBX in the United States, introduced by Rolm Corporation in 1975, used a voice digitization technique similar to that used in T-carrier systems but requiring a transmission rate of 144 kbps [2]. As the cost of conventional coders and decoders (codecs) for transmission applications declines, and as more transmission and switching functions are integrated, the use of higher rate (switching only) voice digitization techniques is disappearing.

Digital storage of voice signals can also be classified according to the conversion bandwidth, which, of course, directly determines the digital storage requirements. At the low end are applications for recorded messages with limited available storage and only minimal quality requirements. One such application is the Speak-and-Spell® learning machine developed by Texas Instruments [3]. This device uses some of the most sophistical voice digitization techniques available, storing up to 200 words with an average of approximately 1000 bits per word. At this level of storage density the equivalent data rate is a little less than 1200 bps. Hence message recordings with limited storage capabilities represent applications using the most sophisticated encoding and decoding techniques. In fact, generation of the recorded words in the Speak-and-Spell® involved considerable human interaction and reprocessing. Thus the technique used to produce these low bit rates cannot, at this time, produce equivalent quality in real time voice applications.

Another example of digital speech storage is the 1A Voice Storage system [4] developed by engineers at Bell Laboratories. This system operates with stored program control switching systems to provide recorded announcements and messages to and from customers. The encoding technique is adaptive delta modulation (discussed later) at 32 kbps.

At the upper end of digital storage applications are high-fidelity recordings of voice and music. Many of the same advantages of digital transmission—as opposed to an analog transmission—also apply to digital recordings. Foremost among these advantages is the ability of defining the fidelity at the time of recording and maintaining the quality indefinitely by periodically copying (regenerating) the digitally stored information before inevitable deterioration produces bit errors. Thus a high-quality (high bit rate) digital recording of Bing Crosby, Elvis Presley, or Luciano Pavarotti (depending on your taste in music) can be saved for posterity. This feat could not be accomplished with analog recordings no matter how well cared for or preserved. Another advantage of digital storage is the ability to use a lower-quality (nonlinear) recording medium. Thus digital record and tape players will become economically attractive in the consumer market as electronics costs continue to drop and lower-cost recording media is mass produced. High-fidelity digital tape systems are already being used in some recording studios, mostly to facilitate flexible high-quality signal processing for development of the final product. These systems use data rates on the order of 400 kbps.

Speech analysis and synthesis is another area of widespread research closely related to voice digitization. In fact, some of the lowest bit rate voice encoders and decoders use certain amounts of analysis and synthesis to digitally represent speech. In its ultimate form, however, analysis and synthesis have unique goals and applications fundamentally different from these of general voice digitization. Basically, these goals are to recognize words or produce machine-generated speech.

One approach to analyzing speech is to process waveforms with the intent of recognizing speech phonemes—the basic units of speech from which spoken words are constructed. Once the phonemes have been identified, they are assigned individual codewords for storage or transmission. A synthesizer can then generate speech by recreating the combinations of phonemes. Analysis of this technique indicates that the information content of speech can be transmitted with a data rate of 50 bps [5]. It must be emphasized, however, that only the information content associated with the words themselves is transmitted—none of the more subjective qualities of speech such as naturalness, voice inflections, accents, speaker recognizability, and so on. Thus such techniques, by themselves, are not applicable to general telephony, which customarily includes qualities other than the message content of spoken words.

Another level of speech analysis involves the actual recognition of spoken words. One such system recently introduced commercially by Nippon Electric Corporation (NEC) of Japan has the capability of recognizing up to 1000 isolated words with an error rate less than 1% [6]. These systems are intended for computer input or voice-activated machine control and generally do not include capabilities for reproducing (synthesizing) the spoken words, except as alphanumeric output. As these recognition techniques improve, particularly for recognizing connected words, voice-activated typewriters (voicewriters) will become available.

Voice digitization techniques can be broadly categorized into two classes: those digitally encoding analog waveforms as faithfully as possible and those processing waveforms to encode only the perceptually significant aspects of speech and hearing processes. The first category is representative of the general problem of analog-to-digital and digital-to-analog conversions and is not restricted to speech digitization. The two most common techniques used to encode a voice waveform are pulse code modulation (PCM) and delta modulation (DM). Except in special cases, digital telephony uses these techniques. Thus when studying the most common digital speech encoding techniques we are, in fact, investigating the more general realm of analog-to-digital conversion.

The second category of speech digitization is concerned primarily with producing very low data rate speech encoders and decoders for narrowband transmission systems or digital storage devices with limited capacity. A device from this special class of techniques is commonly referred to as a "vocoder" (voice coder). Although capable of producing intelligible speech, vocoder techniques generally produce unnatural or synthetic sounding speech. As such, vocoders do not provide adequate quality for general telephony. The following discussions emphasize waveform coding, but the basic principles of vocoders are also included.

3.1 PULSE AMPLITUDE MODULATION

The first step in digitizing an analog waveform is to establish a set of discrete times at which the input signal waveform is sampled. Prevalent digitization techniques are based on the use of periodic, regularly spaced sample times. If the samples occur often enough, the original waveform can be completely recovered from the sample sequence using a low-pass filter to interpolate or "smooth out" between the sample values. These basic concepts are illustrated in Figure 3.1. A representative analog waveform is sampled at a constant sampling frequency $f_s = 1/T$ and reconstructed using a low-pass filter. Notice that the sampling process is equivalent to amplitude modulation of a constant-amplitude pulse train. Hence the technique represented in Figure 3.1 is usually referred to as a pulse amplitude modulation (PAM).

3.1.1 Nyquist Sampling Rate

A classical result in sampling systems was established in 1933 by Harry Nyquist when he derived the minimum sampling frequency required to extract all information in a continuous, time-varying waveform. This result—the Nyquist criterion—is defined by the relation

$$f_s > 2 \cdot BW$$

where f_s = sampling frequency
BW = bandwidth of the input signal

The derivation of this result is indicated in Figure 3.2, which portrays the spectrum of the input signal and the resulting spectrum of the PAM pulse train. The PAM spectrum can be derived by observing that a con-

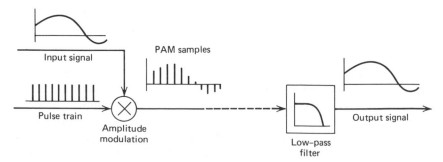

Figure 3.1 Pulse amplitude modulation.

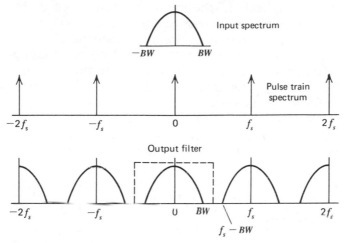

Figure 3.2 Spectrum of pulse amplitude modulated signal.

tinuous train of impulses has a frequency spectrum consisting of discrete terms at multiples of the sampling frequency. The input signal amplitude modulates these terms individually. Thus a double sideband spectrum is produced about each of the discrete frequency terms in the spectrum of the pulse train. The original signal waveform is recovered by a low pass filter designed to remove all but the original signal spectrum. As shown in Figure 3.2, the reconstructive low-pass filter must have a cut-off frequency that lies between BW and $f_s - BW$. Hence, separation is only possible if $f_s - BW$ is greater than BW (i.e., if $f_s > 2 \cdot BW$).

3.1.2 Foldover Distortion

If the input waveform of a PAM system is undersampled ($f_s < 2 \cdot BW$), the original waveform cannot be recovered without distortion. As indicated in Figure 3.3, this output distortion arises because the frequency spectrum centered about the sampling frequency overlaps the original spectrum and cannot be separated from the original spectrum by filter-

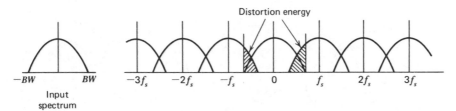

Figure 3.3 Foldover spectrum produced by undersampling an input.

ing. Since it is a duplicate of the input spectrum "folded" back on top of the desired spectrum that causes the distortion, this type of sampling impairment is often referred to as "foldover distortion."

In essence, foldover distortion produces frequency components in the desired frequency band that did not exist in the original waveform. Thus another term for this impairment is "aliasing." Aliasing problems are not confined to speech digitization processes. The potential for aliasing is present in any sample data system. Motion picture taking, for example, is another sampling system that can produce aliasing. A common example occurs when filming moving stagecoaches in old Westerns. Often the sampling process is too slow to keep up with the stagecoach wheel movements, and spurious rotational rates are produced. If the wheel rotates $355°$ between frames, it looks to the eye as if it has moved backwards $5°$.

Figure 3.4 demonstrates an aliasing process occurring in speech if a 5.5 kHz signal is sampled at an 8 kHz rate. Notice that the sample values are identical to those obtained from a 2.5 kHz input signal. Thus after the sampled signal passes through the 4 kHz output filter, a 2.5 kHz signal arises that did not come from the source. This example illustrates that the input must be bandlimited, *before sampling*, to remove frequency terms greater than $f_s/2$, even if these frequency terms are ignored (i.e., are inaudible) at the destination. Thus, a complete PAM system, shown in Figure 3.5, must include a bandlimiting filter before sampling to ensure that no spurious or source-related signals get "folded" back into the desired signal bandwidth. The input filter may also be designed to cut off very low frequencies to remove 60 cycle hum from power lines.

Figure 3.5 shows the signal being recovered by a sample-and-hold circuit that produces a staircase approximation to the sampled waveform. With use of the staircase approximation, the power level of the signal coming out of the reconstructive filter is nearly the same as the

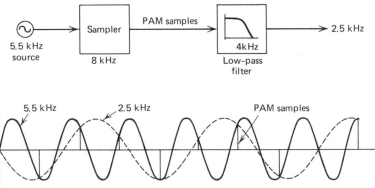

Figure 3.4   Aliasing of 5.5 kHz signal into a 2.5 kHz signal.

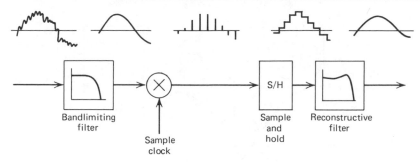

Figure 3.5 End-to-end pulse amplitude modulated system.

level of the sampled input signal. The response of the reconstructive filter, in this case, must be modified somewhat to account for the spectrum of the wider "staircase" samples. (The modification amounts to dividing the "flat" filter spectrum by the spectrum of the finite width pulse. See Appendix C.)

The bandlimiting and reconstructive filters shown in Figure 3.5 are implied to have ideal characteristics.* Since ideal filters are physically unrealizable, a practical implementation must consider the effects of nonideal implementations. Filters with realizable attenuation slopes at the band edge can be used if the input signal is slightly oversampled.

As indicated in Figure 3.2, when the sampling frequency f_s is somewhat greater than twice the bandwidth, the spectral bands are sufficiently separated from each other that filters with gradual roll-off characteristics can be used. As an example, sampled voice systems typically use bandlimiting filters with a 3 dB cutoff around 3.4 kHz and a sampling rate of 8 kHz. Thus the sampled signal is sufficiently attenuated at the overlap frequency of 4 kHz to adequately reduce the energy level of the foldover spectrum. Figure 3.6 shows a filter template designed to meet CCITT recommendations for out-of-band signal rejection in PCM voice coders. Notice that 14 dB of attenuation is provided at 4 kHz.

As mentioned in Chapter 1, the perceived quality of a voice signal is not greatly dependent upon the phase response of the channel (the relative delay of voice frequency components). For this reason the phase responses of the bandlimiting filters in the encoders and the smoothing filters in the decoders are not critical. Nonlinear phase responses in these filters, however, do impact high-rate voiceband data signals when digitized. Hence a somewhat paradoxical situation arises when voiceband data is transmitted over a T-carrier line: the process of converting the voiceband data signal (4800 bps typical maximum) to a virtually

---

*An ideal filter is one with: a frequency-independent time delay (linear phase), no attenuation in the passband (except as might be desired for pulse shaping), an arbitrarily steep cutoff, and infinite attenuation everywhere in the stopband.

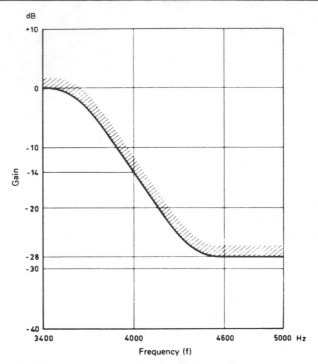

Figure 3.6 Bandlimiting filter template designed to meet CCITT recommendations for PCM voice coders.

error free 64 kbps digital signal causes distortion in the lower-rate data signal. However, the transmission process itself does not add to the signal degradation.

By interleaving the samples from multiple sources, PAM systems can be used to share a transmission facility in a time division multiplex manner. PAM systems are not generally useful over long distances, however, owing to the vulnerability of the individual pulses to noise, distortion, and crosstalk. Instead, for long-distance transmission the PAM samples are converted into a digital format, thereby allowing the use of regenerative repeaters to remove transmission imperfections before errors result. Time division PAM techniques are used in some PBX systems described in Chapter 5.

3.2 PULSE CODE MODULATION

The preceding section describes pulse amplitude modulation, which uses discrete sample times with analog sample amplitudes to extract the information in a continuously varying analog signal. Pulse code modulation (PCM) is an extension of PAM wherein each analog sample value is

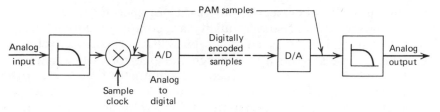

Figure 3.7 Pulse code modulation.

quantized into a discrete value for representation as a digital code word. Thus, as shown in Figure 3.7, a PAM system can be converted into a PCM system by adding an analog-to-digital (A/D) converter at the source and a digital-to-analog (D/A) converter at the destination. Figure 3.8 depicts a typical quantization process in which a set of quantization intervals are associated in a one-to-one fashion with a binary code word. All sample values falling in a particular quantization interval are represented by a single discrete value located at the center of the quantization interval. In this manner the quantization process introduces a certain amount of error or distortion into the signal samples. This error, known as "quantization noise," is minimized by establishing a large number of small quantization intervals. Of course, as the number of quantization intervals increase, so must the number of bits increase to uniquely indentify the quantization intervals.

3.2.1 Quantization Noise

A fundamental aspect of the design and development of an engineering project is the need for analytical measures of systems performance. Only then can a system be objectively measured and its cost effectiveness compared to alternate designs. One of the measures needed by a

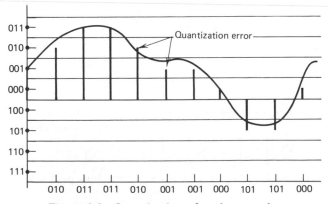

Figure 3.8 Quantization of analog samples.

voice communication engineer is the quality of speech delivered to the listener. Measurements of speech quality are complicated by subjective attributes of speech as perceived by a typical listener. One subjective aspect of noise or distortion on a speech signal involves the frequency content, or spectrum, of the disturbance in conjunction with the power level. These effects of noise as a function of frequency are discussed in Chapter 1 with the introduction of C-message and psophometric weighting.

Successive quantization errors of a PCM encoder are generally assumed to be distributed randomly and uncorrelated to each other. Thus the cummulative effect of quantization errors in a PCM system can be treated as additive noise with a subjective effect that is similar to bandlimited white noise. Figure 3.9 shows the quantization noise as a function of signal amplitude for a coder with uniform quantization intervals. Notice that if the signal has enough time to change in amplitude by several quantization intervals, the quantization errors are independent. If the signal is oversampled, that is sampled much higher than the Nyquist rate, successive samples are likely to fall in the same interval causing a loss of independence in the quantization errors.

The quantization error or distortion created by digitizing an analog signal is customarily expressed as an average noise power relative to the average signal power. Thus the signal-to-quantizing noise ratio SQR (also called a signal-to-distortion ratio or a signal-to-noise ratio)

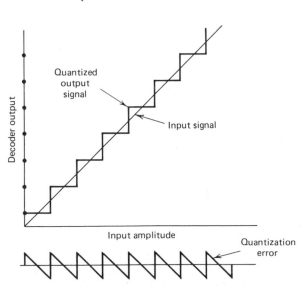

Figure 3.9 Quantization error as a function of amplitude over a range of quantization intervals.

can be determined as

$$SQR = \frac{E\{x^2(t)\}}{E\{[y(t) - x(t)]^2\}} \tag{3.1}$$

where $E\{\cdot\}$ = expectation or averaging
$x(t)$ = the analog input signal
$y(t)$ = the decoded output signal

In determining the expected value of the quantization noise, three observations are necessary:

1 The error $y(t) - x(t)$ is limited in amplitude to $q/2$ where q is the height of the quantization interval. (Decoded output samples are ideally positioned at the middle of a quantization interval.)
2 A sample value is equally likely to fall anywhere within a quantization interval—implying a uniform probability density of amplitude $1/q$.
3 Signal amplitudes are assumed to be confined to the maximum range of the coder. If a sample value exceeds the range of the highest quantization interval, overload distortion (also called peak limiting) occurs.

If we assume (for convenience) a resistance level of 1 ohm, the average quantization noise power is determined in Appendix A as:

$$\text{quantization noise power} = \frac{q^2}{12} \tag{3.2}$$

If all quantization intervals have equal lengths (uniform quantization), the quantization noise is independent of the sample values and the signal-to-quantization ratio is determined as*

$$SQR \text{ (dB)} = 10 \log_{10} \left(\frac{v^2}{q^2/12}\right)$$

$$= 10.8 + 20 \log_{10} \left(\frac{v}{q}\right) \tag{3.3}$$

*The signal-to-quantization noise ratios commonly compare unfiltered decoder outputs to unfiltered quantization errors. In actual practice, the decoder output filter reduces the power level of both the signal and the noise. The noise power experiences a greater reduction than the signal power, since the noise samples have a wider spectrum than the voice samples. Thus filtered signal-to-noise ratios are usually higher than the values calculated here by 1 to 2 dB.

where v is the rms amplitude of the input. In particular, for a sinewave input the signal-to-quantizing noise ratio produced by uniform quantization is

$$SQR \ (dB) = 10 \ \log_{10} \left(\frac{A^2/2}{q^2/12} \right)$$

$$= 7.78 + 20 \ \log_{10}(A/q) \tag{3.4}$$

where A is the peak amplitude of the sinewave.

EXAMPLE 3-1

A sine wave with a 1 V maximum amplitude is to be digitized with a minimum signal-to-quantization noise ratio of 30 dB. How many uniformly spaced quantization intervals are needed, and how many bits are needed to encode each sample?

Solution. Using Equation 3.4, the maximum size of a quantization interval is determined as

$$q = (1) \ 10^{-(30-7.78)/20}$$

$$= 0.078 \ V$$

Thus 13 quantization intervals are needed for each polarity for a total of 26 intervals in all. The number of bits required to encode each sample is determined as

$$N = \log_2 (26)$$

$$= 4.7$$

$$= 5 \ \text{bits per sample}$$

When measuring quantization noise power, the spectral content is often weighted in the same manner as noise in an analog circuit. Unfortunately, spectrally weighted noise measurements do not always reflect the true perceptual quality of a voice encoder/decoder. If the spectral distribution of the quantization noise more or less follows the spectral content of the speech waveform, the noise is masked by the speech and is much less noticeable than noise uncorrelated to the speech [7]. On the other hand, if the quantization process produces energy at voice-band frequencies other than those contained in particular sounds, they are more noticeable.

High-quality PCM encoders produce quantization noise that is evenly distributed across voice frequencies and independent of the encoded waveforms. Thus quantization noise ratios defined in Equation 3.4, are good measures of PCM performance. In some of the encoders discussed later (vocoders in particular) quantization noise power is not very useful. Reference 8 describes other measures of encoder speech quality providing better correlations to quality as perceived by a listener.

3.2.2 Idle Channel Noise

Examination of Equations 3.3 and 3.4 reveals that the SQR is small for small sample values. In fact, as shown in Figure 3.10, the noise may actually be greater than the signal when sample values are in the first quantization interval. This effect is particularly bothersome during speech pauses and is known as idle channel noise. Figure 3.11 depicts one method of minimizing idle channel noise in PCM systems by establishing a quantization interval that straddles the origin. In this case all sample values in the central quantization interval are decoded as a constant zero output. PCM systems of this type use an odd number of quantization intervals since the encoding ranges of positive and negative signals are usually equal.

The quantization characteristics required to produce the output waveforms shown in Figures 3.10 and 3.11 are shown in Figures 3.12 and 3.13, respectively. The first characteristic (mid-riser) cannot produce a zero output level. The second characteristic (mid-tread) is obviously preferred since very low signals are decoded into a constant, zero-level

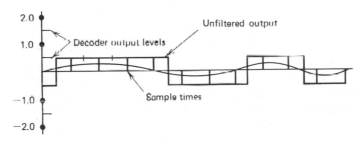

Figure 3.10 Idle channel noise produced by mid-riser quantization.

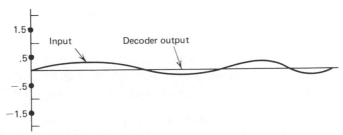

Figure 3.11 Elimination of idle channel noise by mid-tread quantization.

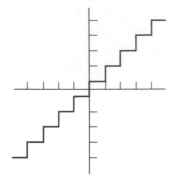

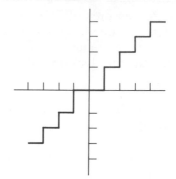

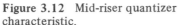

Figure 3.12 Mid-riser quantizer   Figure 3.13 Mid-tread quantizer
characteristic.   characteristic.

output. However, if the signal amplitude is comparable to the size of
the quantization interval, or if a dc bias exists in the encoder, idle channel noise is a problem with mid-tread quantization also.

As mentioned in Chapter 1, noise occurring during speech pauses is
more objectionable than noise with equivalent power levels during
speech. Thus idle channel noise is specified in absolute terms separate
from quantization noise which is specified relative to the signal level.
For example, Bell system D3 channel bank specifications list the maximum idle channel noise as 23 dBrnc0 [9].

3.2.3 Uniformly Encoded PCM

An encoder using equal length quantization intervals for all samples
produces code words linearly related to the analog sample values.
That is, the numerical equivalent of each code word is proportional to
the quantized sample value it represents. In this manner a uniform PCM
system uses a conventional analog-to-digital converter to generate the
binary sample codes. The number of bits required for each sample is
determined by the maximum acceptable noise power. Minimum digitized voice quality requires a signal-to-noise ratio in excess of 26 dB
[10]. For a uniform PCM system to achieve a signal-to-quantization
ratio of 26 dB, Equation 3.4 indicates that $q_{max} = 0.123A$. For equal
positive and negative signal excursions (encoding from $-A$ to A), this
result indicates that just over 16 quantization intervals or 4 bits per
sample are required.*

In addition to providing adequate quality for small signals, a telephone system must be capable of transmitting a large range of signal
amplitudes. A typical minimum dynamic range is 30 dB [10]. Thus

---

*This SQR objective is for minimum acceptable performance and assumes all
degradations occur in a single encoder. If additional signal impairments occur (such
as multiple analog-to-digital conversions), the encoder must use more bits to provide noise margin for other elements in the network.

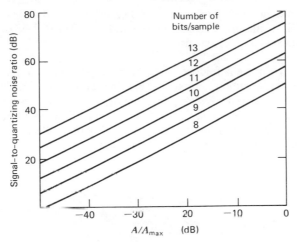

Figure 3.14   Signal-to-quantizing noise of uniform PCM coding.

signal values as large as 31 times A must be encoded without exceeding the range of quantization intervals. Assuming equally spaced quantization intervals for uniform coding, the total number of intervals is determined as 496, which requires 9 bit code words.*

The performance of an n-bit uniform PCM system is determined by observing that

$$q = \frac{2 A_{max}}{2^n} \qquad (3.5)$$

where A_{max} is the maximum (unoverloaded) amplitude.

Substituting Equation 3.5 into Equation 3.4 produces the PCM performance equation for uniform coding:

$$SQR = 1.76 + 6.02n + 20 \log_{10}\left(\frac{A}{A_{max}}\right) \qquad (3.6)$$

The first two terms of Equation 3.6 provide the SQR when encoding a full range sine wave. The last term indicates a loss in SQR when encoding a lower level signal. These relationships are presented in Figure 3.14 which shows the SQR of uniform PCM system as a function of the number of bits per sample and the magnitude of an input sine wave.

EXAMPLE 3.2

What is the minimum bit rate that a uniform PCM encoder must provide to encode a high fidelity audio signal with a dynamic range of 40

---

*This result is derived with the assumption of minimum performance requirements. Higher performance objectives (less quantization noise and greater dynamic range) require as many as 13 bits per sample for uniform PCM systems.

dB? Assume the fidelity requirements dictate passage of a 10 kHz bandwidth with a minimum signal-to-noise ratio of 50 dB. For simplicity,
assume sinusoidal input signals.

Solution. To prevent foldover distortion, the sampling rate must be at
least 20 kHz. Assuming an excess sampling factor comparable to that
used in D-type channel banks (4000/3400), we choose a sampling rate
of 24 kHz as a compromise for a practical bandlimiting filter. By observing that a full amplitude signal is encoded with an SQR of 40 + 50 =
90 dB, we can use Equation 3.6 to determine the number of bits n required to encode each sample.

$$n = \frac{(90 - 1.76)}{6.02}$$

$$= 15 \text{ bits}$$

Thus the required bit rate is:

$$(15)(24) = 360 \text{ kbps}$$

3.2.4 Companding

In a uniform PCM system the size of every quantization interval is
determined by the SQR requirement of the lowest signal level to be
encoded. Larger signals are also encoded with the same quantization
interval. As indicated in Equation 3.6 and Figure 3.14, the SQR increases with the signal amplitude A. For example, a 26 dB SQR for
small signals and a 30 dB dynamic range produces a 56 dB SQR for a
maximum amplitude signal. In this manner a uniform PCM system provides unneeded quality for large signals. Moreover, the large signals are
the least likely to occur. For these reasons the code space in a uniform
PCM system is very inefficiently utilized.

A more efficient coding procedure is achieved if the quantization
intervals are not uniform but allowed to increase with the sample value.
When quantization intervals are directly proportional to the sample
value, the SQR is constant for all signal levels. With this technique
fewer bits per sample provide a specified SQR for small signals and an
adequate dynamic range for large signals. When the quantization intervals are not uniform, a nonlinear relationship exists between the code
words and the samples they represent.

Historically, the nonlinear function was first implemented on analog
signals using nonlinear devices such as specially designed diodes [11].
The basic process is shown in Figure 3.15 where the analog input sample is first compressed and then quantized with uniform quantization
intervals. The effect of the compression operation is shown in Figure

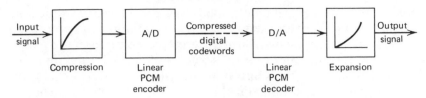

Figure 3.15 Companded PCM with analog compression and expansion.

3.16. Notice that successively larger input signal intervals are compressed into constant length quantization intervals. Thus the larger the sample value, the more it is compressed before encoding. As shown in Figure 3.15, a nonuniform PCM decoder expands the compressed value using an inverse compression characteristic to recover the original sample value. The process of first compressing and then expanding a signal is referred to as "companding." When digitizing, companding amounts to assigning small quantization intervals to small samples and large quantization intervals to large samples.

Various compression-expansion characteristics can be chosen to implement a compandor. By increasing the amount of compression, we increase the dynamic range at the expense of the signal-to-noise ratio for large amplitude signals. One family of compression characteristics used in North America and Japan is the μ-law characteristic defined as:

$$F_\mu(x) = \text{sgn}(x) \frac{\ln (1 + \mu|x|)}{\ln (1 + \mu)} \tag{3.7}$$

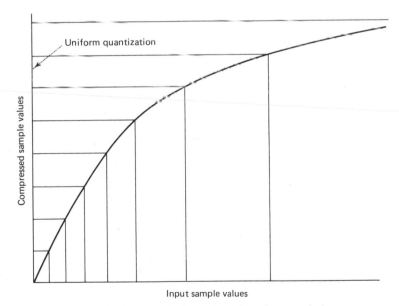

Figure 3.16 Typical compression characteristic.

where x is the input signal amplitude ($-1 \leqslant x \leqslant 1$), sgn($x$) is the polarity of x, and μ is a parameter used to define the amount of compression.

Because of the mathematical nature of the compression curve, companded PCM is sometimes referred to as log-PCM. A logarithm compression curve is ideal in the sense that quantization intervals and, hence, quantization noise is proportional to the sample amplitude. The inverse or expansion characteristic for a μ-law compandor is defined as

$$F_\mu^{-1}(y) = \text{sgn}(y)\left(\frac{1}{\mu}\right)[(1+\mu)^{|y|} - 1] \tag{3.8}$$

where y is the compressed value $= F_\mu(x)$ ($-1 \leqslant y \leqslant 1$), sgn($y$) is the polarity of y, and μ is the companding parameter.

The first T-carrier systems in the United States used D1 channel banks [12] which approximated Equation 3.7 for $\mu = 100$. The compression and expansion functions were implemented with the specially biased diodes mentioned previously. Figure 3.17 depicts a block diagram of a D1 channel bank. Notice that the time division multiplexing and demultiplexing functions are implemented on analog PAM samples. Thus the companding and encoding/decoding functions are shared by all 24 voice channels. The ability to share this relatively expensive equipment was one of the reasons that PCM was originally chosen as the means of digitally encoding speech. Recent developments for large scale integrated circuit PCM codecs have diminished the need to share this equipment. Thus newer systems can use per channel codecs and provide more flexibility in implementing various system sizes. When

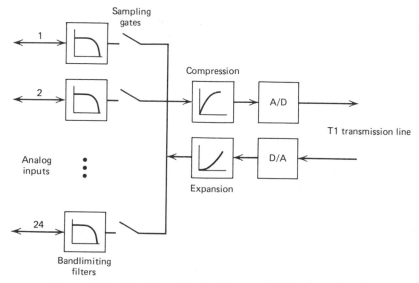

Figure 3.17   Functional block diagram of D1 channel bank.

most of the cost of a channel bank is in the common equipment, as in the original channel banks, less than fully implemented systems are overly expensive.

Each sample produced by a D1 channel bank is encoded into 7 bits: 1 polarity bit and 6 compressed magnitude bits. In addition, 1 signaling bit is added to each channel to produce an 8 bit code word for each time slot. Since the sampling rate is 8 kHz, a 64 kbps channel results. Even though the D1 channel banks have been superceded by newer channel banks utilizing a different coding format, the 64 kbps channel rate has been retained as a standard.

3.2.5   Easily Digitally Linearizable Coding

The success of the first T1 systems for interoffice exchange area transmission paved the way for further use of digital time division multiplex systems. As it became apparent that digital transmission was useful within the toll network, it also became clear that the form of PCM encoding used in the D1 channel banks was inadequate. In contrast to the exchange area, an end-to-end connection through the toll network may conceivably involve as many as nine tandem connections.* Since digital switching was not in existence at the time that T-carrier systems for the toll network were being developed, each of these tandem connections implied an extra digital-to-analog and analog-to-digital conversion. Thus the quality of each conversion had to be improved to maintain the desired end-to-end quality. The D2 channel bank [13] was therefore developed with improved digitized voice quality.

When the D2 channel bank was being developed digital switching was recognized as a coming technology–implying that channel banks would be paired on a dynamic basis as opposed to the one-to-one basis in T-carrier systems. Thus a greater degree of uniformity in companding characteristics would be required to allow pairing of channel banks on a nationwide basis. The main features incorporated into the D2 channel bank (and ensuing channel banks such as D3 and D4) to achieve the improved quality and standardization are:

1   Eight bits per PCM code word.
2   Incorporation of the companding functions directly into the encoder and decoder.
3   A new companding characteristic ($\mu 255$).

The D1 channel banks use 1 bit per time slot for signaling and 7 bits for voice. Thus a signaling rate of 8 kbps is established, which is more than necessary for basic voice service. In order to provide a higher data

---

*Nine tandem connections have never been known to occur.

rate for voice, signaling between D2, D3, D4 channel banks is inserted into the least significant bit position of 8 bit code words in every sixth frame. Thus every sixth $\mu255$ PCM code word contains only 7 bits of voice information—implying that the effective number of bits per sample is actually $7\frac{5}{6}$ bits instead of 8. When common channel inter-office signaling is established between two switching offices, however, the corresponding T-carrier systems no longer need to carry signaling information on a per channel basis, and a full 8 bits of voice can be transmitted in every channel of every frame.

The compression and expansion characteristics of the D1 channel banks were implemented separately from the encoders and decoders. The D2 channel bank incorporates the companding operations into the encoders and decoders themselves. In these channel banks a resistor array is used to establish a sequence of nonuniformly spaced decision thresholds. A sample value is encoded by successively comparing the input value to the sequence of decision thresholds until the appropriate quantization interval is located. The digital output becomes whatever code is used to represent the particular quantization interval. (See Appendix B for a detailed description of the direct encoding procedure used in the D2 channel banks.) By incorporating the companding functions directly into the encoders and decoders, D2 channel banks avoid certain problems associated with parameter variability and temperature sensitivity of companding diodes in D1 channel banks.

The D2 channel banks also provide improved performance in terms of the effect of channel errors on the decoded outputs. Of paramount concern in a PCM system is the effect of a channel error in the most significant bit position of a code word. Bit errors in other positions of a code word are much less noticeable to a listener. A channel error in the most significant bit of a code word produced by a D1 channel bank causes an output error that is always equal to one half of the entire coding range of the coder. The D2 channel bank, on the other hand, uses a sign-magnitude coding format. With this format, a channel error in the polarity bit causes an output error that is equal to twice the sample magnitude (i.e. the polarity is inverted). In the worst case this error may correspond to the entire encoding range. Maximum amplitude samples are relatively rare, however, so most channel errors in a D2 coding format produce outputs with error magnitudes less than one-half the coding range. Thus, on average, the sign-magnitude coding format of the D2 channel bank provides superior performance in the presence of channel errors [13].

In addition to a need for improved voice quality, it also became apparent that as more of the network began using digital techniques it would be necessary, or at least desirable, to implement many signal processing functions directly on digital signals and not convert them to an analog format for processing. Most signal processing functions (such

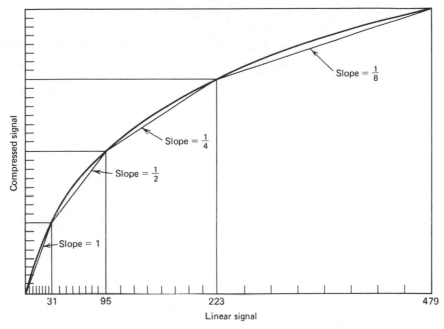

Figure 3.18 First four segments of straight line approximation to $\mu255$ compression curve.

as attenuating a signal or adding signals together) involve linear operations. Thus before processing a log-PCM voice signal it is necessary to convert the compressed transmission format into a linear format.

To simplify the conversion process, the particular companding characteristic with $\mu = 255$ was chosen. This companding characteristic has the property of being closely approximated by a set of eight straight-line segments. Furthermore, the slope of each successive segment is exactly one-half the slope of the previous segment. The first four segments of a $\mu255$ approximation are shown in Figure 3.18. The overall result is that the larger quantization intervals have lengths that are binary multiples of all smaller quantization intervals. Because of this property a compressed code word is easily expanded into a linear representation. Similarly, the linear representation is easily converted into a compressed representation.* In fact, some recent companding techniques use digital compression of linear code words instead of directly compressed encoding, as in the D2 channel banks. These techniques use a linear encoder with a relatively large number of bits to cover the entire dynamic range of the signal. As described in Appendix B, the

---

*The inexorable advance of semiconductor technology has obviated much of the ingenuity that went into selecting EDL coding formats. Brute force, table look up conversion between codes using read only memories (ROMs) is now the most cost-effective approach.

least significant bits of large sample values are discarded when compressing the code. The number of insignificant bits deleted is encoded into a special field included in the compressed code format. In this manner digital companding is analogous to expressing a number in scientific notation.

As shown in Figure 3.18, each major segment of the piecewise linear approximation is divided into equally sized quantization intervals. For 8 bit code words the number of quantization intervals per segment is 16. Thus an 8 bit $\mu255$ code word is composed of: 1 polarity bit, 3 bits to identify a major segment, and 4 bits for identifying a quantizing interval within a segment. Table 3.1 lists the major segment endpoints, the quantization intervals, and the corresponding segment and quantization interval codes.

The quantization intervals and decoded sample values in Table 3.1 have been expressed in terms of a maximum amplitude signal of 8159 so that all segment endpoints and decoder outputs are integers. Notice that the quantizing step is doubled in each of eight successive linear segments. It is this property that facilitates the conversion to and from a linear format. A complete encoding table is provided in Appendix B, along with a detailed description of the linear conversion process.

The straight-line approximation of the $\mu255$ companding curve is sometimes referred to as a 15 segment approximation. The 15 segments arise because, although there are eight positive segments and eight negative segments, the two segments nearest the origin are colinear and therefore can be considered as one. When viewed in this manner, the middle segment contains 31 quantization intervals with one segment straddling the origin (from -1 to $+1$ in Table 3.1). Code words for this middle quantization interval arise as a positive value less than $+1$ or a negative value greater than -1. There are, in effect, a positive zero and a negative zero. As represented in Table 3.1, these values are encoded as 00000000 and 10000000, respectively. However, the D2 channel bank inverts all code words for transmission. The smaller amplitude signals, with mostly "0" segment codes, are most probable and would therefore cause less than 50% pulses on the transmission line. The density of pulses is increased by inversion of the transmitted data, which improves the timing and clock recovery performance of the receiving circuitry in the regenerative repeaters. Thus the actual transmitted code words corresponding to a positive zero and a negative zero are respectively, 11111111 and 01111111, indicating strong timing content for the line signal of an idle channel.

In the interest of ensuring clock synchronization in the T1 repeaters, D2, D3, and D4 channel banks alter the transmitted data in one other way. As indicated in Table 3.1, a maximum amplitude negative signal is all 1's, which would normally be converted to all 0's for transmission. Instead, for the all 0's code word only, the second least significant bit

TABLE 3.1   ENCODING/DECODING TABLE FOR $\mu255$ PCM[a]

Input Amplitude Range	Step Size	Segment Code S	Quantization Code Q	Code Value	Decoder Amplitude
0–1	1		0000	0	0
1–3			0001	1	2
3–5	2	000	0010	2	4
:			:	:	:
29–31			1111	15	30
31–35			0000	16	33
:	4	001	:	:	:
91–95			1111	31	93
95–103			0000	32	99
:	8	010	:	:	:
215–223			1111	47	219
223–239			0000	48	231
:	16	011	:	:	:
463–479			1111	63	471
479–511			0000	64	495
:	32	100	:	:	:
959–991			1111	79	975
991–1055			0000	80	1023
:	64	101	:	:	:
1951–2015			1111	95	1983
2015–2143			0000	96	2079
:	128	110	:	:	:
3935–4063			1111	111	3999
4063–4319			0000	112	4191
:	256	111	:	:	:
7903–8159			1111	127	8031

[a]This table displays magnitude encoding only. Polarity bits are assigned as "0" for positive and "1" for negative. In transmission all bits are inverted.

is set to 1 so that 00000010 is transmitted. In effect, an encoding error is produced to preclude an all 0's code word. Fortunately, maximum amplitude samples are extremely unlikely so that no significant degradation occurs. (If the least significant bit were forced to a 1, a smaller decoding error would result. However, in every sixth frame this bit is "stolen" for signaling purposes and therefore is occasionally set to 0 independently of the code word. To ensure that "all-zero" code words are never transmitted, the second least significant bit is forced to a 1 when necessary.)

EXAMPLE 3.3:

Determine the sequence of code words for a D2 (also D3, D4) channel bank format representing a half maximum power 1 kHz digital signal.

Solution. Since the sampling rate of the standard $\mu 255$ PCM channel bank is 8 kHz, a sequence of eight samples can be repeated in cyclic fashion to generate the 1 kHz waveform. For convenience the phases of the samples are chosen to begin at $22.5°$. Thus the eight samples correspond to $22.5°$, $67.5°$, $112.5°$, $157.5°$, $202.5°$, $247.5°$, $292.5°$, and $337.5°$. With these phases, only two different magnitudes corresponding to $22.5°$ and $67.5°$ are required. The maximum amplitude of a half maximum power sine wave is $(0.707) 8159 = 5768$. Thus the two amplitudes contained in the sample sequence are

$$(5768) \sin (22.5°) = 2207$$

$$(5768) \sin (67.5°) = 5329$$

Using the encoding table in Appendix B, we determine the codes for these sample magnitudes to be 1100001 and 1110100, respectively. The sequence of eight samples can now be established as:

Polarity	Segment	Quantization		
0	1 1 0	0 0 0 1	$22.5°$
0	1 1 1	0 1 0 0	$67.5°$
0	1 1 1	0 1 0 0	$112.5°$
0	1 1 0	0 0 0 1	$157.5°$
1	1 1 0	0 0 0 1	$202.5°$
1	1 1 1	0 1 0 0	$247.5°$
1	1 1 1	0 1 0 0	$292.5°$
1	1 1 0	0 0 0 1	$337.5°$

Note. This sequence defines a 1 kHz test signal at a power level of 1 mW at the transmission level point (0dBm0). However, the actual transmitted data pattern is the complement of that provided above. Because only two amplitude samples are required to produce the test tone, this tone does not test all encoding/decoding circuitry. In general, a 1004 Hz tone is a better test tone since it is not harmonically related to a 8000 Hz sampling rate and will therefore exercise all encoder and decoder levels.

Performance of a $\mu 255$ PCM Encoder

As mentioned, the main motivation for developing a successor to the D1 channel bank was to provide better speech quality for digital toll network transmission links. The signal-to-quantizing noise ratio for a maximum amplitude sine wave in the first segment of a $\mu255$ codec is determined easily from Equation 3.4 as:

$$\text{SQR } (A = 31) = 7.78 + 20 \log_{10} \left(\frac{31}{2} \right)$$

$$= 31.6 \text{ dB}$$

The signal-to-quantizing noise ratios for larger amplitude sinusoids are not as easy to calculate since the lengths of the quantization intervals vary with the sample size. Thus a general calculation of the quantizing noise power involves finding the expected value of the power of the quantization errors:

$$\text{noise power} = \tfrac{1}{12} \sum_{i=0}^{7} p_i q_i^2 \tag{3.9}$$

where p_i = probability of a sample in the ith segment
$\qquad q_i$ = quantization size for segment i
$\qquad = 2^{i+1}$ for segmented $\mu255$ coding

Using Equation 3.9, we determine the signal-to-quantizing noise power for a full range sinusoid as:

$$\text{SQR } (A = 8159) = 39.3 \text{ dB}$$

For comparison, if all quantization intervals had the maximum length of 256 as in the upper segment, Equation 3.4 provides an SQR of 37.8 dB. The difference of only 1.5 dB reflects the fact that a full scale sine wave spends 67% of the time in the upper segment where the quantization intervals are maximum (i.e. $p_7 = 0.67$). A voice signal, on the other hand, has a higher peak-to-average ratio than a sinewave. The average quantization error is smaller, but so is the average signal power. Hence the SQR is approximately the same.

The dynamic range of a segmented PCM encoder is determined as the signal power difference between a low-level signal occupying the entire range of the first segment and a high-level signal extending to the limits of the coder. Thus the dynamic range of a segmented 255 coder is determined as:

$$\text{DR } (A = 31 \text{ to } 8159) = 20 \log_{10} \left(\tfrac{8159}{31} \right) = 48.4 \text{ dB}$$

In summary, an 8 bit $\mu255$ PCM codec provides a theoretical signal-to-quantizing noise ratio greater than 30 dB across a dynamic range of 48 dB. For comparison, Equation 3.4 or Figure 3.14 reveals that a linear PCM encoder/decoder requires 13 bits for equivalent performance. (The extra quality of a linear coder at high signal levels is unneeded.)

The theoretical performance of an 8 bit segmented $\mu255$ coder is

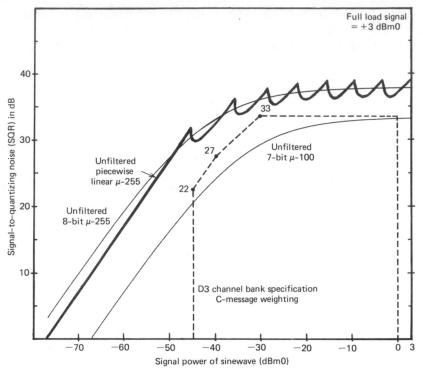

Figure 3.19   Signal-to-quantizing noise of μ-law coding with sinewave inputs.

shown in Figure 3.19 as a function of the amplitude of a sine wave input. Also shown is the theoretical performance of an unsegmented $\mu255$ coder and a 7 bit $\mu100$ coder used in the D1 channel bank. Notice that the 8 bit coders provide about 5 dB improvement over the 7 bit coder for high-level signals and even more improvement for low-level signals. The performance shown for the 8 bit coders does not include the effect of using only 7 bits for voice coding in every sixth frame. When this effect is included, the 8 bit coders lose 1.76 dB in performance.

The scalloped effect of the segmented coder occurs because the quantization intervals change abruptly at the segment endpoints instead of continuously as in analog companding.* Also as shown in Figure 3.19, note the required performance of a D3 channel bank as specified by the Bell System [9]. This specification assumes all noise measurements are made using C-message weighting. C-message weighting reduces the effective noise level by 2 dB and therefore improves the SQR. Thus an ideal 8 bit $\mu255$ coder actually exceeds the specification by more than that

_____

*A finer scalloped effect also exists within individual quantization intervals because the quantization noise is zero at the center and peaks at the edges of the interval (see Figure 3.9).

shown in Figure 3.19. When the least significant bit of every sixth frame is used for signaling, however, the SQR is reduced by a comparable 1.76 dB.

A-Law Companding

The companding characteristic recommended by CCITT is referred to as an A-law characteristic. This characteristic has the same basic features and implementation advantages as does the μ-law characteristic. In particular, the A-law characteristic can also be well approximated by straight-line segments to facilitate direct or digital companding, and can be easily converted to and from a linear format. The normalized A-law compression characteristic is defined as:

$$F_A(x) = \text{sgn}(x) \left(\frac{A|x|}{1 + \ln(A)} \right) \quad 0 \leqslant |x| \leqslant \frac{1}{A}$$

$$= \text{sgn}(x) \left(\frac{1 + \ln|Ax|}{1 + \ln(A)} \right) \quad \frac{1}{A} \leqslant |x| \leqslant 1 \tag{3.10}$$

The inverse or expansion characteristic is defined as:

$$F_A^{-1}(y) = \text{sgn}(y) \frac{|y|[1 + \ln(A)]}{A} \quad 0 \leqslant |y| \leqslant \frac{1}{1 + \ln(A)}$$

$$= \text{sgn}(y) \frac{(e^{|y|[1+\ln(A)]-1})}{A} \quad \frac{1}{1 + \ln(A)} \leqslant |y| \leqslant 1 \tag{3.11}$$

where $y = F_A(x)$

Notice that the first portion of the A-law characteristic is linear by definition. The remaining portion of the characteristic ($1/A \leqslant |x| \leqslant 1$) can be closely approximated by linear segments in a fashion similar to the μ-law approximation. All in all, there are eight positive and eight negative segments. The first two segments of each polarity (four in all) are colinear and therefore are sometimes considered as one straight-line segment. Thus the segmented approximation of the A-law characteristic is sometimes referred to as a "13 segment approximation." For ease in describing the coding algorithms of the segmented companding characteristic, however, a 16 segment representation is used—just as in the case of the segmented μ-law characteristic.

The segment endpoints, quantization intervals, and corresponding codes for an 8 bit segmented A-law characteristic are shown in Table 3.2. The values are scaled to a maximum value of 4096 for integral representations. Figure 3.20 displays the theoretical performance of the A-law approximation where it is compared to the performance of a μ-law approximation presented in Figure 3.19. Notice that the A-law characteristic provides a slightly larger dynamic range. However, the

TABLE 3.2 SEGMENTED A-LAW ENCODING/DECODING TABLE

Input Amplitude Range	Step Size	Segment Code S	Quantization Code Q	Code Value	Decoder Amplitude
0–2			0000	0	1
2–4		000	0001	1	3
:			:	:	:
:					
30–32	2		1111	15	31
32–34			0000	16	33
:		001	:	:	:
:			:	:	:
62–64			1111	31	63
64–68			0000	32	66
:	4	010	:	:	:
124–128			1111	47	126
128–136			0000	48	132
:	8	011	:	:	:
248–256			1111	63	252
256–272			0000	64	264
:	16	100	:	:	:
496–512			1111	79	504
512–544			0000	80	528
:	32	101	:	:	:
992–1024			1111	95	1008
1024–1088			0000	96	1056
:	64	110	:	:	:
1984–2048			1111	111	2016
2048–2176			0000	112	2112
:	128	111	:	:	:
3968–4096			1111	127	4032

A-law characteristic is inferior to the μ-law characteristic in terms of small signal quality (idle channel noise). The difference in small signal performance occurs because the minimum step size of the A-law standard is 2/4096 whereas the minimum step size of μ-law is 2/8159. Furthermore, notice that the A-law approximation does not define a zero-level output for the first quantization interval (i.e. uses a mid-riser quantizer). However, the difference between mid-riser and mid-tread performance at 64 kbps is imperceptible [14].

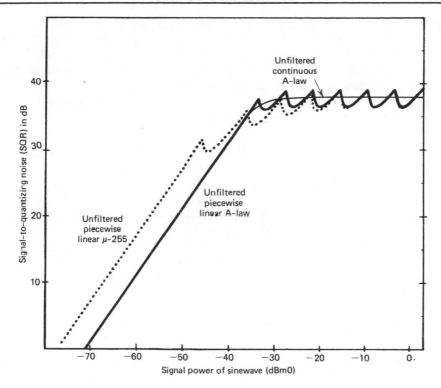

Figure 3.20   Signal-to-quantizing noise of A-law PCM coding with sinewave inputs.

3.2.6   Adaptive Gain Control

Another method of reducing the number of bits required in a PCM codeword (without increasing quantization noise) is to restrict the dynamic range of the input to the encoder. Since the actual power level of the voice cannot be controlled, automatic gain control (AGC) can be used to adjust all inputs to the encoder to a standard power level. Telephone companies have not generally used automatic gain control on voice signals, partly because of tradition and partly because of requirements. One drawback to full range AGC on a voice circuit is the implication that all speech would be received at a single power level. Thus it would be fruitless to ask someone to talk louder to overcome a noisy circuit or a listener's hearing deficiencies. Furthermore, background sounds at a speaker's telephone would be amplified to the average speech power during pauses. Thus a thresholding or squelch circuit would be needed at the source to preserve pauses.

These basic shortcomings of AGC can be overcome if the source terminal encodes and transmits the gain adjustment factor to the destination. Then the receiving terminal adjusts the standard power level (coming out of the decoder) to the original power level of the analog input. In this manner encoders and decoders are concerned only with limited range signals, while source and destination terminals are adapted to track changes in the input signal power level. Since voice-signal power levels tend to change gradually, the gain measurements do not have to occur too often, and the gain factors do not represent significant transmission overhead. Meanwhile, waveform coding is confined to a limited amplitude range so that fewer bits per sample can be used. Figure 3.21 shows a basic block diagram of an adaptive gain control system for PCM encoding.

There are two basic modes of operation for adaptive gain control depending on how gain factors are measured and to which speech segments the factors are applied. One mode of operation, as implied in Figure 3.21, involves measuring the power level of one segment of speech and using that information to establish a gain factor for ensuing speech segments. Obviously, this mode of operation relies on gradually changing power levels. This mode of operation is sometimes referred to as "backward estimation."

Another mode of operation involves measuring the power level of a speech segment and using the gain factor thus derived to adapt the encoder to the same segment. This approach, referred to as "forward estimation," has the obvious advantage that the encoder and decoder use gain factors specifically related to the speech segments from which they are derived. The disadvantage is that each speech segment must be delayed while the gain factor is being determined. This disadvantage is becoming less objectionable, however, owing to new technologies such as charge-coupled devices for analog delay or digital memory and shift registers for digital delay.

Adaptive gain control with explicit transmission of gain factors is not without shortcomings. First, when the periodic gain information is inserted into the transmitted bit stream, some means of framing the bit stream into blocks is needed so gain information can be distinguished

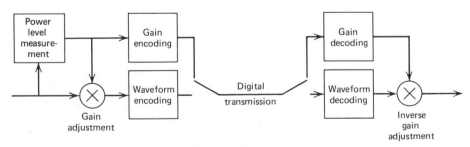

Figure 3.21   Adaptive gain control.

from waveform coding. Second, periodic insertion of gain information disrupts information flow causing higher transmitter clock rates that are often inconveniently related to the waveform sample clock. Third, correct reception of gain factors is usually critical to voice quality—indicating a need to redundantly encode gain information.

Reference 15 describes a modified form of PCM using forward estimation of gain factors that is referred to as "Nearly Instantaneously Companded PCM." The need for transmitting speech segments in blocks is not a disadvantage in the application mentioned (mobile telephone) because repetitive bursts with error checking are used as a means of overcoming short-lived multipath fading.

Adaptive gain control with explicit transmission of gain information has not been used extensively for digital voice transmission. One exception is the Bell System's subscriber loop multiplexer [16], which is now obsolete. A form of gain control using backward estimation to derive the gain information from waveform coding has been used quite successfully in most delta modulation systems. Rather than adjusting the power levels of the signals directly, as indicated in Figure 3.21, these systems perform the equivalent function of adjusting the step sizes in the encoders and decoders. These techniques are described more fully in Section 3.5.3.

3.3 SPEECH REDUNDANCIES

As mentioned in the previous sections, a conventional PCM system encodes each sample of the input waveform independently from all other samples. Thus a PCM system is inherently capable of encoding an arbitrarily random waveform whose maximum frequency component does not exceed one-half the sampling rate. Analyses of speech waveforms, however, indicates there is considerable redundancy from one sample to the next. In fact, as reported in Reference 7 the correlation coefficient (a measure of predictability) between adjacent 8 kHz samples is generally .85 or higher. Hence the redundancy in conventional PCM codes suggests significant savings in transmission bandwidths are possible through more efficient coding techniques. All of the digitization techniques described in the rest of this chapter are tailored, in one degree or another, to the characteristics of speech signals with the intent of reducing the bit rate.

In addition to the correlation existing between adjacent samples of a speech waveform, several other levels of redundancy can be exploited to reduce encoded bit rates. Table 3.3 lists these redundancies. Not included are higher-level redundancies related to context dependent interpretations of speech sounds (phonemes), words, and sentences.

TABLE 3.3   SPEECH REDUNDANCIES

TIME DOMAIN REDUNDANCIES

1 Nonuniform amplitude distributions
2 Sample-to-sample correlations
3 Cycle-to-cycle correlations (periodicity)
4 Pitch-interval to pitch-interval correlations
5 Inactivity factors (speech pauses)

FREQUENCY DOMAIN REDUNDANCIES

6 Nonuniform long term spectral densities
7 Sound specific short term spectral densities

These topics are not covered because techniques that analyze speech waveforms to extract only information content eliminate subjective qualities essential to general telephony.

3.3.1   Nonuniform Amplitude Distributions

As mentioned in the introduction to companding, lower amplitude sample values are more common than higher amplitude sample values. Most low-level samples occur as a result of speech pauses in a conversation. Beyond this, however, the power levels of active speech signals also tend to occur at the lower end of the encoding range. The companding procedures described in the previous section provide slightly inferior quality (i.e., lower signal-to-noise ratios) for small signals compared to large signals. Thus the average quality of PCM speech could be improved by further shortening of lower-level quantization intervals and increasing of upper-level quantization intervals. The amount of improvement realized by such a technique, however, probably would not justify the additional complexities, particularly if the linearizability of contemporary compandors was compromised.

The most beneficial approach to processing signal amplitudes in order to reduce encoder bit rates involves some form of adaptive gain control as discussed earlier. On a long-term basis, a single speaker confines his or her power levels to a narrower range than that of the population of all speakers. Over a shorter term, individual speech segments (syllables) maintain a fairly constant power level in their duration. Since a syllable lasts for approximately 30 ms, a 64 kbps digitized voice signal typically produces 1920 bits of data between changes in power level. Thus power level transmission overhead is insignificant in terms of bandwidth requirements. The "nearly instantaneously companded" PCM system mentioned previously [15] provides a bit rate reduction of 30% with respect to conventional PCM.

3.3.2 Sample-to-Sample Correlation

The high correlation factor of .85 mentioned in Section 3.3 indicates that any significant attempt to reduce transmission rates must exploit the correlation between adjacent samples. In fact, at 8 kHz sampling rates, significant correlations also exist for samples two to three samples apart. Naturally, samples become even more correlated if the sampling rate is increased.

The simplest way to exploit sample-to-sample redundancies in speech is to encode only the differences between adjacent samples. The difference measurements are then accumulated in a decoder to recover the signal. In essence these systems encode the slope or derivative of a signal at the source and recover the signal by integrating at the destination. Digitization algorithms of this type are discussed at length in later sections.

3.3.3 Cycle-to-Cycle Correlations

Although a speech signal requires the entire 300 to 3400 Hz bandwidth provided by a telephone channel, at any particular instant in time certain sounds may be composed of only a few frequencies within the band. When only a few underlying frequencies exist in a sound, the waveform exhibits strong correlations over numerous samples corresponding to several cycles of an oscillation. The cyclic nature of a voiced sound is evident in the time waveform shown in Figure 3.22. Encoders exploiting the cycle-to-cycle redundancies in speech are markedly more complicated than those concerned only with removing the redundancy in adjacent samples. In fact, these encoders more or less represent a transition from the relatively high-rate, natural-sounding waveform encoders to the relatively low-rate, synthetic-sounding vocoders.

Figure 3.22 Time waveform of voiced sound.

3.3.4 Pitch Interval to Pitch Interval Correlations

Human speech sounds are often categorized as being generated in one of two basic ways. The first category of sounds encompasses "voiced" sounds which arise as a result of vibrations in the vocal cords. Each vibration allows a puff of air to flow from the lungs into the vocal tract. The interval between puffs of air exciting the vocal tract is referred to as the pitch interval or more simply, the rate of excitation is the pitch. Generally speaking, voiced sounds arise in the generation of vowels and the latter portions of some consonants. An example of a time waveform for a voiced sound is shown in Figure 3.22.

The second category of sounds includes the fricatives or "unvoiced" sounds. Fricatives occur as a result of continuous air flowing from the lungs and passing through a vocal tract constricted at some point to generate air turbulence (friction). Unvoiced sounds correspond to certain consonants such as f, j, s, x. An example of a time waveform of an unvoiced sound is shown in Figure 3.23. Notice that an unvoiced sound has a much more random waveform than a voiced sound.

As indicated in Figure 3.22, a voiced sound not only exhibits the cycle-to-cycle redundancies mentioned in Section 3.3.3, but the waveform also displays a longer term repetitive pattern corresponding to the duration of a pitch interval. Thus one of the most efficient ways of encoding the voiced portions of speech is to encode one pitch interval waveform and use that encoding as a template for each successive pitch interval in the same sound. Pitch intervals typically last from 5 ms to 20 ms for men and from 2.5 ms to 10 ms for women. Since a typical voiced sound lasts for approximately 100 ms, there may be as many as 20 to 40 pitch intervals in a single sound. Although pitch interval encoding can provide significant reductions in bit rates, the pitch is sometimes very difficult to detect. (Not all voiced sounds produce a readily identifiable pitch interval as in Figure 3.22.) If the pitch gets encoded erroneously, strange sounds result.

An interesting aspect of pitch interval encoding is that it provides a means of speeding up speech while maintaining intelligibility. By deleting some percentage of pitch intervals from each sound (phoneme) the rate of sound generation is effectively increased in a manner analogous to more rapid word formation. The pitch of the sounds remains unchanged. In contrast, if the rate of reconstruction is merely increased, all frequencies including the pitch increase proportionately. Moderate

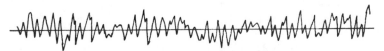

Figure 3.23 Time waveform of unvoiced sound.

speedups produce obvious distortion while greater speedups become unintelligible. Devices designed to simulate faster word formation have demonstrated that we are capable of assimilating spoken information much faster than we normally generate it.

3.3.5 Inactivity Factors

Analyses of telephone conversations have indicated that a party is typically active about 40% of a call duration. Most inactivity occurs as a result of one person listening while the other is talking. Thus a conventional (circuit-switched) full-duplex, connection is significantly underutilized. Time assignment speech interpolation (TASI) described in Chapter 1, is a technique to improve channel utilization on expensive analog links. Digital speech interpolation (DSI) is a term used to refer to a digital circuit counterpart of TASI. In essence DSI involves: sensing speech activity, seizing a channel, digitally encoding and transmitting the utterances, and releasing the channel at the completion of each speech segment.

Digital speech interpolation is obviously applicable to digital speech storage systems where the duration of a pause can be encoded more efficiently than the pause itself. In recorded messages, however, the pauses are normally short since a "half-duplex" conversation is not taking place. DSI is being proposed for numerous transmission systems discussed in Chapter 8.

3.3.6 Nonuniform Long-Term Spectral Densities

The time domain redundancies described in the preceding sections exhibit certain characteristics in the frequency domain that can be judiciously processed to reduce the encoded bit rate. Frequency domain redundancies are not independent of the redundancies in the time domain. Frequency domain techniques merely offer an alternate approach to analyzing and processing the redundancies.

A totally random or unpredictable signal in the time domain produces a frequency spectrum that is flat across the bandwidth of interest. Thus a signal that produces uncorrelated time domain samples makes maximum use of its bandwidth. On the other hand, a nonuniform spectral density represents inefficient use of the bandwidth and is indicative of redundancy in the waveform. Figure 3.24 shows the long-term spectral density of speech signals averaged across two populations: men and women [17]. Notice that the upper portions of the 3 kHz bandwidth passed by the telephone network have significantly reduced power levels. The lower power levels at higher frequencies are a direct consequence of the time domain sample-to-sample correlations discussed previously. Large amplitude signals cannot change rapidly

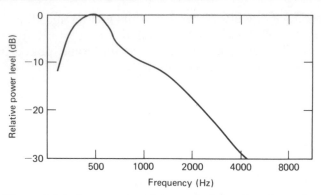

Figure 3.24 Long-term power spectral density of speech.

because, on the average, they are predominantly made up of lower-frequency components.

A frequency domain approach to more efficient coding involves flattening the spectrum before encoding the signal. The flattening process can be accomplished by passing the signal through a high-pass filter to emphasize the higher frequencies before sampling. The original waveform is recovered by passing the decoded signal through a filter with a complementary, low-pass characteristic. An important aspect of this process is that a high-pass filter exhibits time domain characteristics similar to a differentiator and a low-pass filter has time domain characteristics analogous to an integrator. Thus, the spectrum flattening process essentially means the slope of the signal is encoded at the source, and the signal is recovered by integrating at the destination—the basic procedure described previously for sample-to-sample redundancy removal in the time domain.

In studying Figure 3.24 it is natural to think that the remarkably low levels of signal energy at the higher frequencies (2 to 3.4 kHz) means that more bandwidth is being allocated to a voice signal than is really necessary. The error in such a conclusion, however, lies in the distinction between energy content and information content of the voice frequency spectrum. As any beginning computer programmer soon learns, the meaning of a program variable can be retained even though it is shortened by deleting all of the vowels. In speech the vowels require most of the energy and occupy primarily the lower portion of the frequency band. The consonants, on the other hand, contain most of the information but use much less power and generally higher frequencies. Hence merely reproducing a high percentage of the original speech energy is an inadequate goal for a digital speech transmission or storage system.

3.3.7 Short-Term Spectral Densities

The speech spectrums shown in Figure 3.24 represent long-term averages of the spectral densities. Over shorter periods of time the spectral densities vary considerably and exhibit sound specific structures with energy peaks (resonances) at some frequencies and energy valleys at others. The frequencies at which the resonances occur are called formant frequencies, or simply formants. Voiced speech sounds typically contain three to four identifiable formants. These features of the short-term spectral density are illustrated in the spectogram of Figure 3.25. A spectogram is a display of speech spectral energy as a function of time and frequency. The horizontal axis represents time, the vertical axis represents frequency, and the shadings represent energy levels. Thus the darker portions in Figure 3.25 indicate relatively high energy levels (formants) at particular instants in time.

Frequency domain voice coders provide improved coding efficiencies by encoding the most important components of the spectrum on a dynamic basis. As the sounds change different portions (formants) of the frequency band are encoded. The period between formant updates is typically 10 to 20 ms. Instead of using periodic spectrum measurements, some higher quality vocoders continuously track gradual changes in the spectral density at a higher rate. Frequency domain vocoders often provide lower bit rates than the time domain coders, but typically produce more unnatural sounding speech.

3.4 DIFFERENTIAL PULSE CODE MODULATION

Differential pulse code modulation (DPCM) is designed specifically to take advantage of the sample-to-sample redundancies in a typical speech waveform. Since the range of sample differences is less than the range of individual amplitude samples, fewer bits are needed to encode difference samples. The sampling rate is often the same as for a comparable PCM system. Thus the bandlimiting filter in the encoder and the smoothing filter in the decoder are basically identical to those used in conventional PCM systems.

The simplest means of generating the difference samples for a DPCM coder is to store the previous input sample directly in a sample-and-hold circuit and use an analog subtractor to measure the change. The change in the signal is then quantized and encoded for transmission. The DPCM structure shown in Figure 3.26 is more complicated, however, because the previous input value is reconstructed by a feedback loop that integrates the encoded sample differences. In essence, the feedback signal is an estimate of the input signal as obtained by integrating the encoded

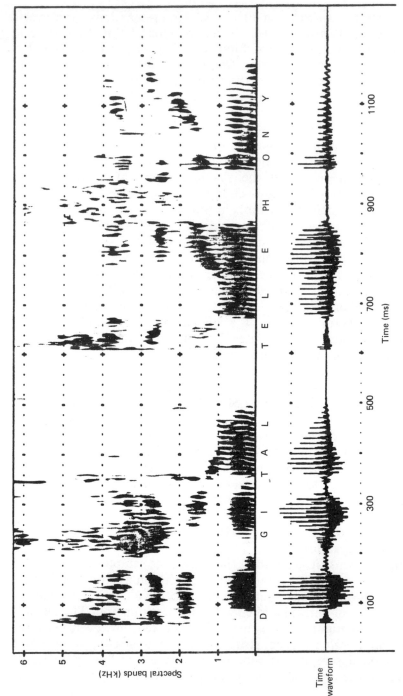

Figure 3.25 Spectogram of phrase: "digital telephony." (Courtesy of Robert L. Davis of Texas Instruments Inc.)

120

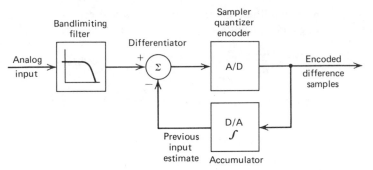

Figure 3.26 Functional block diagram of differential PCM.

sample differences. Thus the feedback signal is obtained in the same manner used to reconstruct the waveform in the decoder.

The advantage of the feedback implementation is that quantization errors do not accumulate indefinitely. If the feedback signal drifts from the input signal, as a result of an accumulation of quantization errors, the next encoding of the difference signal automatically compensates for the drift. In a system without feedback the output produced by a decoder at the other end of the connection might accumulate quantization errors without bound.

As in PCM systems, the analog-to-digital conversion process can be uniform or companded. Some DPCM systems also use adaptive techniques to adjust the quantization step size in accordance with the average power level of the signal. (See Reference 18 for an overview of various techniques.) These adaptive techniques are often referred to as syllabic companding, in accordance with the time interval between gain adjustments. Syllabic companding is discussed in conjunction with delta modulation systems, where it is most often used.

EXAMPLE 3.4

Speech digitization techniques are sometimes measured for quality by use of a 800 Hz sine wave as a representative test signal. Assuming a uniform PCM system is available to encode the sine wave across a given dynamic range, determine how many bits per sample can be saved by using a uniform DPCM system.

Solution. In essence, the solution is obtained by determining how much smaller the dynamic range of the difference signal is in comparison to the dynamic range of the signal amplitude. Assume the maximum amplitude of the sine wave is A, so that

$$x(t) = A \sin (2\pi \cdot 800t)$$

The maximum amplitude of the difference signal can be obtained by differentiating and multiplying by the time interval between samples:

$$\frac{dx}{dt} = A \cdot (2\pi) \cdot (800) \cdot \cos (2\pi \cdot 800t)$$

$$|\Delta x(t)|_{\max} = A \cdot (2\pi) \cdot (800) \cdot \left(\frac{1}{8000}\right) = 0.628A$$

The savings in bits per sample can be determined as:

$$\log_2 \left(\frac{1}{0.628}\right) = 0.67 \text{ bits}$$

Example 3.4 demonstrates that a DPCM system can use $\frac{2}{3}$ bit per sample less than a PCM system with the same quality. Typically DPCM systems provide a full 1 bit reduction in code word size. The larger savings is achieved because, on average, speech waveforms have a lower slope than an 800 Hz tone. (See Figure 3.24.)

3.4.1   DPCM Implementations

Differential PCM encoders and decoders can be implemented in a variety of ways depending on how the signal processing functions are partitioned between analog and digital circuitry. At one extreme the differencing and integration functions can be implemented with analog circuitry, while at the other extreme all signal processing can be implemented digitally using conventional PCM samples as input. Figure 3.27 shows block diagrams of three different implementations with differing amounts of digital signal processing.

Fig. 3.27a depicts a system using analog differencing and integration. Analog-to-digital (A/D) conversion is performed on the difference signal, and conversion from digital-to-analog (D/A) for the feedback loop is immediately performed on the limited-range difference code. Analog summation and storage in a sample-and-hold (S/H) circuit is used to provide integration.

Figure 3.27b shows a system that performs the integration function digitally. Instead of immediately converting the difference code back to analog for feedback, the difference code is summed and stored in a data register to generate a digital representation of the previous input sample. A full scale D/A converter is then used to produce the analog feedback signal for differencing. Notice that the D/A converters in Figure 3.27b must provide full amplitude range conversion whereas the D/A converters in Figure 3.27a convert the more limited difference signals.

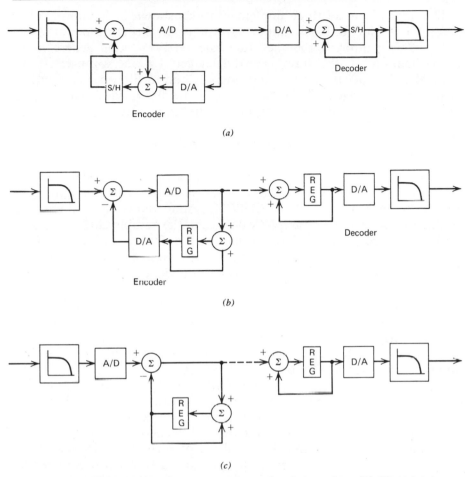

Figure 3.27 DPCM implementations. (*a*) Analog integration. (*b*) Digital integration. (*c*) Digital differencing.

Figure 3.27*c* shows a system where all signal processing is performed by digital logic circuits. The A/D converter produces full amplitude range sample codes which are compared to digitally generated approximations of the previous amplitude code. Notice that the A/D converter in this case must encode the entire dynamic range of the input whereas the A/D converters in the other two versions operate on only the difference signals.

The advantages of the digital implementations lie in the robustness and reproducibility of digital circuitry and its applicability to large scale integrated circuit (LSI) implementations. For implementations with standard components, the analog versions are obviously simpler. Furthermore, the cost of the analog-to-digital conversion equipment decreases as the conversion range decreases.

The decoders in all three implementations shown in Figure 3.27 are exactly like the feedback implementations in the corresponding encoder. This reinforces the fact that the feedback loop generates an approximation of the input signal (delayed by one sample). If no channel errors occur, the decoder output (before filtering) is identical to the feedback signal. Thus the closer the feedback signal matches the input, the closer the decoder output matches the input.

3.4.2 Higher Order Prediction

A more general viewpoint of a DPCM encoder considers it a special case of a linear predictor with encoding and transmission of the prediction error. The feedback signal of a DPCM system represents a first-order prediction of the next sample value, and the sample difference is a prediction error. Under this viewpoint the DPCM concept can be extended to incorporate more than one past sample value into the prediction circuitry. Thus the additional redundancy available from all previous samples can be weighted and summed to produce a better estimate of the next input sample. With a better estimate, the range of the prediction error decreases to allow encoding with fewer bits. For systems with constant predictor coefficients, results have shown that most of the realizable improvement occurs when using only the last three sample values. The basic implementation of a linear predictive coder using the last three sample values is shown in Figure 3.28. This implementation uses analog integration as shown in Figure 3.27a for standard DPCM. Of course, implementations utilizing a greater amount of digital signal processing, as in Figures 3.27b and c, are possible.

As mentioned in Section 3.4, analysis of differential PCM systems with first-order predication typically provides a 1 bit per sample reduction in code length relative to PCM systems with equivalent performance. Extended DPCM systems utilizing third-order prediction can provide reductions of $1\frac{1}{2}$ to 2 bits per sample [19]. Thus a standard

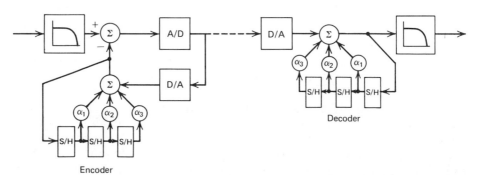

Figure 3.28 Extension of DPCM to third-order prediction.

DPCM system can provide 64 kbps PCM quality at 56 kbps, and third-order linear prediction can provide comparable quality at 48 kbps. However, subjective evaluations often indicate that somewhat higher bit rates are needed to match 64 kbps PCM quality.

Although the DPCM techniques described here can provide worthwhile reductions in transmission bit rates, they have not been used extensively in public telephony for two reasons. First, the 64 kbps PCM systems are well entrenched and accepted as being able to provide the desired quality. Second, delta modulation, described in the next section, is a special case of DPCM that provides comparable quality and is much simpler to implement. As LSI implementations are developed for the more sophisticated coding algorithms, the significance of the implementation differences will diminish.

3.5 DELTA MODULATION

Delta modulation (DM) is another digitization technique that specifically exploits the sample-to-sample redundancy in a speech waveform. In fact, DM can be considered as a special case of DPCM using only 1 bit per sample of the difference signal. The single bit specifies merely the polarity of the difference sample and thereby indicates whether the signal has increased or decreased since the last sample. An approximation to the input waveform is constructed in the feedback path by stepping up one quantization level when the difference is positive ("one") and stepping down when the difference is negative ("zero"). In this way the input is encoded as a sequence of "ups" and "downs" in a manner resembling a staircase. Figure 3.29 shows a DM approximation of a typical waveform. Notice that the feedback signal continues to step in one direction until it crosses the input at which time the feedback step reverses direction until the input is crossed again. Thus when tracking the input signal the DM output "bounces" back and forth across the input waveform allowing the input to be accurately reconstructed by a smoothing filter.

Since each encoded sample contains a relatively small amount of information (1 bit), delta modulation systems require a higher sampling

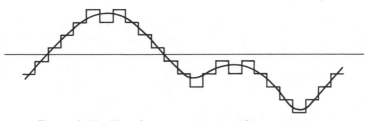

Figure 3.29 Waveform encoding by delta modulation.

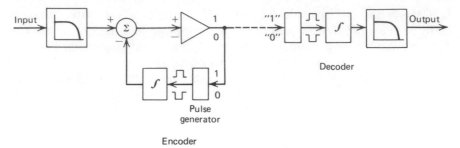

Figure 3.30   Delta modulation encoder and decoder.

rate than PCM or multibit DPCM systems. In fact, the sampling rate is necessarily much higher than the minimum (Nyquist) sampling rate of twice the bandwidth. From another viewpoint, "oversampling" is needed to achieve better prediction from one sample to the next.

Delta modulation has attracted considerable interest as a method of digitizing various types of analog signals. One of the main attractions of DM is its simplicity. Figure 3.30 shows a basic implementation of a DM encoder and decoder. Notice that the A/D conversion function is provided by a simple comparator. A positive difference voltage produces a 1, and a negative difference voltage produces a 0. Correspondingly, the D/A function in the feedback path, and in the decoder, is provided by a two-polarity pulse generator. In the simplest form the integrator can consist of nothing more than a capacitor to accumulate the charge from the pulse generator.

In addition to these obvious implementation simplicities, a delta modulator also allows the use of relatively simple filters for bandlimiting the input and smoothing the output [20]. As discussed in Section 3.1, the spectrum produced by a sampling process consists of replicas of the sampled spectrum centered at multiples of the sampling frequency. The relatively high sampling rate of a delta modulator produces a wider separation of these spectrums, and, hence, foldover distortion is prevented with less stringent roll-off requirements for the input filter.

3.5.1   Slope Overload

The conceptual operation of a delta modulator shown in Figure 3.29 indicates that the encoded waveform is never much more than a step size away from the input signal. Sometimes, however, a delta modulator may not be able to keep up with rapid changes in the input signal and thus fall more than a step size behind. When this happens the delta modulator is said to be experiencing "slope overload." A slope overload condition is shown in Figure 3.31.

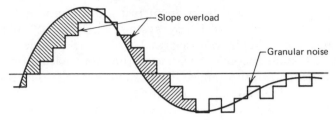

Figure 3.31   Slope overload and granular noise of delta modulation system.

Basically, slope overload occurs when the rate of change of the input exceeds the maximum rate of change that can be generated by the feedback loop. Since the maximum rate of change in the feedback loop is merely the step size times the sampling rate, a slope overload condition occurs if:

$$\left| \frac{dx(t)}{dt} \right| > q \cdot f_s$$

where $x(t)$ = input signal
      q = step size,   (3.12)
      f_s = sampling frequency

The design of a delta modulator necessarily involves a trade-off between two types of distortion: the more or less random quantization noise, sometimes referred to as granular noise, and the slope overload noise. As indicated in Figure 3.31, granular noise is a predominant consideration for slowly changing signals, whereas slope overload is dominant during rapidly changing signals. Obviously, granular noise is small if step sizes are small, but small step sizes increase the likelihood of slope overload. The optimum step size in terms of minimizing the total of granular and slope overload noise has been considered by Abate [21].

The perceptual effects of slope overload on the quality of a speech signal are significantly different from the perceptual effects produced by granular noise. As indicated in Figure 3.31, the slope overload noise reaches its peaks just before the encoded signal reaches its peaks. Hence, slope overload noise has strong components identical in frequency and approximately in phase with a major component of the input. Distortion that is correlated in this manner to the speech signal is effectively "masked" by the speech energy and therefore is less noticeable than "uncorrelated" distortion. In fact, overload noise is much less objectionable to a listener than random or granular noise at an equivalent power level [22]. Hence, from the point of view of perceived speech quality, the optimum mix of granular and slope overload noise is difficult to determine.

Slope overload is not a limition of just a delta modulation system, but an inherent problem with any system, such as DPCM in general, that encodes the difference in a signal from one sample to the next. A difference system encodes the slope of the input with a finite number of bits and hence a finite range. If the slope exceeds that range, slope overload occurs. In contrast, a conventional PCM system is not limited by the rate of change of the input, only by the maximum encodable amplitude. Notice that a differential system can encode signals with arbitrarily large amplitudes, as long as the large amplitudes are attained gradually.

3.5.2 Linear Delta Modulation

The simplest delta modulator uses a constant step size for all signal levels and is therefore referred to as a uniform or linear delta modulator (LDM). The design of a linear delta modulator is fundamentally concerned with selecting a fixed step size and a sampling rate to satisfy two criteria. First, the signal-to-granular noise ratio must be some minimum value for the lowest level signal to be encoded. Second, the signal-to-slope overload distortion ratio must be some minimum for the highest level signal to be encoded.

A convenient procedure for satisfying the slope overload criterion is to use Equation 3.12 and design the system so that slope overload is just on the verge of occurring at the highest level of input. Thus the step size for the highest level signal is not optimum in a perceptual sense or even in terms of minimizing the sum of granular noise and overload distortion. Nevertheless, when only granular noise is present, a delta modulator's SQR performance can be appropriately compared to the SQR performance of a PCM system. Comparing the two encoding methods under other conditions, for example when overloads occur in either system, requires listener evaluations.

The expected energy level of a single (unoverloaded) quantization error in a delta modulation system is determined from Equation 3.2 as $q^2/12$ where q is the step size. However, owing to the relatively high sampling rate of a DM system, the frequency spectrum of the quantization errors is much wider than the frequency spectrum of quantization errors in a PCM system. Thus the output filter of a DM decoder removes a higher percentage of the quantization noise than does the output filter of a PCM decoder. For sampling rates that are more than six times the maximum frequency of the input filter, the output quantization noise is essentially proportional to the ratio of the filter cutoff frequency to the sampling rate. Thus the quantization noise power of a linear delta modulator can be approximated as:

$$\text{LDM noise power} = K \left(\frac{f_c}{f_s} \right) q^2 \qquad (3.13)$$

where f_c = cutoff frequency of the output filter
    f_s = sampling frequency
    q = step size
    $K = 0.32$

The value of K in Equation 3.13 has been determined by numerous researchers producing such values as: 0.18 [23], 0.32 [24], 0.51 [25], and 0.67 [26]. The value 0.32 is chosen here because it appears to be the most commonly used and is the original value derived by de Jager [24] in his pioneering work on delta modulation. Many varieties of delta modulators have been proposed, leading to numerous other expressions for the quantization noise power.

Using a sine wave of frequency f as an input signal, Equations 3.12 and 3.13 can be combined to provide the following commonly used expression for the signal-to-quantizing noise ratio of a linear delta modulator:

$$SQR = \frac{(A^2/2)}{K(f_c/f_s)(2\pi f A/f_s)^2}$$

$$= (0.04)\frac{f_s^3}{f_c f^2}$$

(3.14)

where the input is a sine wave: $A\sin(2\pi ft)$, and the step size is chosen to prevent slope overload (Equation 3.12).

When using Equation 3.14, some important assumptions in its derivation must be kept in mind. First, Equation 3.14 pertains only to granular noise. Second, the equation uses a step size matched to the slope overload of the particular sine wave input. If the step size implied in Equation 3.14 is used on an input with the same amplitude but higher frequency, slope overload would occur. Furthermore, lower-level signals experience lower SQRs if the same step size is used. The most misleading aspect of Equation 3.14 is that it implies that a linear delta modulator provides a higher SQR for a low-frequency signal than for a high-frequency signal. The implication is true only if the step size is adjusted to a smaller value for the lower frequency.

Equation 3.14 is plotted in Figure 3.32 for a number of different input frequencies. Notice that the SQR improves at a rate of 9 dB per octave (per doubling) of the sampling frequency. In contrast, a PCM system SQR improves 6 dB with each additional bit. Thus PCM systems provide better quality at high data rates, but a delta modulator provides better performance at low bit rates. The crossover point is typically around 40 kbps.

Using an 800 Hz sine wave as a test signal, we determine from Equation 3.14 or Figure 3.32 indicates that linear delta modulation requires a sampling frequency of 28 kHz to provide a 26 dB SQR. The implied

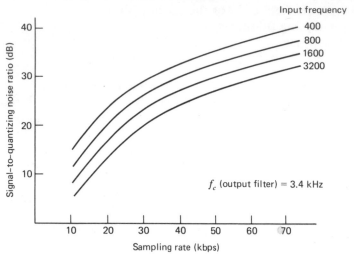

Figure 3.32   Signal-to-quantizing noise of linear delta modulation.

data rate of 28 kbps should be compared with the 4 bits per sample or 32 kbps linear PCM system determined previously for the same SQR.

Equation 3.14 only specifies the SQR for a sine wave with a fixed amplitude. The following relationship expresses the sampling frequency (bit rate) required for a particular SQR and also includes a dynamic range factor R:

$$f_s = \{25(R)\,(f^2)\,(f_c)\,(\text{SQR})\}^{1/3} \tag{3.15}$$

where R = dynamic range $(A_{max}/A_{min})^2$
   f = input frequency
   f_c = output filter cutoff
SQR = desired minimum signal-to-quantizing noise ratio

Equation 3.15 holds under the assumption that the step size is selected for no slope overload on the highest-level signal:

$$q = 2\pi A_{max}\left(\frac{f}{f_s}\right) \tag{3.16}$$

Again, Equations 3.15 and 3.16 represent a conservative design procedure for a delta modulator since they do not allow slope overload. They are useful, however, because they are consistent with the un-overloaded PCM designs presented earlier.

EXAMPLE 3.5

Using an 800 Hz sine wave as an input signal, determine the step size and sampling rate of a linear delta modulator to provide voice encoding with an SQR of 398 (26 dB) and a dynamic range of 1000 (30 dB). Assume an output filter cutoff at 3.4 kHz.

Solution. Equations 3.15 and 3.16 provide the following values for the sampling frequency and step size, respectively:

$$f_s = \{25(1000)(800)^2(3400)(398)\}^{1/3}$$

$$= 279 \text{ kHz}$$

$$q = 2\pi A_{max}(800/279000)$$

$$= 0.018 A_{max}$$

$$= 0.56 A_{min}$$

The 279 kbps data rate should be compared with the 9 bits per sample and 72 kbps data rate obtained earlier for uniform PCM.

The step size value obtained in Example 3.5 is just large enough to prevent overload for a maximum-level 800 Hz input. Thus a higher-frequency signal with the same amplitude would overload the encoder. Examination of the long-term spectral density of speech in Figure 3.24, however, shows that 800 Hz is near the maximum amplitude compo-nent and higher frequencies have significantly less power on average. Thus the single step size is approximately matched for no slope over-load at all frequencies. In fact, Figure 3.24 shows that the amplitude of a 2400 Hz component is typically down by a factor of about 5.6 (15 dB) when it only needs to be down by a factor of 3 to have a slope lower than the 800 Hz signal. Hence the 800 Hz input adequately determines the necessary step size in terms of slope overload.

A rather startling result presented in Example 3.5 is the relatively coarse step size (0.56 A_{min}) used for encoding the minimum amplitude sine wave. Despite this coarseness, the delta modulator provides a 26 dB SQR by virtue of the large amount of encoding noise that is removed by the output filter.* Figure 3.33 depicts a portion of a coarsely en-coded 800 Hz sine wave using the sampling rate and step size obtained in Example 3.5. Also shown is the "instantaneous" average of the en-coded signal which, for all practical purposes, is an equivalent input to the output filter. Notice that the high sampling rate provides very fine resolution as to when the signal crosses a quantization boundary. Thus a delta modulator encodes a signal with coarse quantization in ampli-tude but fine quantization in time. In contrast, a PCM system uses relatively fine resolution in amplitude but coarse quantization in time.

The basic drawback of linear delta modulation, like uniform PCM, is that low-level signals and high-level signals are both encoded with the

---

*As discussed later, practical considerations in the implementation of this encoder create an excessive amount of idle channel noise.

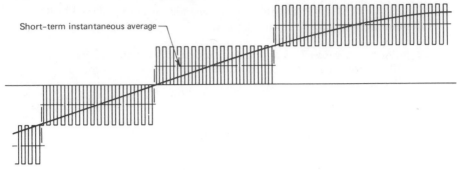

Short-term instantaneous average

Figure 3.33 800 Hz tone encoded by delta modulation with a sampling rate of 279 kHz.

same step size. Thus, high-level signals are encoded with excess quality at an expense in data rate. The solution, of course, is to vary the step size in some manner dependent on the magnitude of the input slope. Commonly, the step size is adjusted gradually based on a short-term average of the slope amplitude.

A significant consideration in the implementation of a delta modulator is maintaining equal positive and negative steps in the encoder and decoder. If a step size imbalance exists in the decoder, the reconstructed waveform accumulates a dc offset that may lead to circuit saturations. As discussed in the next section, modest step size imbalances in the decoder are relatively easy to accommodate. On the other hand, step size imbalances in the encoder are more troublesome and therefore require close attention.

When a delta modulator encodes the signal of an idle channel (speech pause), the output data ideally alternates between a 1 and a 0. When alternation does occur, the output filter of the decoder eliminates the relatively high-frequency sampling noise. However, if an imbalance in step sizes exists, as shown in Figure 3.34, two 1's or two 0's occasionally occur together. Since double 0's or double 1's occur relatively infrequently, they produce low-frequency distortion which is not filtered out and idle channel noise results. Idle channel noise is a

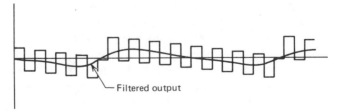

Filtered output

Figure 3.34 Effect of imbalance in positive and negative step sizes in a delta modulator with a zero level input.

particular problem in delta modulators because of the relatively large quantization intervals (see Example 3.5). One means of balancing the step sizes in the encoder is to subtract an offset obtained by low-pass filtering (integrating) the feedback signal over a long time period [20].

3.5.3 Syllabic Companding

A significant attribute of the companding techniques described for PCM systems is that they are designed to encompass instantaneously the entire dynamic range of the coder on a sample-to-sample basis. Thus μ-law and A-law companding is sometimes referred to as instantaneous companding. As mentioned previously, however, there is power level redundancy in a speech signal by virtue of the fact that during intervals of 100 or more 8 kHz samples, the power level remains fairly constant. Hence adaptive gain control provides more efficient coding than does instantaneous companding. When the gain control of the system is adapted on a periodic basis that more or less corresponds to the rate of syllable generation, these adaptive techniques are referred to as syllabic companding.

Syllabic companding was first developed for use on noisy analog circuits to improve the idle channel noise [27]. As shown in Figure 3.35 the power level of low-level syllables is increased (compressing the dynamic range) for transmission but attenuated upon reception (expanding the dynamic range). The process of attenuating the received signal restores the low-power syllable to its original level, but attenuates any noise arising on the transmission link. Thus the SNR for transmission link noise is improved for low-level signals. The amount of amplification applied at the source is dependent on the short-term (syllabic)

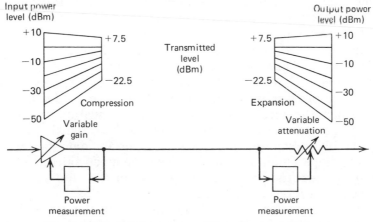

Figure 3.35 Syllabic companding of an analog signal.

power level of the signal. Similarly, the compensating attenuation applied at the receiving terminal is determined from the short-term power level of the received signal.

Syllabic companding on digital systems provides the same basic improvement in signal-to-quantizing noise ratios as it does on noisy analog transmission links. When the digital encoders and decoders are considered as part of the transmission link, the process of amplifying low-level signals before encoding and attenuating them after decoding effectively reduces the quantization noise with no net change in signal level. In practice syllabic companding, as implemented in digitized voice terminals, does not amplify the signal at the source and attenuate it at the destination. Instead, an equivalent process of controlling the step sizes in the encoder and decoder is used. As far as the transmitted bit stream is concerned, it makes no difference if the signal is amplified and encoded with fixed quantization or if the signal is unmodified but encoded with smaller quantization intervals. Thus syllabic compandors in digital voice terminals typically reduce the quantization intervals when encoding and decoding low-power syllables but increase the quantization intervals for high-power syllables.

Although syllabic companding can be used in conjunction with any type of voice coding, the technique has been applied most often to differential systems, in general, and delta modulation systems, in particular. In many of the applications, the adaption time has been reduced to 5 or 10 ms which is somewhat shorter than the duration of a typical syllable (approximately 30 ms). The technique is still generally referred to as syllabic companding, however, to distinguish it from the instantaneous variety.

In order to adjust the step sizes in the decoder in synchronism with adjustments made in the encoder, some means must be established to communicate the step size information from the source to the destination. One method explicitly transmits the step size information as auxiliary information [16]. As mentioned when discussing forward gain control, this approach has several shortcomings. A generally more useful approach is to derive the step size information from the transmitted bit stream. Thus, in the absence of channel errors, the decoder and the feedback circuitry in the encoder operate on the same information. This procedure is analogous to syllabic companded analog systems in which the receiver determines its attenuation requirements from the short-term power level of the received signal. In a digital system the bit stream is monitored for certain data patterns that indicate the slope amplitude of the signal being encoded. Indications of high slope amplitude initiate an increase in the step size, whereas indications of low levels cause a decrease.

Determining the step size information from the transmitted bit stream overcomes the basic shortcomings of auxiliary encoding mentioned previously. Specifically, since there is no explicit transmission of step size, the transmission of sampled speech information is never interrupted, and the speech sample rate is equal to the transmission rate. Also, the bit stream does not have to be framed to identify step size information separately from the waveform coding. Furthermore, if the step size adjustments are made on a more gradual basis, the individual increments are small enough that occasional incorrect adjustments in the receiver caused by channel errors are not critical. However, on transmission links with very high error rates (one error in a hundred bits or so), better decoded voice quality can be obtained if the step size is transmitted explicitly and redundantly encoded for error correction [28].

3.5.4 Adaptive Delta Modulation

Numerous researchers have proposed and studied a large variety of algorithms for adapting the step size of a delta modulator. Basically, all algorithms increase the step size when the onset of slope overload is detected and reduce the step size as the input slope decreases. Some of the algorithms measure the input slope directly and transmit explicit step size information. Other algorithms derive the step size information from the transmitted bit stream for both the encoder and decoder. The rate of adaption is sometimes instantaneous in that adjustments are made on a sample-to-sample basis [16] but more often syllabic in that significant changes in the step size occur only once every 10 ms or so. All of the basic algorithms appear to provide approximately the same voice quality [19], but some algorithms have certain features making them more attractive than others in specific applications. In particular, the algorithms vary in terms of their channel error sensitivity, transmission format, interconnectability with other coders, and amenability to digital signal processing.

This section describes the basic operation of one particular type of adaptive-delta modulation (ADM) generally referred to as continuously variable slope delta modulation (CVSD). The basic CVSD encoding algorithm was first described by Greefkes and de Jager in 1968 [29]. A subsequent design by Greefkes and Riemens [20, 30] incorporated what is referred to as: delta modulation with digitally controlled companding. This particular type of adaptive delta modulation is chosen because it is simple and embodies many generally desirable features. Furthermore, the basic algorithm is available from several manufacturers as a single integrated circuit.

Digitally controlled companding, as the name implies, derives its step size information from the transmitted bit stream. As shown in Figure 3.36, the adaption logic monitors the transmitted data to detect the occurrence of four successive ones or four successive zeros.* A string

---

*The number of successive ones or zeros causing a step size increase is reduced to three on lower-rate delta modulators to improve the response time.

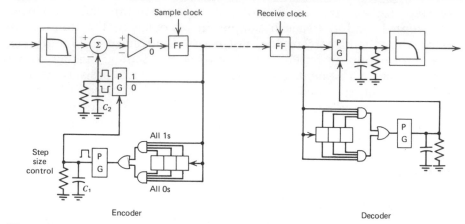

Encoder   Decoder

Figure 3.36 Continuously variable slope delta modulation with digitally controlled companding.

of ones indicates that the feedback signal is probably not rising as fast as the input, while a string of zeros indicates the feedback is probably not falling as fast as the input. In either case, all ones or all zeros implies that slope overload is occurring and the step size should be increased. Thus the all ones and all zeros signals are "or" ed together to control a pulse generator. During overload the pulse generator is enabled so that the step size voltage, as stored on capacitor C_1, is increased.

The CVSD system shown in Figure 3.36 makes no explicit measurements to determine if the step size is too large and therefore should be reduced. Instead, a resistor is connected to capacitor C_1 to allow the step voltage to decay with time. Thus, in the absence of explicit increases, the step size automatically becomes smaller through exponential decay. Eventually the step size decays to a minimum value or to a point at which overload is detected, causing the step size to increase again.

The resistor-capacitor combination is sometimes referred to as a "leaky integrator" implying that the capacitor integrates the input pulses but the resistor allows the accumulated step size voltage to leak off. Besides providing an automatic way to reduce the step size, the leaky integrator also eliminates the long-term effects of channel errors in the receiving terminal. If "perfect" integration were used, an erroneous increment in the step size would be retained indefinitely. With leaky integration, however, the duration of an erroneous increment (or lack of increment) is limited to a few time constants of the leaky integrator.

Notice that leaky integration is also used to reconstruct the input signal at the output of the decoder and, correspondingly, in the feedback path of the encoder. Thus an offset in the output of the receiving terminal, caused by a channel error, naturally decays in time and is effectively eliminated within a few time constants of the second leaky integrator. Elimination of an offset in the output signal is not so important from a listener's point of view since a fixed offset is inaudible, but is useful for preventing saturation in the decoder. If "perfect" integration were used, a large excess of positive or negative errors would eventually cause the electronics in the decoder to saturate and distort the voice waveform. In addition, leaky integrators provide compensation for small imbalances in positive and negative step sizes that would otherwise lead to saturation also.

Figure 3.37 shows the performance of two CVSD codecs as a function of the input level of a 1 kHz test tone. The performance curves were obtained from specification sheets for Motorola's MC3417 and MC3418 CVSD codecs. The 16 kbps codec (MC3417) provides what is referred to as general communications quality. The 37.7 kbps codec (MC3418) provides commercial telephone quality. Notice, however, that the 37.7 kbps performance does not meet the D3 channel bank requirements of 33 dB minimum signal-to-noise ratio (SQR) across 30 dB dynamic range. Recall that the D3 specification arose out of a need for higher quality because of the possibility of multiple conversions. Thus, although a CVSD at 37.7 kbps provides acceptable end-to-end quality, the data rate must be increased for tandem coding applications. The idle channel noise of the CVSD is also inferior if step size imbalances exist.

The performance curves shown in Figure 3.37 do not completely reflect the output quality when applied to an input speech signal. Since the 1 kHz tone is a steady state input, it does not test the adaptive aspects of the codec. In particular, adaptive delta modulators are generally noted for a certain lack of "crispness" in the speech output.

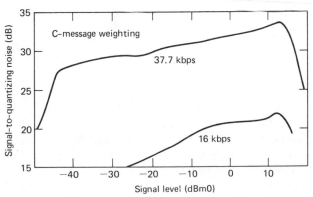

Figure 3.37   Signal-to-quantizing noise ratios of CVSD.

The lack of crispness occurs at the beginning of words and relatively strong syllables that experience temporary slope overloads.

In conclusion, it should be mentioned again that numerous varieties of delta modulation algorithms have been proposed and implemented. A CVSD codec is described here because it is currently the most popular type and represents a very basic form of delta modulation.

3.6 ADAPTIVE PREDICTIVE CODING

The encoding algorithms discussed previously, PCM and delta modulation, are relatively simple to implement but require considerably more transmission bandwidth than the analog signals they encode. Beginning in this section, digitization techniques are discussed that encode voice signals with significantly lower bit rates. However, the complexity of these encoders and decoders is much greater.

It was mentioned previously that extending the DPCM concept to use several past samples in predicting the present input sample provides better performance than conventional DPCM (first-order prediction). An adaptive predictive coder (APC) extends this concept to include prediction from one cycle to the next, or from one pitch interval to the next [31]. Thus the complexity and amount of signal delay required in the feedback loop (and the decoder) is significantly increased. The benefit is an average prediction gain of about 13 dB for voiced speech [7]. Owing to their more random structure, unvoiced inputs cannot be predicted as accurately. Fortunately, exact reproduction of an unvoiced sound is not as important as it is for voiced sounds. (One random waveform is perceptually the same as another random waveform).

In addition to greatly increasing the number of past samples used for prediction, an APC coder also extends the DPCM concept by using adaptive prediction circuitry instead of fixed prediction circuitry. In other words, the weighting coefficients assigned to each past sample are adjustable. An adaptive technique provides improved performance because the various voiced sounds have distinctly different repetitive patterns. In order to adapt the prediction circuitry as desired, an APC encoder first determines the amount of delay to be used in the prediction (the pitch interval), and then the weighting coefficients for the delayed sample values. The predictor coefficients are typically recalculated and retransmitted once every 5 to 20 ms. For a method of determining these parameters see References 31 and 32.

A block diagram of one form of an adaptive predictive coder is shown in Figure 3.38. Notice that there are basically two feedback paths: one to adapt the quantizer (make gain adjustments) and one to predict the next sample value. The information for adapting the quantizer is derived from the transmitted data stream (backwards estima-

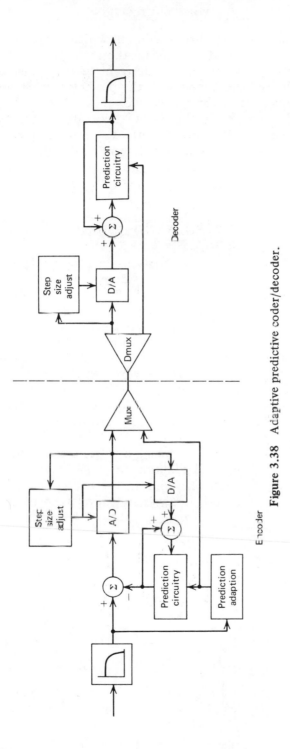

Figure 3.38 Adaptive predictive coder/decoder.

139

tion). APC structures are also possible in which adaptive quantization is either not used or based directly on the input (forward estimation). When forward estimation is used, the gain factors must be explicitly encoded and transmitted along with the predictor coefficients and the waveform coding (difference signals).

Jayant [18] has reported that an APC encoder of this type can provide as much as 20 dB improvement over a PCM system. The large amount of computation required to determine the pitch period and the predictor coefficients is becoming less of a drawback as integrated circuit technology continues to advance.

3.7 SUBBAND CODING

A subband coder is one form of coder using a frequency domain analysis of the input signal instead of a time domain analysis as in previously described coders. As shown in Figure 3.39, the coder first divides the input spectrum into separate bands using a bank of bandpass filters. The signal passing through each of the relatively narrow subbands is individually encoded with separate adaptive PCM (APCM) encoders. If instantaneously companded PCM encoding were used, a single encoder could be shared by all subbands. However, in the interest of minimizing the bit rate, an adaptive coder is used for each subband. After each subband is encoded, the individual bit streams are multiplexed for transmission to the decoder where they are demultiplexed, decoded, and combined to reconstruct the input.

Separately encoding each subband is advantageous for several reasons. First, by using separate adaption for each band, the quantization step sizes can be adjusted according to the energy level in each band. Those bands with relatively high energy levels can be encoded with

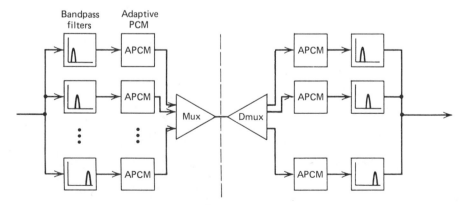

Figure 3.39 Sub-band coder.

relatively coarse quantization. In this manner the spectrum of the quantization noise is matched to the short-term spectrum of the signal. This property is very desirable in a perceptual sense because it allows the speech signal to mask the quantization noise. (The human ear perceives speech by measuring the short-term energy level of individual frequency bands. Hence, relatively low noise in a band with no speech energy is perceptually more significant than greater noise in a band with significant speech energy.)

A second advantage of subband coding is that the bit rate (quality) assigned to each band can be optimized according to the perceptual importance of each individual band. In particular, a relatively large number of bits per sample can be used for low frequencies where it is important to preserve the pitch and formant structure of voiced sounds. At higher frequencies, however, fewer bits per sample can be used because noise-like fricatives do not require comparable quality in reproduction.

As reported in Reference 7, subband coders provide significant bit rate reductions compared to the more common and simpler coding algorithms: adaptive delta modulation and adaptive differential PCM. Specifically, subband coders at 16 kbps are reported to be perceptually equivalent to ADPCM coders at 22 kbps. A subband coder at 9.6 kbps is reported to be equivalent to an ADM coder at 19.5 kpbs. An extensive description and performance analysis of subband coding is available in Reference 33.

3.8 VOCODERS

For the most part, the encoding/decoding algorithms described previously have been concerned primarily with reproducing the input waveform as accurately as possible. Thus they assume little or no knowledge of the nature of the signal they process and are basically applicable to any signal occurring in a voice channel. Exceptions occur when subband coding and adaptive predictive coding are designed for relatively low bit rates (20 kbps or less). At these bit rates the encoders have been closely tailored to the statistics of a speech signal and cannot provide comparable quality for other signals. Differential systems, such as DPCM and delta modulation, also exhibit a certain amount of speech specific properties by virtue of their high-frequency encoding deficiencies (slope overload).

The digitization procedures described in this section very specifically encode speech signals and speech signals only. For this reason these techniques are referred to collectively as "vocoders," an acronym for voice coders. Since these techniques are designed specifically for voice signals, they are not applicable to the public telephone network in which other analog signals (such as modem signals) must be accommodated. Furthermore, vocoders typically produce unnatural or synthetic sounding speech.

The basic goal of a vocoder is to encode only the perceptually important aspects of speech with fewer bits than the more general waveform encoders. Thus they can be used in limited bandwidth applications where the other techniques cannot.

Some of the main applications for vocoders are: recorded (e.g. "wrong number") messages, encrypted voice transmission over analog telephone circuits, computer output, and educational games. A particularly interesting use of one type of vocoder, linear predicative coding, has arisen recently for deriving multiple voice channels over a single voice frequency leased line. Using a well conditioned leased line to obtain a 9600 bps circuit, four 2400 bps voice signals can be time division multiplexed into a single leased line. Hence, in this case, digitization actually decreases the bandwidth of a voice signal. Although these systems provide intelligible voice, the overall quality is below telephone standards.

This section describes three of the most basic vocoding techniques: the channel vocoder, the formant vocoder, and the linear predictive coder. Many other forms and variations of vocoders have been proposed and studied. For a discussion of some of the other techniques and an extensive bibliography on the subject, see Reference 7.

A fundamental requirement for maintaining good speech quality involves preserving the short-term power spectrum of the signal. The phase relationship between individual frequency components is perceptually much less important. One of the best examples of the ear's insensitivity to phase is demonstrated when two notes are played simultaneously, or nearly simultaneously, on a piano. The composite sound, as perceived by a listener, is seemingly no different if one note is struck slightly later than the other. In essence, the ear senses how much energy is present at various frequencies in the speech spectrum but does not sense the phase relationship between individual frequency components.

The effect of a phase shift in one component of a composite signal is shown in Figure 3.40. The first composite waveform is produced when two individual frequency components have identical starting phases.

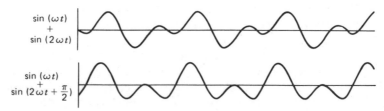

Figure 3.40 Effect of phase shift in the superposition of two tones.

The second composite waveform occurs when the two frequency terms have starting phases shifted by 90° with respect to each other. Notice that the composite waveforms are markedly different even though the difference is imperceptible to the ear. For these reasons the time waveform produced by a vocoder generally bears little resemblance to the original input waveform. Instead, the emphasis of a vocoder is in reproducing the short-term power spectrum of the input.

3.8.1 Channel Vocoder

Channel vocoders were first developed in 1928 by Homer Dudley [34]. Dudley's original implementation compressed speech waveforms into an analog signal with a total bandwidth of about 300 Hz. Based on the original concept, digital channel vocoders have been developed operating in the range of 1 to 2 kbps.

A major part of the encoding process of a channel vocoder involves determining the short-term signal spectrum as a function of time. As indicated in Figure 3.41, a bank of bandpass filters is used to separate the speech energy into subbands that are full wave rectified and filtered to determine relative power levels. The individual power levels are encoded and transmitted to the destination. Notice that this much of a channel vocoder is very similar to the subband coder discussed previously. A subband coder, however, typically uses wider bandpass filters, which necessitate sampling the subband waveforms more often (determining a waveform instead of just a power level). Since a subband coder encodes waveforms, it also includes phase information that is ignored by a channel vocoder.

In addition to measuring the signal spectrum, modern channel vocoders also determine the nature of speech excitation (voice or unvoiced) and the pitch frequency of voiced sounds. The excitation measurements are used to synthesize the speech signal in the decoder by passing an appropriately selected source signal through a frequency domain model of the vocal tract transfer function. Voiced excitation is simulated by a pulse generator using a repetition rate equal to the measured pitch period. Unvoiced excitation is simulated by a noise generator. Owing to the synthesized nature of the excitation, this form of a vocoder is sometimes referred to as a pitch excited vocoder.

As indicated in Figure 3.41, a decoder implements a vocal tract transfer function as a bank of bandpass filters whose input power levels are determined by respective subband power levels in the encoder. Thus outputs of each bandpass filter in the decoder correspond to outputs of respective bandpass filters in the encoder. Superposing the individual bands recreates, in a spectral sense, the original signal.

Many variations in the basic channel vocoder have been developed involving the nature of the excitation and the means of encoding the power levels. Recent advances in digital technology have introduced the use of digital signal processing to determine the input spectrum by way of Fourier transform algorithms in lieu of the bank of analog filters. All forms of vocoders that measure the power spectral density are sometimes referred to collectively as "spectrum channel vocoders" to distinguish them from time domain vocoders such as a linear predictive coder to be discussed later.

The most difficult aspect of most vocoder realizations involves determining the pitch of voiced sounds. Furthermore, certain sounds are not clearly classifiable as purely voiced or purely unvoiced. Thus a desirable extension of the basic vocoder involves more accurate characterization of the excitation. Without accurate excitation information, vocoder output quality is quite poor and often dependent on both the speaker and the particular sounds being spoken. Some of the more advanced channel vocoders have produced highly intelligible, although somewhat synthetic sounding, speech at 2400 bps [19].

3.8.2 Formant Vocoder

As indicated in the spectogram of Figure 3.25, the short-term spectral density of speech is rarely distributed across the entire voice band (200 to 3400 Hz). Instead, speech energy tends to be concentrated at three or four peaks called formants. A formant vocoder determines the location and amplitude of these spectral peaks and transmits this information instead of the entire spectrum envelope. Thus a formant vocoder produces lower bit rates by encoding only the most significant short-term components in the speech spectrum.

The most important requirement for achieving useful speech from a formant vocoder involves accurately tracking changes in the formants. Once this is accomplished, a formant vocoder can provide intelligible speech at less than 1000 bps [7].

3.8.3 Linear Predictive Coding

A linear predictive coder (LPC) is a popular vocoder that extracts perceptually significant features of speech directly from a time waveform rather than from frequency spectra, as does a channel vocoder and formant vocoder. Fundamentally, an LPC analyzes a speech waveform to produce a time varying model of the vocal tract excitation and transfer function. A synthesizer in the receiving terminal recreates the speech by passing the specified excitation through a mathematical model of the vocal tract. By periodically updating the parameters of the model and the specification of the excitation, the synthesizer adapts to changes in either. During any one specification interval, however, the vocal tract is assumed to represent a linear time-invariant

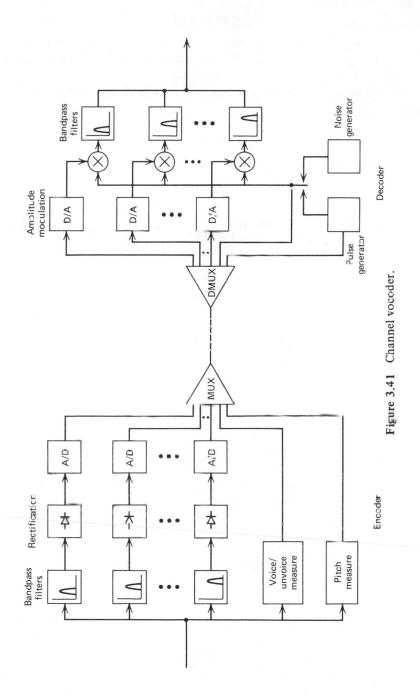

Figure 3.41 Channel vocoder.

process. A block diagram of the basic model for speech generation is shown in Figure 3.42. Thus Figure 3.42 is also a model of an LPC decoder/synthesizer.

The equation of the vocal tract model shown in Figure 3.42 is defined as follows:

$$y(n) = \sum_{k=1}^{p} a_k y(n - k) + Gx(n) \qquad (3.17)$$

where $y(n)$ = nth output sample
$\qquad a_k$ = kth predictor coefficient
$\qquad G$ = gain factor
$\qquad x(n)$ = input at sample time n
$\qquad p$ = order of the model

Notice that the speech output in Equation 3.17 is represented as the present input value plus a linear combination of the previous p outputs of the vocal tract. The model is adaptive in that the encoder periodically determines a new set of predictor coefficients corresponding to successive speech segments. In this manner linear predictive coding is similar to adaptive DPCM or APC coding. The basic difference, however, lies in the method of determining the prediction coefficients and the fact that an LPC does not measure and encode difference waveforms or error signals. Instead, the error signals are minimized in a mean squared sense when the predictor coefficients are determined. The ability to avoid encoding the predictor errors comes from the fact that an LPC uses prediction parameters based on the actual input segments to which the parameters are applied (forward estimation). In contrast, the predictive coders mentioned previously base their prediction on past measurements only (backward estimation). Rapid changes in the vocal tract or excitation cause models based on past measurements to be more inaccurate.

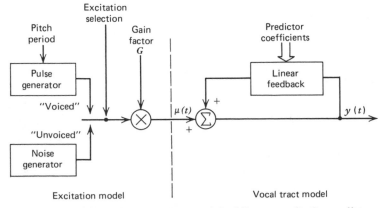

Figure 3.42   Speech generation model of linear predictive coding.

The information that an LPC encoder/analyzer determines and transmits to the decoder/synthesizer consists of:

1 Nature of excitation (voiced or unvoiced).
2 Pitch period (for voiced excitation).
3 Gain factor.
4 Predictor coefficients (parameters of vocal tract model).

The nature of the excitation is determined, as in other vocoders, by determining whether or not strong periodic components exist in the waveform. Pitch is determined by measuring periodicity when it exists. In addition to measuring pitch with techniques similar to those used by other vocoders, an LPC encoder/analyzer has particular properties that aid in pitch determination [35].

The predictor coefficients can be determined using one of several different computational procedures [35]. All procedures use actual waveform samples as the desired outputs of the synthesizer. Using these sample values, a set of p linear equations in p unknown coefficients is produced. Thus the coefficients are determined by inverting a p-by-p matrix. Since the order of p may vary from 6 to 12, depending on the speech quality desired, straightforward matrix inversion represents a formidable amount of computation. Depending on specific assumptions made in the model, however, the matrices have special properties that greatly simplify the solution of the equations.

Although linear predictive coders process time domain representations of speech, their operation is such that they provide good estimates of the peaks of the speech spectrum. Furthermore, an LPC is capable of effectively tracking gradual changes in the spectrum envelope. The overall result is that LPCs provide more natural sounding speech than the purely frequency domain based vocoders [19]. Most LPC research has concentrated on encoding speech in the range of 1.2 to 2.4 kbps.

3.9 ENCODER/DECODER SELECTION CONSIDERATIONS

This chapter has described several techniques for digitizing voice signals and has indicated that many other variations and types are possible. This last section summarizes the main considerations to be made when choosing a particular codec or digitization algorithm for implementation. Foremost among the factors to be considered are: speech quality, transmission rate, tolerance of transmission errors, coding format and data patterns, signal processing requirements, timing requirements, and implementation cost.

Naturally, the relative importance of these factors depends on the application: all applications have certain minimum quality requirements.

Beyond these requirements, the most important characteristics of a transmission codec are its cost, bit rate, and its performance in the presence of channel errors. On the other hand, an encoder for a digital switch is concerned mostly with implementation costs. In integrated transmission and switching systems, transmission considerations dominate—particularly since the cost of an LSI transmission codec is not much more than the cost of an LSI switching codec.

A codec for digital voice storage is mostly concerned with storage requirements (bit rate) but allows considerable flexibility in data format. Thus some factors, such as sensitivity to channel errors and coding formats, that might influence codec selection for transmission applications are immaterial for storage applications.

3.9.1. Speech Quality

As mentioned at the beginning of this chapter, characterizations of speech quality involve two general considerations: listener acceptability and intelligibility. Assessment of both factors necessarily involves listener evaluations. Listener acceptability factors include: naturalness, speaker recognition, and perception of noise or distortion. One method of evaluating acceptability is to have a group of listeners rate various speech segments in terms of overall subjective quality. The acceptability of an encoder/decoder pair is determined as the percentage of listeners rating the quality of the speech segments as adequate. A generally more consistent procedure requires that listeners indicate a preference between the outputs of two different encoder/decoder pairs, or between one pair and unprocessed speech. Generally speaking, only higher-rate waveform encoders such as PCM, DPCM, or DM provide high acceptability scores. However, continued research using vocoder techniques at intermediate bit rates (10 to 20 kbps) may produce higher quality, medium-rate voice coders in the future.

Speech researchers have long sought an objective measurement of speech quality that relates to subjective evaluations. A signal-to-quantizing noise ratio is a simple attempt to provide such an evaluation. However, as indicated when discussing delta modulation, slope overload noise is a subjectively less disturbing than granular (uncorrelated) noise with a much lower power level. Furthermore, SQR measurements are meaningless with respect to most vocoders since these digitization algorithms make no attempt to preserve source waveforms—only the perceptually significant factors of speech. Thus an SQR measurement has limited relevance when comparing coders with different noise characteristics. Nevertheless, SQR measurements are often used as a comparative measure for waveform coders. A more useful and more complicated measure of speech quality is the degree of preservation of the short-term amplitude spectrum of the signal [7].

Another problem with objectively determining the quality of speech produced by an encoder/decoder pair involves selecting an appropriate input. Sine waves are often used because of their convenience. However, a sine wave differs from a typical speech waveform in several respects. First, speech typically contains several strong frequency components (formants) at any one time, but some encoding algorithms encode different frequencies with differing quality. Second, a speech waveform has a higher peak-to-average ratio than a sine wave. Thus overload or clipping is more likely to occur on a speech waveform. Third, speech activity is intermittent. Hence, the transient response (adaption speed) must be analyzed in addition to steady state performance.

Intelligibility tests require listeners to recognize specially designed utterances consisting of isolated syllables and words or whole phrases and sentences. These tests are obviously more objective than other quality assessments since they do not ask for preferential evaluations on the part of the listeners. However, they are still subject to the nature of the test material (word or sound familiarity, accents, etc.) and the capabilities of the listeners. To improve the objectivity of these tests, standards for speech material have been established, and small groups of trained listeners are recommended [36]. Encoding techniques that score well in quality preference tests score well in intelligibility tests. Thus intelligibility tests are used mostly for low bit rate vocoders where intelligibility may be the only achievable and necessary criterion.

In addition to voice quality, it may be necessary to select or design a codec on the basis of how well it preserves the essential characteristics of other analog signals such as signaling tones, facsimile waveforms, or voiceband data signals (modem signals). Since existing telephone networks contain significant amounts of analog equipment, only analog signals can be serviced universally. Thus a codec embedded in a hybrid network must adequately digitize these nonvoice signals. High-rate PCM and ADM codecs provide acceptable and approximately equal quality for most modem signals [37]. However, low-rate coders, by virtue of their voice-specific designs, cannot be expected to provide good quality for signals with different characteristics. For a comprehensive theoretical treatment of voiceband data signal digitization see Reference [38].

With respect to low bit rate (speech specific) coders, it may also be necessary to evaluate their performance in the presence of background noise at the source. Since narrowband coders are tailored to certain characteristics of the speech process, they may be incapable of processing combinations of voice and background sounds.

Another, more subtle, consideration of codec quality is how well it performs in tandem with other codecs already in use in a network. High-quality waveform codecs generally operate in tandem with no more degradation than the predictable increase in quantization noise caused by multiple encodings. As already mentioned, the 64 kbps $\mu255$ PCM codecs were chosen in light of the possibility of up to nine tandem encodings occurring in a long-distance connection. In contrast, some low bit rate vocoders do not tandem well with themselves and particularly not in combinations. The end-to-end degradation may be worse than expected from each codec individually. Furthermore, the amount of degradation may depend on which codec processes the speech signal first [14, 39].

3.9.2 Transmission Rates

In discussing the transmission rates of various types of codecs, Flanagan, et al. [7] has described three general categories of quality: toll quality, communications quality, and synthetic quality. Toll quality is defined as being equivalent to 56 kbps log-PCM (D1 channel banks).* Communications quality refers to systems with noticeable degradations but good intelligibility and at least some naturalness. Synthetic quality loosely refers to systems that provide intelligibility and little else. The following table, also obtained from Reference 7, lists the minimum data rate for each technique achieving the indicated quality.[†]

As can be seen in Table 3.4, the differential systems, ADM and ADPCM, have significant bandwidth advantages over log-PCM. The more specialized subband and adaptive predictive coders provide even greater savings in bandwidth but are significantly more complicated than the first three. The vocoder techniques included in Table 3.4 produce speech with unique peculiarities. Their quality and that of low bit rate waveform coders is often dependent on both the speaker and the material being spoken.

3.9.3 Tolerance of Transmission Errors

Of the coding algorithms discussed in this chapter, delta modulation generally provides the best performance in the presence of random, independent channel errors. In fact, even with error rates as high as 10%, a CVSD codec provides better than 90% intelligibility [20]. The threshold of channel error rate perceptibility with a CVSD codec is approximately 10^{-3}. That is, error probabilities below .001 are generally unnoticeable. The relative immunity of delta modulation codecs to channel errors stems from the fact that a single channel error produces a decoder error magnitude of only one quantization level. Fur-

---

*North American toll quality is actually defined according to the D3 channel bank specification [9] which requires 64 kbps log-PCM when multiple conversions are possible.

[†]This table is included mostly for qualitative comparisons with single encodings. The quality categories and bit rates are not proven or universally accepted.

TABLE 3.4 DATA RATE REQUIREMENTS OF VARIOUS ENCODERS

Coder	Minimum Data Rate (kbps)		
	Toll Quality	Communications Quality	Synthetic Quality
log-PCM	56	36	
Adaptive delta modulation	40	24	
Adaptive DPCM	32	16	
Subband coder	24	9.6	
Adaptive predictive	16	7.2	
Channel vocoder			2.4
LPC			2.4
Formant vocoder			0.5

thermore, in a CVSD system, most channel errors do not produce erroneous adjustments in the step size. When erroneous adjustments do occur, they decay with time.

In contrast to delta modulation systems, PCM systems place more decoding significance on some bits than on others. When an error in the most significant bit occurs, a relatively large error spike is produced. The result is that PCM, and other systems with critically significant bits, have a lower threshold of perceptibility for channel errors. However, different coding formats for the PCM samples can be used to improve the sensitivity to error rates. As mentioned earlier, the coding format for D2 channel banks was changed to a sign-magnitude format from the "ones" complement format of the D1 channel bank. An error in the most significant bit of a "ones" complement encoding always causes an output error equal to half the maximum range of the coder. An error in the most significant bit of a sign-magnitude code causes a decoder error spike equal to twice the amplitude of the sample. Since most sample values of a speech waveform are below half maximum amplitude, the expected error power of the D2 format is considerably lower. The threshold of perceptability for μ255 PCM channel errors is about 10^{-4}. Within the Bell System, an error rate of 10^{-3} is the threshold for declaring a PCM radio channel (or group of channels) out of service [40].

The channel error sensitivity of vocoders is approximately equal to that of PCM systems. The performance with respect to channel errors can improve significantly by redundantly encoding the most critical information in the bit stream. If errors are bursty, the error rate advantage of delta modulation is diminished relative to other systems because additional errors in a word or block of the other systems may be incrementally insignificant. (If the polarity bit of a PCM code word is in error, errors in the lower significant bits are immaterial.)

3.9.4   Coding Formats and Data Patterns

Coding formats are of concern primarily in transmission and switching applications where the existence of a word or block oriented structure imposes framing requirements in the data stream. Data patterns become a consideration in some transmission systems that must restrict certain sequences in order to minimize interference from repetitive patterns, ensure timing recovery in regenerative repeaters, or reserve certain codes for transmission line control.

Framing requirements arise in PCM, DPCM, and all vocoders because the bit stream is arranged in words or blocks in which specific bits require specific interpretations. On the other hand, some delta modulation systems (CVSD with digitally controlled companding, for example) have no framing requirements whatsoever. Every bit in the bit stream is generated and processed in identical fashion. The absence of framing requirements is most useful in single channel systems with a constant (synchronous) transmission rate. In other systems the need for word or block framing may not be a drawback. For example, a T1 system automatically provides word framing for each channel when the 193 bit, 24 channel multiplex frame is established. In addition, systems using activity dependent transmission (DSI) automatically establish transmission blocks. In these cases, encoding algorithms with periodic gain factors or parameter adjustments can be inserted into predefined fields in the transmission blocks with no additional framing overhead.

Another consideration that may arise in transmission applications involves the possibility that an encoder might produce a data pattern incompatible with the transmission medium. For example, a T1 line cannot support a continuous string of zeros. The D1 channel bank (and all other subsequent channel banks used by the Bell System) precludes generation of an all zeros code word. Although this operation purposely induces an encoding error, it is necessary to ensure a minimum density of pulses on a T1 line.

3.9.5   Signal Processing Requirements

As mentioned earlier, digital signal processing of an encoded speech signal can be advantageous for such functions as adding gain or attenuation, conferencing, sensing speech activity, or converting from one encoding algorithm to another. Instantaneously companded PCM (both $\mu255$ and A-law) was chosen with the specific intent of simplifying digital processing operations. Most other encoders, however, have been designed with no specific attention given to digital processing requirements. Thus the signal processing functions may require conversion to analog or a relatively complicated digital simulation of an analog process. A number of adaptive delta modulation schemes have been proposed with convenient properties for various processing functions, particularly conversion to log-PCM formats. See, for example, References 41 and 42.

3.9.6 Timing Requirements

As mentioned at the beginning of this chapter, the first step in digitizing a voice signal involves sampling the waveform at regularly spaced time intervals. It is desirable that the encoder and decoder sample clocks be easily derived from the line clock. This situation exists for T1 systems where the frame rate is identical to the sample rate. The 8 kHz sample clocks for each of 24 channels can be derived from the line clock by a modulo 193 counter (1544 kbps/193 = 8 kHz).

A more difficult situation occurs when a delta modulator is interfaced to a T1 line. If the same channel rate of 64 kbps is used, the delta modulator must produce eight evenly spaced samples during each frame (instead of a single sample per frame as with a PCM coder). The 64 kHz sample clock is not derivable directly from the line clock since the sample times do not coincide with line clock transitions. Instead, the 64 kHz sample clock must be derived by synchronizing every eighth transition of a 64 kHz oscillator to every one hundred ninety-third transition of the 1.544 MHz line clock. The basic problem is demonstrated in Figure 3.43, which shows the fundamental timing requirements of the two systems.

Timing complications for delta modulation with T1 lines arise because the one hundred ninety third bit that is inserted for framing purposes. Without the framing bit the line rate would be 1.536 Mbps, and the 8 kHz sample clocks could be obtained by dividing the line rate by 24 (a modulo 24 counter). Other framing procedures that do not require an extra bit are also discussed in Chapter 7. One of these is the CCITT transmission standard that dedicates one of 32 channels for

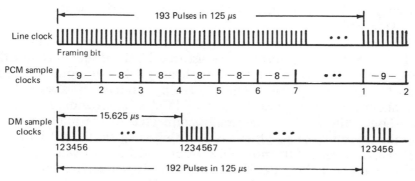

Figure 3.43 Sample clock requirements of PCM and DM as derived from T1 line clock.

framing. In this case a 64 kHz sample rate can be derived from the 2.048 MHz line rate with a modulo 32 counter.

Notice that other coders, particularly those that transmit auxiliary information such as gain factors or predictor coefficients, may have complicated sample clock derivations irrespective of framing procedures on the transmission links. In many of these systems, however, the terminal cost may be such that additional clock synchronization circuitry is incidental compared to the cost of encoding at a lower bit rate.

3.9.7   Implementation Cost

The cost of implementation is the last consideration, not because it is the least important, but because it is usually the most important. When considering an encoding algorithm available as an integrated circuit, operational advantages of any other comparable algorithm are probably academic. At the present time integrated circuit implementations of μ255 PCM, A-law PCM, and CVSD codecs are available from several integrated circuit manufacturers. Hence, even though other algorithms with comparable inherent complexity, such as ADPCM, can provide better quality or lower bit rates, their implementation cost from individual components precludes their use in all but special applications. In the area of low bit rate coding no encoders have yet been announced, but LSI linear predictive synthesizers have been developed by Texas Instruments [3] and Nippon Telegraph and Telephone of Japan [43]. The synthesizers are used for voice response units such as Texas Instruments' Speak-and-Spell.®

In regard to adaptive delta modulation versus log-PCM codecs, ADM codecs are simpler and will always be less expensive than log-PCM codecs, particularly when filter costs are included. However, the cost differential for integrated circuit implementations tends to be insignificant compared to such other considerations as performance, error rate sensitivity, and so on.

The universal acceptance of log-PCM by the telephone companies for T-carrier systems seems to have assured PCM's dominance in other applications—even when the other applications do not require compatibility to T-carrier systems. The marketing strategy of several digital PBX manufacturers stresses the use of PCM as an encoding technique even though the PBXs are normally connected to the public network with analog trunks or to other PBXs with analog tie lines. Digital connectivity is coming, but most PBX users will have to wait until their local end office is "digitized" or until new digital services are offered such as switched dataphone digital service [44].

REFERENCES

1 "Proposal Concerning A System for Digital Coding of Sound Signals," Document CMTT/186-E, Study Group 10A-1, January 1977.

2 J. M. Kasson, "The Rolm Computerized Branch Exchange: An Advanced Digital PBX," *Computer Magazine*, June 1979, p 24.

3 Richard Wiggins, "Low Cost Voice Response Systems Based on Speech Synthesis," SAE Technical Paper Series 800197, Society of Automotive Engineers, Inc., 1980.

4 R. G. Cornell, "The 1A Voice Storage System," *National Electronics Conference Proceedings*, 1979, pp 399–402.

5 J. L. Flanagan, *Speech Analysis, Synthesis, and Perception*, 2nd ed. Springer-Verlag, New York, 1972.

6 S. Tsuruta, H. Sakoe, and S. Chiba, "DP-100 Connected Speech Recognition System," *Intelcom '79 Proceedings*, Horizon House, 1979.

7 J. Flanagan, M. Schroeder, B. Atal, R. Crochiere, N. Jayant, and J. Tribolet, "Speech Coding," *IEEE Transactions on Communications*, April 1979, pp 710–737.

8 B. J. McDermott, C. Scagliola, and D. J. Goodman, "Perceptual and Objective Evaluation of Speech Processed by Adaptive Differential PCM," *Bell System Technical Journal*, May–June 1978, pp 1597–1618.

9 "The D3 Channel Bank Compatibility Specification—Issue 3," Technical Advisory No. 32, American Telephone and Telegraph Company, October 1977.

10 M. B. Akgun, "Transmission Objectives for a Subscriber Loop System Digital Voice Codec," *International Symposium-Subscriber Loops and Services*, Ottawa, 1974, pp 7.1.1–7.1.7.

11 H. Nann, H. M. Straube, and C. P. Villars, "A Companded Coder for an Experimental PCM Terminal," *Bell System Technical Journal*, January 1962, pp 173–226.

12 K. E. Fultz and D. B. Penick, "The T1 Carrier System," *Bell System Technical Journal*, September 1965, pp 1405–1451.

13 H. H. Henning and J. W. Pan, "D2 Channel Bank: System Aspects," *Bell System Technical Journal*, October 1972, pp 1641–1657.

14 W. R. Daumer and J. R. Cauanaugh, "A Subjective Comparison of Selected Digital Codecs for Speech," *Bell System Technical Journal*, November 1978, pp 3119–3165.

15 D. L. Duttweiler and D. G. Messerschmitt, "Nearly Instantaneous Companding and Time Diversity as Applied to Mobile Radio Transmission," *International Communications Conference*, 1975, pp 40-12 to 40-15.

16 I. M. McNair, Jr., "Digital Multiplexer for Expanded Rural Service," *Bell Labs Record*, March 1972, pp 80–86.

17 H. K. Dunn and S. D. White, "Statistical Measurements on Conversational Speech," *Journal of Acoustic Society of America*, January 1940, pp 278–288.

18 N. S. Jayant, "Digital Coding of Speech Waveforms: PCM, DPCM, and DM Quantizers," *Proceedings of IEEE*, May, 1974, pp 611–632.

19 J. W. Bayless, S. J. Campanella, and A. J. Goldberg, "Voice Signals: Bit by Bit," *IEEE Spectrum*, October 1973, pp 28–34.

20 J. A. Greefkes and K. Riemens, "Code Modulation with Digitally Controlled Companding for Speech Transmission," *Phillips Technical Review*, Vol. 31, 1970, pp 335–353.

21 J. E. Abate, "Linear and Adaptive Delta Modulation," *Proceedings of IEEE*, March 1967, pp 298–308.

22 N. S. Jayant and A. E. Rosenberg, "The Preference of Slope Overload to Granularity in the Delta Modulation of Speech," *Bell System Technical Journal*, December 1971, pp 3117–3125.

23 A. Tomozawa and H. Kaneko, "Companded Delta Modulation for Telephone Transmission," *IEEE Transactions on Communications Technology*, February 1968, pp 149–157.

24 F. deJager, "Delta Modulation, A Method of PCM Transmission Using the 1-Unit Code," *Phillips Research Reports*, No. 7, 1962, pp 422–466.

25 J. A. Betts, *Signal Processing Modulation and Noise*, American Elsevier, New York, 1971.

26 A. Lender, "Delta Modulation Study," Internal IT&T Fed. Lab. Report 1565.

27 Members of Technical Staff, *Transmission Systems for Communications*, Bell Telephone Laboratories, 4th ed., 1971.

28 N. S. Jayant, "Step-Size Transmitting Differential Coders for Mobile Telephony," *Bell System Technical Journal*, November 1975, pp 1557–1581.

29 J. A. Greefkes and F. de Jager, "Continuous Delta Modulation," *Phillips Research Reports*, Vol. 23, 1968, pp 233–246.

30 J. A. Greefkes, "A Digitally Companded Delta Modulation Modem for Speech Transmission," *Proceedings of IEEE International Communications*, 1970, pp 7-33 to 7-48.

31 B. S. Atal and M. R. Schroeder, "Adaptive Predictive Coding of Speech Signals," *Bell System Technical Journal*, October 1970, pp 1973–1986.

32 P. Noll, "On Predictive Quantizing Schemes," *Bell System Technical Journal*, May–June 1978, pp 1499–1532.

33 R. E. Crochiere, "An Analysis of 16 kb/s Sub-Band Coder Performance, Dynamic Range, Tandem Connections, and Channel Errors," *Bell System Technical Journal*, October 1978, pp 2927–2951.

34 H. Dudley, "Remaking Speech," *Journal of the Acoustic Society of America*, October 1939, pp 169–177.

35 L. R. Rabiner and R. W. Schafer, *Digital Processing of Speech Signals*, Prentice-Hall, Englewood Cliffs, New Jersey, 1978.

36 *IEEE Recommended Practice for Speech Quality Measurements*, IEEE Standards Publication No. 297, IEEE, 345 E. 47th St., New York, N.Y.

37 P. J. May, C. J. Zarone, and K. Ozone, "Voice Band Data Modem Performance Over Companded Delta Modulation Channels," *Proceedings of IEEE International Conference on Communications*, 1975, pp 40-16 to 40-21.

38 J. B. O'Neal Jr., "Waveform Encoding of Voiceband Data Signals," *Proceedings of the IEEE*, February 1980, pp 232–247.

39 D. J. Goodman, C. Scagliola, R. E. Crochiere, L. R. Rabiner, and J. Goodman, "Objective and Subjective Performance of Tandem Connections of Waveform Coders with an LPC Vocoder," *Bell System Technical Journal*, March 1979, pp 601–628.

40 T. S. Giuffrida, "Measurements of the Effects of Propagation on Digital Radio Systems Equipped with Space Diversity and Adaptive Equalization," *International Communications Conference*, 1979, pp 48.1.1–48.1.6.

41 D. J. Goodman and J. L. Flanagan, "Direct Digital Conversion between Linear and Adaptive Delta Modulation Formats," *Proceedings of IEEE International Conference on Communications*, 1971.

42 D. J. Goodman, "The Application of Delta Modulation to Analog-to-PCM Encoding," *Bell Systems Technical Journal*, February 1969, pp 321–343.

43 "Japanese Develop Speech Synthesizer on a Chip," *Electronics*, May 24, 1979, pp 69.

44 Irwin Dorros, "ISDN," *IEEE Communications Magazine*, March 1981, pp 16–19.

PROBLEMS

3.1 Assume a signal consists of three tones: one at 1 kHz, one at 10 kHz, and one at 21 kHz. What tones will be present at the output of a PAM decoder if the sampling rate is 12 kHz and the input is unfiltered? (Assume the output filter cutoff frequency is 6 kHz.)

3.2 Derive an expression for the average quantization noise power that occurs when the decoder output samples are offset from the center of a quantization interval by a distance equal to 25% of the interval. (The output values are at the 75% point instead of the 50% point.) How much degradation in decibels does this offset represent? (Assuming uncorrelated offsets.)

3.3 How much does the signal-to-noise ratio of a uniform PCM encoder improve when 1 bit is added to the code word?

3.4 A black and white television signal has a bandwidth of about 4.2 MHz. What bit rate is required if this signal is to be digitized with uniform PCM at an SQR of 30 dB? Use a sampling-rate to Nyquist-rate ratio comparable to that used for PCM voice encoding.

3.5 How much dynamic range is provided by a uniform PCM encoder with 12 bits per sample and a minimum SQR of 33 dB?

3.6 What is the signal-to-quantizing noise ratio produced by a segmented $\mu 255$ PCM coder when the input signal is a full-range triangular wave? (Assume the repetition frequency is low enough that the bandlimiting filter does not change the waveform significantly.)

3.7 Given a sample value of 420 mV for a $\mu255$ PCM encoder capable of encoding a maximum level of 2 V, determine each of the following:
   (a) The compressed $\mu255$ codeword.
   (b) The linear representation of the compressed code.
   (c) The A-law code obtained by converting from the $\mu255$ code.
   (d) The μ-law code obtained by converting back from the A-law code.

3.8 Given the following $\mu255$ code words, determine the code word that represents the linear sum: (0 110 1001), (1 011 0111).

3.9 Generate an encoding table (i.e., list the quantization intervals and corresponding codes) for the magnitude of a piecewise linear code with segment slopes: $1, \frac{1}{2}, \frac{1}{4}, \frac{1}{8}$. Assume four equally spaced intervals are in each segment. Assume all intervals in the first segment are of equal length (as in A-law coding).

3.10 What is the signal-to-noise ratio of a full amplitude sample for the coder of Problem 3.9?

3.11 What is the signal-to-noise ratio of a maximum amplitude sample in the first linear segment of the preceding coder?

3.12 What is the dynamic range implied by Problems 3.10 and 3.11?

3.13 For the encoder in Problem 3.9, how many bits are required for uniform encoding of the same dynamic range and same minimum quantization interval?

3.14 A uniform PCM system is defined to encode signals in the range of -8159 to $+8159$ in quantization intervals of length 2. (The quantization interval at the origin extends from -1 to $+1$.) Signals are encoded in sign-magnitude format with a polarity bit = 1 denoting a negative signal.
   (a) How many bits are required to encode the full range of signals?
   (b) How many unused codes are there?
   (c) Determine the quantization noise, noise power, and signal-to-noise ratio (in decibels) of each of the following sample values: 30.2, 123.2, -2336.4, 8080.9.

3.15 Repeat part c in Problem 3.14 for piecewise linear $\mu255$ PCM.

3.16 Given two A-law piecewise linear codewords (00110110 and 00101100) determine their linear representations, add them, and convert back to compressed representation.

3.17 A D3 channel bank uses "robbed digit" signaling wherein the least significant bit of every sixth frame is stolen for signaling. Determine the relative increase in overall quantization noise produced by this process under the following conditions:

(a) The decoder merely treats the signaling bit as a voice bit and decodes the PCM sample accordingly.

(b) The decoder knows which bits are signaling bits and generates an output sample corresponding to the middle of the double-length quantization interval defined by the seven available bits. (This is the actual technique specified for D3 channel bank decoders.)

3.18 If two bits per sample are added to a PCM bit stream, how much can the dynamic range be increased if the quantization intervals are adjusted to improve the SQR by 3 dB?

CHAPTER FOUR

DIGITAL TRANSMISSION AND MULTIPLEXING

A fundamental consideration in the design of digital transmission systems is the selection of a finite set of discrete electrical waveforms for encoding the information. In the realm of digital communications theory these discrete waveforms are referred to as signals. The same terminology is used in this chapter with the understanding that signals in the present context refer to internal waveforms (pulses) of a transmission system and not the control information (signaling) used to set up and monitor connections within a network. In communications theory terminology "signal processing" refers to filtering, shaping, and transformations of electrical waveforms—not the interpretation of control signals by the processor of a switching machine.

A second aspect of digital transmission involves defining certain time relationships between the individual transmission signals. The source terminal transmits the individual signals using predefined time relationships so the receiving terminal can recognize each discrete signal as it arrives. Invariably the establishment of a proper time base at the receiver requires transmission capacity above that needed for the digital information itself. Over relatively short distances (as within a switching system or a computer complex) the timing information (clock) is usually distributed separately from the information bearing signals. Over long distances, however, it is more economical to incorporate the timing information into the signal format itself. In either case the timing information requires channel capacity in terms of bandwidth, data rate, or code space.

This chapter discusses the most common digital signaling techniques for wire-line transmission systems. Thus the subject of this chapter is often referred to as line coding. These techniques are generally applicable to any transmission system transmitting digital signals directly in the form of pulses (such as coaxial cable or optical fiber). The fact that these techniques include low-frequency components in their transmitted spectrum leads to their also being called low-pass or baseband transmission systems. In Chapter 6 we discuss bandpass transmission

160

systems, that is, radio systems that require modulation and carrier frequency transmission.

The following discussions concentrate on system and application level considerations of digital transmission. Analytic details of pulse transmission systems are not emphasized since material of this nature is available in all works on digital communications theory. (See the bibliography.) Some fundamentals of pulse transmission systems are presented in Appendix C where many of the equations presented in this chapter are derived.

4.1 PULSE TRANSMISSION

All digital transmission systems are designed around some particular form of pulse response. Even carrier systems must ultimately produce specific pulse shapes at the detection circuitry of the receiver. As a first step, consider the perfectly square pulse shown in Figure 4.1. The frequency spectrum corresponding to the rectangular pulse is derived in Appendix A and shown in Figure 4.2. It is commonly referred to as a $\sin(x)/x$ response:

$$F(\omega) = (T) \frac{\sin(\omega T/2)}{\omega T/2} \tag{4.1}$$

where $\omega = 2\pi f$ is the radian frequency, and T is the duration of a signal interval.

Notice that Figure 4.2 also provides the percentage of total spectrum power at various bandwidths. As indicated, 90% of the signal energy is contained within the first spectral null at $f = 1/T$. The high percentage of energy within this band indicates that the signal can be confined to a bandwidth of $1/T$ and still pass a good approximation to the ideal waveform. In theory, if only the sample values at the middle of each signal interval are to be preserved, the bandwidth can be confined to $1/2T$. From this fact the maximum signaling rate in a specified band-

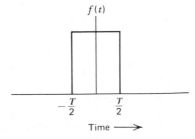

Figure 4.1   Definition of a square pulse.

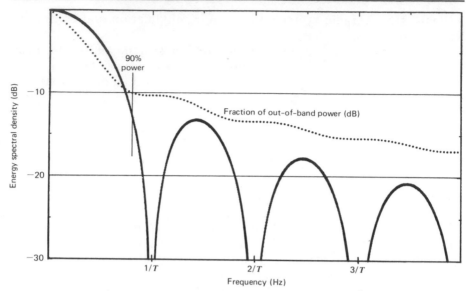

Figure 4.2 Spectrum of square pulse with duration T.

width is determined as:

$$R_{max} = 2 \cdot BW \tag{4.2}$$

where R is the signaling rate = $1/T$, and BW is the available bandwidth.

Equation 4.2 states a fundamental result from communications theory credited to Harry Nyquist: the maximum signaling rate achievable through a low-pass bandwidth with no intersymbol interference is equal to twice the bandwidth. This rate R_{max} is sometimes referred to as the "Nyquist rate."

Although discrete, square-shaped pulses are easiest to visualize, preservation of the square shape requires wide bandwidths and is therefore undesirable. A more typical shape for a single pulse is shown in Figure 4.3. The ringing on both sides of the main part of the pulse is a necessary accompaniment to a channel with a limited bandwidth. Normally, a digital transmission link is excited with square pulses (or modulated equivalents thereof), but bandlimiting filters and the transmission medium itself combine to produce a response like the one shown. Figure 4.3 shows pulse output in negative time so the center of the pulse occurs at $t = 0$. Actually, the duration of the preringing is limited to the delay of the filters and equalizers.

An important feature of the pulse response shown in Figure 4.3 is that, despite the ringing, a pulse can be transmitted once every T seconds and be detected at the receiver without interference from adjacent pulses. Obviously, the sample time must coincide with the zero crossings of the adjacent pulses. Pulse responses like the one shown in

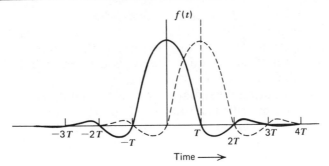

Figure 4.3   Typical pulse response of a bandlimited channel.

Figure 4.3 can be achieved in channel bandwidths approaching the minimum (Nyquist) bandwidth equal to one-half of the signaling rate. Appendix C describes pulse transmission design in more detail.

Intersymbol Interference

As the signaling rate of a digital transmission link approaches the maximum rate for a given bandwidth, both the channel design and the sample times become more critical. Small perturbations in the channel response or the sample times produce nonzero overlap at the sample times called intersymbol interference. The main causes of intersymbol interference are:

1   Timing inaccuracies.
2   Insufficient bandwidth.
3   Amplitude distortion.
4   Phase distortion.

Timing Inaccuracies

Timing inaccuracies occurring in either the transmitter or the receiver produce intersymbol interference. In the transmitter, timing inaccuracies cause intersymbol interference if the rate of transmission does not conform to the ringing frequency designed into the channel. Timing inaccuracies of this type are insignificant unless extremely sharp filter cutoffs are used while signaling at the Nyquist rate.

Since timing in the receiver is derived from noisy and possibly distorted receive signals, inaccurate sample timing is more likely than inaccurate transmitter timing. Sensitivity to timing errors is small if the transmission rate is well below the Nyquist rate. (E.g. if the transmission rate is equal to or less than the channel bandwidth, as opposed to being equal to the theoretical maximum rate of twice the bandwidth. See Appendix C.)

Insufficient Bandwidth

The ringing frequency shown in Figure 4.3 is exactly equal to the theoretical minimum bandwidth of the channel. If the bandwidth is reduced further, the ringing frequency is reduced and intersymbol interference necessarily results.

Some systems purposely signal at a rate exceeding the Nyquist rate, but do so with prescribed amounts of intersymbol interference accounted for in the receiver. These systems are commonly referred to as partial response systems—so called because the channel does not fully respond to an input during the time of a single pulse. The most common forms of partial response systems are discussed in a later section.

Amplitude Distortion

Digital transmission systems invariably require filters to bandlimit transmit spectrums and to reject noise and interference in receivers. Overall, the filters are designed to produce a specific pulse response. When a transmission medium with predetermined characteristics is used, these characteristics can be included in the overall filter design. However, the frequency response of the channel cannot always be predicted adequately. A departure from the desired frequency response causes pulse distortions (reduced peak amplitudes and improper ringing frequencies) in the time domain. Compensation for irregularities in the frequency response of the channel is referred to as amplitude equalization.

Phase Distortion

When viewed in the frequency domain, a pulse is represented as the superposition of frequency components with specific amplitude and phase relationships. If the relative amplitudes of the frequency components are altered, amplitude distortion results as discussed. If the phase relationships of the components are altered, phase distortion occurs. Basically, phase distortion results when the frequency components of a signal experience differing amounts of delay in the transmission link (see Chapter 1). For a good tutorial on equalization, including a description of an automatic equalizer for data transmission, see Reference 1.

4.2 ASYNCHRONOUS VERSUS SYNCHRONOUS TRANSMISSION

There are two basic modes of digital transmission involving two fundamentally different techniques for establishing a time base (sample clock) in the receiving terminal of a digital transmission link. The first of these techniques is asynchronous transmission, which involves separate transmissions of groups of bits or characters. Within an individual group a specific predefined time interval is used for each discrete signal. However, the transmission times of the groups are unrelated to each other. Thus the sample clock in the receiving terminal is reestablished for reception of each group. With the second technique, called synchronous transmission, digital signals are sent continuously at a constant rate. Hence the receiving terminal must establish and maintain a sample clock that is synchronized to the incoming data for an indefinite period of time.

4.2.1   Asynchronous Transmission

Between transmissions an asynchronous line is in an inactive or idle state. The beginning of each transmission group is signified by a start bit. The width of the start bit is measured and its midpoint determined. Succeeding information bits are sampled at a nominal rate beginning at the middle of the second bit interval. Following the information symbols, one or more stop bits are transmitted to allow the line to return to the inactive state.* Figure 4.4 shows an asynchronous mode of operation commonly used for low-speed data communication.

As shown in Figure 4.4, the detection of each information bit is accomplished by ideally sampling the input waveform at the middle of each signal interval. In practice, sample times depart from the ideal depending on how much the start bit is corrupted by noise and distortion. Since the sample time for each information bit is derived from a single start bit, asynchronous systems do not perform well in high-noise environments. Of course, more than one start bit could be used to improve the accuracy of the starting phase of the sample clock, but this would complicate the receiver and add more overhead for transmission of timing information.

Timing errors also arise if the nominal rate of the sample clock in the receiver is different from the nominal rate of transmission at the source. Even though the start bit might define the proper starting phase for the sample clock, an offset in the clock frequency of the receiver causes each successive sample time to drift farther from the center of the respective signal intervals. Since the very use of the term "asynchronous" implies a free-running clock in the receiver, a certain amount of drift is inevitable in all asynchronous systems. The maximum length of each symbol group or character is determined by the limits of the initial phase inaccuracies and the maximum expected frequency difference between the transmitter and receiver clocks.

The main attraction of asynchronous transmission is the ease with which it determines the sample times in the receiver. In addition, asyn-

---

*Originally, stop bits were inserted to allow electromechanical equipment enough time to reset before the next character arrived. With electronic equipment the only purpose of stop bits is to allow a start bit to always be a space (logic "0").

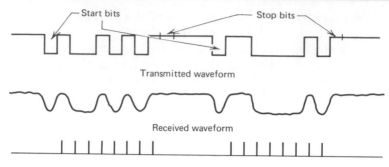

Figure 4.4 Asynchronous transmission.

chronous transmission is inherently flexible in the range of average data rates that can be accommodated. For high rates, one character after another is transmitted. Lower data rates are automatically accommodated by increasing the idle time between characters. In contrast, a synchronous receiver must track changes in a transmitter rate before it can sample the incoming signals properly. Normally the receive clock of a synchronous system can be adjusted only quite slowly and only over a narrow range. Hence an asynchronous system is more naturally suited to applications where the data rate varies.

Synchronous transmission systems can support variable information rates, but the task of adjusting the information rate falls upon higher level processes (line protocols) that insert null codes into the bit stream. The null codes are used as filler when a source has nothing to send. This form of transmission is sometimes referred to as "isochronous." An isochronous mode of operation is required whenever a synchronous line carries data from an asynchronous source.

Another attractive feature of asynchronous transmission is that the mode of operation automatically provides character framing. Thus if PCM encoded voice is transmitted asynchronously, the start and stop bits define the character boundaries automatically. With synchronous transmission, additional provisions must be established for character framing.

The major drawbacks of asynchronous transmission are: its poor performance in terms of error rates on noisy lines and the relatively high overhead of transmission capacity for timing. Thus asynchronous transmission is most appropriate for applications where implementation costs dominate signal-to-noise ratio and bandwidth considerations. Asynchronous transmission is used in voiceband data sets (modems) for transmission rates up to 1200 bps. For digital telephony, something similar to asynchronous transmission is being considered for two-wire digital subscriber loops [2, 3, 4]. These systems provide a full duplex (four-wire) circuit by transmitting bursts of data alternately in each direction on a single pair of wires. Thus these systems are sometimes referred to as "ping-pong" transmission systems, they are not truly asynchronous since each transmission in each direction occurs at prescribed times, possibly eliminating the need for start and stop bits.

4.2.2 Synchronous Transmission

T1 lines, and all other interoffice digital transmission links currently used in the U.S. telephone network, use synchronous transmission exclusively. Thus the line coding format for these systems must incorporate special considerations to ensure that each regenerative repeater or receiver can synchronize a local sample clock to the incoming signaling rate. Generally speaking, the synchronization requirements imply that a certain minimum density of signal transitions is required to provide continuous indication of signaling boundaries. Often, purely random data cannot be relied upon to provide a sufficient number of transitions. In these cases certain provisions must be made to insert artificial transitions into the transmitted waveforms. Although these extra transitions imply a certain amount of transmission overhead, the loss in capacity can be relatively small. Following are descriptions of five techniques for ensuring the existence of signal transitions:

1   Source code restriction.
2   Dedicated timing bits.
3   Bit insertion.
4   Data scrambling.
5   Forced bit errors.

A sixth technique—inserting the transitions into the signal waveforms themselves—is discussed in Section 4.3 under "line coding."

Source Code Restriction

One means of ensuring a sufficient number of signal transitions is to restrict the code set or data patterns of the source so that long transition-free data sequences do not occur. PCM channel banks in the North American telephone network preclude all-zeros codewords since the line code of T1 lines produces no transitions for zeros. In the case of 8 bit PCM code words, the exclusion of a single code word represents a loss in transmission capacity of only one part in 256.*

Relying on source coding to ensure sufficient transitions in the line code has one significant drawback: the transmission link cannot be used

*From an information theory point of view the loss in voice information is even lower since the probability of occurrence of the all-zeros code word is much less than 1 in 256.

for new applications where the source does not exclude the unwanted data patterns. For this reason, the total capacity of a T1 line can not be used for random data.

Dedicated Timing Bits

As an alternative to excluding transition-free data patterns, the line itself can periodically insert transition bearing bits into the data stream. These bits are inserted at regular intervals, independently of the source data, to ensure the existence of a minimum number of signal transitions. Thus some fraction of the channel capacity is dedicated to timing bits.

As an example, Dataphone® Digital Service (DDS) offered by the Bell System for data communications over T1 lines, provides a maximum capacity of 56 kbps for each channel. Thus only 7 of the 8 bits in each time slot of each channel are available for the user. Among other functions, the unused bit in each time slot provides an assurance that all 8 bits of a time slot are not zero. As discussed in a later section, a more recent digital transmission system (the T2 system) uses a line coding format guaranteeing sufficient transitions independently of the source codes and therefore does not require a timing bit for random data. Notice that a dedicated timing bit is essentially the same procedure used to establish timing for asynchronous transmission. In a synchronous receiver, however, a sample clock is obtained by averaging the timing information over a large number of timing transitions, not just one.

Bit Insertion

In the preceding DDS example 1 bit of every 8 bits in a time slot is dedicated to ensuring sufficient timing information in the bit stream. Another possibility for precluding unwanted line patterns is to use "bit insertion" only when necessary. As an example, the source data over a T1 line could be monitored for all zeros in the first 7 bits of a time slot. Whenever the zeros occur, a one could be inserted into the data stream as the eighth bit of the time slot. At the other end of the line the one following seven zeros is removed automatically. Each insertion delays the source data by the time of 1 bit, but otherwise the full capacity of the channel is available.

This procedure is directly analogous to the "zero bit insertion" algorithm used in IBM's Synchronous Data Link Control (SDLC) for data communications. In these protocols (and other similar bit-oriented protocols such as CCITT's High Level Data Link Control (HDLC) or American National Standards Institute's Advanced Data Communications Control Procedure ADDCP), a specific data pattern called a "flag" is used to indicate the end of a data block. The transmitter must be precluded from inadvertently sending the flag as part of the user data. The means of preventing inadvertent flags is to insert a 0 following a string of five 1's in the user data. Since a flag contains a string of six 1's, zero bit insertion effectively precludes unintended flag transmissions. The receiving node of an SDLC data link removes any 0 following five 1's. The receipt of six 1's, however, can only be part of a legitimate flag (01111110).

Although a bit insertion algorithm allows for more efficient use of a channel than dedicated timing bits, the insertion procedure has a number of drawbacks. First, this process causes the source data to be delayed every time an insertion is made. Hence a continuous, real-time transmission application (such as voice) experiences "jitter" in the arrival rate of the source information. Although jitter can be minimized by smoothing the arrival rate with data buffers at the destination, the jitter can never be completely removed. Second, the bit insertion process causes any character structure in the user's data to become unrelated to the time slot structure of a time division multiplexed transmission link. Thus if user data consists of 8 bit characters (like PCM voice samples), character boundaries can not be maintained with respect to 8 bit time slots in a T1 transmission link.

Data Scrambling

Many digital transmission systems use data scramblers to randomize the data patterns on their transmission links. Although these data scramblers are similar to those used for encryption, the fundamental purpose of these scramblers is to prevent the transmission of repetitive data patterns, and not to encrypt the traffic. Repetitive data patterns generate line spectra that can be significantly more degrading from an interference point of view than continuously distributed spectra produced by random data patterns. Voiceband data modems, for example, are allowed to operate at higher power levels if they include scramblers to randomize the data traffic. Also, digital radio systems are required by the FCC to not transmit line spectra, which essentially means that repetitive data patterns must be excluded.

Even when not required, data scramblers are useful in transforming data sequences with low transition densities into sequences with strong timing components. Scrambling is not used on lower rate T-carrier systems (T1 and T2), but it is used on the 274 Mbps T4M coaxial transmission links of the Bell System [5].

Data scramblers (with equal input and output bit rates) do not prevent long strings of zeros in an absolute sense. They merely ensure that relatively short repetition patterns are transformed to randomized traffic with a minimum density of transitions. If purely random input data is scrambled, the output data is also purely random and therefore has a certain statistical probability of producing arbitrarily long strings of zeros. The probability of a random occurrence, however, is acceptable when compared to the probability of nonrandom sequences representing speech pauses or idle data terminals. To determine which seemingly random data sequence produces all zeros at the output of a scrambler, apply all zeros to the corresponding descrambler.

The T4M coaxial transmission system uses a data scrambler as the basic means of producing adequate timing information. This system can tolerate much longer strings of zeros because the timing recovery circuits in the regenerative repeaters use phase-locked-loops that maintain timing over relatively long periods of time. In contrast, T1 systems recover timing with tuned circuits that resonate at the desired clock frequency (1.544 Mbps) when excited by a received pulse. Because the tuned circuits have lower effective Qs than a phase-locked-loop, the oscillations drift from the proper frequency and die out more rapidly. Hence T1 receivers cannot tolerate as long a string of zeros as can T4M receivers.

Forced Bit Errors

A fifth method of maintaining sufficient timing information in the line signals involves having the transmission terminal at the source force an occasional bit error in order to interrupt a long, transition-free data pattern. If the transition-free sequences are long enough and uncommon enough, the intentional bit errors might be less frequent than random channel errors on the digital transmission link. Thus the intentional errors may not represent a significant degradation over and above that already existing. Nevertheless, forced bit errors are not generally recommended as part of a line coding procedure but are mentioned in the interest of completeness.

As mentioned previously, North American channel banks force a bit error in the second least significant bit of an all-zeros transmission code to ensure sufficient signal transitions. An important aspect of this procedure is that it is performed by the source where the significance of the bit error is known. If the transmission link itself inserted the bit errors, the effects would not be as controllable—particularly when a variety of traffic types are being serviced.

A more subtle problem with forced transmission errors arises if the digital transmission link is used for automatic repeat request (ARQ) data transmission. An ARQ data communication link is designed to provide error-free transmission, despite random channel errors, by inserting redundancy into the data stream and checking the received data for error-free reception. If errors are detected, a retransmission is requested. When the errors are not random, but forced by the transmission link, the ARQ system will become frustrated if it ever encounters the restricted sequence, no matter how unlikely it is.* Once again, if

---

*If the transmission link uses a scrambler, the unlikely sequence will not be repeated.

forced errors are used they should be incorporated into the source as part of the source code restriction process and not as a function of the transmission link.

4.3 LINE CODING

In the preceding section various techniques for establishing timing information are described in general terms. The choice of any particular technique is dependent on the specific line code in use. This section describes the most common line codes used for digital transmission and indicates what additional steps, if any, are needed to maintain synchronization between the transmitter and receiver. Some line coding techniques provide adequate timing information themselves and do not require any of the procedures discussed previously. In addition to synchronization requirements, other considerations for choosing a line code are: the spectrum of the line code and the available bandwidth (particularly at low frequencies), noise and interference levels, synchronization acquisition times, performance monitoring, and implementation requirements.

4.3.1 Level Encoding

Conceptually, the simplest form of line coding uses a different signal level to encode each discrete symbol transmitted. Within a computer system the most common form of coding is an on/off code using a 3 V level for a 1 and near 0 V for a 0. Over a transmission link, however, it is more efficient in terms of power to encode binary data with an equivalent difference in levels but symmetrically balanced about 0 V. For example, the average power required for equally likely +3 and 0 V encodings is 4.5 W (assuming 1 ohm resistance). With +1.5 and −1.5 V encodings, the same error distance is achieved with half the power requirements (2.25 W). Communications engineers commonly refer to the unbalanced code as a unipolar code and the balanced code as a polar code. A representative sequence of binary data and its corresponding balanced and unbalanced level encoding is shown in Figure 4.5. Notice that the level of each signal is maintained for the duration of a signal interval. For this reason the balanced (polar) encoding is also referred to as a nonreturn-to-zero (NRZ) code.*

As indicated in Figure 4.5, an NRZ signal contains no transitions for long strings of ones or zeros. Hence one of the procedures described

---

*Some communications theorists refer to a balanced two-level code as a "bipolar code." The North American telephone industry, however, uses the term bipolar to refer to a three-level code described in the next section.

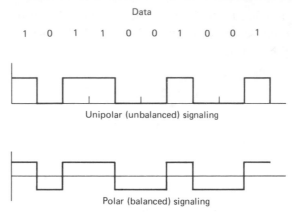

Figure 4.5 Unipolar and polar (NRZ) line codes.

previously to ensure timing transitions must be employed if NRZ encoding is used on a synchronous transmission link.

An NRZ line code is a pulse transmission system wherein the pulse (before filtering) lasts for the duration of a signaling interval T. Hence the frequency spectrum (assuming random data) of an NRZ code is the $\sin(x)/x$ spectrum of Equation 4.1 and shown in Figure 4.2. As indicated, the frequency spectrum is significantly nonzero at zero frequency (dc). Most wire-line transmission links, however, do not pass dc signals by virtue of their being ac coupled with transformers or capacitors to eliminate dc ground loops. Furthermore, some systems purposely remove dc components from the signal to allow line powering of repeaters or to facilitate single sideband transmission. The elimination of the low-frequency components in the waveform causes long strings of ones or zeros to decay gradually in amplitude. Hence not only would a receiver lose timing information during these strings, but it would also lose its amplitude reference for optimally discriminating between a 1 level and a 0 level. The effect of low-frequency cutoff, called dc wander, is shown in Figure 4.6 for a typical transmission sequence. Notice that following the long string of ones, the output of the link is such that 1 to 0 errors are more likely than 0 to 1 errors. Similarly, long strings of zero increase the likelihood of a 0 to 1 error. This problem arises not only for long strings of ones or zeros, but when-

Figure 4.6 DC wander of NRZ signal.

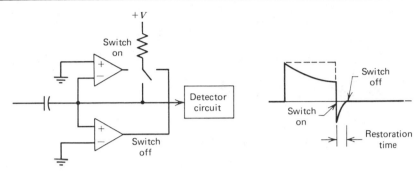

Figure 4.7 DC restoration for unipolar pulses.

ever there is an imbalance in the number of ones and zeros. Hence periodic timing pulses are not sufficient to remove dc wander.

The existence of low frequencies in a random data signal is the basic reason why modems are needed for data communications over the analog telephone network. (Analog telephone circuits don't pass direct current either.) It is also the reason that NRZ coding is not often used for long-distance transmission. Dc wander is not unique to data transmission systems. It is a phenomenon that must be reconciled in television receivers, radar receivers, or radiation detectors.

One technique of offsetting dc wander is referred to as dc or baseline restoration [6]. As illustrated in Figure 4.7, dc restoration basically involves passing received pulses through a capacitor, detecting them, and then removing the charge on the capacitor before the next pulse arrives. Charge on the capacitor is removed by driving the voltage to a specific threshold (0 V in Figure 4.7) and then removing the driving voltage when the threshold is reached. Since all charge on the capacitor is removed after each pulse, the baseline or decision reference level is constant at the beginning of a signal interval. An obvious disadvantage of this technique is that the signal input must have zero amplitude or be disabled during the reset time.

A generally more useful technique for overcoming baseline wander is to use decision feedback, also called quantized feedback equalization [7]. In contrast to dc restoration which drives the capacitor voltage to a constant, predetermined level, quantized feedback compensates for dc wander by locally generating the unreceived low-pass response and adding it to the received signal. To accomplish this, the original data stream is reconstructed. As shown in Figure 4.8, the reconstructed data stream is passed through a low-pass filter that generates a pulse equal to the tail or droop characteristic of the channel. The feedback signal adds to the received signal to eliminate the droop (intersymbol interference). Using a frequency domain analysis, the feedback response is complimentary to the channel response [7]. Quantized feedback is used in T4M repeaters of the Bell System [5].

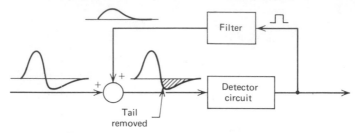

Figure 4.8 Decision feedback restoration.

4.3.2 Bipolar Coding

The dc restoration techniques mentioned in the preceding section simplify pulse detection by creating a low-pass pulse response in the receiver. There are numerous line codes that are specifically designed to not contain dc energy and thereby be unaffected by dc removal. Bipolar coding solves the dc wander problem by using three levels to encode binary data. Specifically, a logic 0 is encoded with zero voltage while a logic 1 is alternately encoded with positive and negative voltages. Hence the average voltage level is maintained at zero—to eliminate dc components in the signal spectrum. Since bipolar coding uses alternate polarity pulses for encoding logic 1's, it is also referred to as "alternate mark inversion" (AMI).*

Bipolar coding is the basic line coding procedure used by T1 lines in the telephone network. Rather than use full period pulses, however, T1 lines use a 50% duty cycle pulse to encode each logic 1. Fifty percent duty cycle pulses (Figure 4.9) were selected to simplify timing recovery in the regenerative repeaters of a T1 line [8]. The power spectrum of a bipolar code is obtained from [9] as:

$$S(\omega) = \frac{2p(1-p)}{T} |G(\omega)|^2 \frac{1 - \cos \omega T}{1 - 2(2p-1)\cos \omega T + (2p-1)^2} \quad (4.3)$$

where p is the probability of a 1, and $G(\omega)$ is the spectrum of an individual pulse.

$$G(\omega) = \left(\frac{T}{2}\right) \frac{\sin(\omega T/4)}{\omega T/4} \quad \text{for 50\% duty cycle pulses.}$$

Equation 4.3 is plotted in Figure 4.10 for various values of p. For purely random data, $p = \frac{1}{2}$. Recall, however, that source coding for $\mu255$ PCM channel banks produces more 1's than 0's in the interest of establishing a strong clock signal. Hence the appropriate value of p for

*A mark is a term arising from telegraphy to refer to the active or 1 state of a level encoded transmission line.

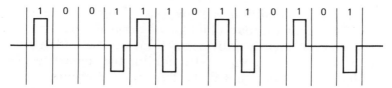

Figure 4.9 Bipolar coding.

a T1 voice line is normally somewhat larger than .5 and depends on the amplitude of the voice signal. Low-level signals that remain in the first encoding segment produce a value of p equal approximately to .65. On the other hand, full-scale sine waves produce a value for p that is somewhat below .5 since most of the samples occur near maximum amplitude.

Because a bipolar code uses alternating polarities for encoding 1's, strings of "1"s have strong timing components. However, a string of 0's contains no timing information and therefore must be precluded by the source. The specifications for T1 line repeaters states that the repeaters will maintain timing as long as no string of greater than 15 zeros is allowed to occur [10]. A string of 14 zeros is the longest that can occur if during any single 8 bit time slot at least one 1 is present.

Code Space Redundancy

In essence, bipolar coding uses a ternary code space, but only two of the levels during any particular signal interval. Hence bipolar coding eliminates dc wander with an inefficient and redundant use of the code space. The redundancy in the waveform also provides other benefits. The most important additional benefit is the opportunity to monitor the quality of the line with no knowledge of the nature of the traffic being transmitted. Since pulses on the line are supposed to alternate in polarity, the detection of two successive pulses of one polarity implies

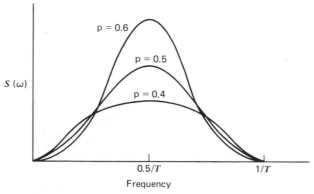

Figure 4.10 Spectral density of bipolar coding.

an error. This error condition is known as a "bipolar violation." No single error can occur without a bipolar violation also occurring. Hence the bipolar code inherently provides a form of line code parity. The terminals of T1 lines are designed to monitor the frequency of occurrence of bipolar violations, and if the frequency of occurrence exceeds some threshold, an alarm is set.

In T-carrier systems, bipolar violations are used merely to detect channel errors. By adding some rather sophisticated detection circuitry, the same redundancy can be used for correcting errors in addition to detecting them. Whenever a bipolar violation is detected, an error has occurred in one of the bits between and including the pulses indicating the violation. Either a pulse should be a 0, or an intervening 0 should have been a pulse of the opposite polarity. By examining the actual sample values more closely, a decision can be made as to where the error was most likely to have occurred. The bit with a sample value closest to its decision threshold is the most likely bit in error. This technique belongs to a general class of decision algorithms for redundant signals called maximum likelihood or Viterbi decoders [11]. Notice that this method of error correction requires storage of pulse amplitudes. If decision values only are stored, error correction can not be achieved (only error detection). At the present time the implementation cost of these error correction algorithms precludes their use in commercial applications.

An additional application of the unused code space in bipolar coding is to purposely insert bipolar violations to signify special situations such as time division multiplex framing marks, alarm conditions, or special codes to increase the timing content of the line signals. Since bipolar violations are not normally part of the source data, these special situations are easily recognized. Of course, the ability to monitor the quality of the line is compromised when bipolar violations occur for reasons other than channel errors.

4.3.3 Binary N Zero Substitution

A major limitation of bipolar coding is its dependence on a minimum density of 1's in the source code to maintain timing at the regenerative repeaters. Even when strings of zeros greater than 14 are precluded by the source, a low density of pulses on the line increases timing jitter and therefore produces higher error rates. Binary N zero substitution (BNZS) [12] augments a basic bipolar code by replacing all strings of N zeros with a special N length code containing several pulses that purposely produce bipolar violations. Thus the density of pulses is increased while the original data is obtained by recognizing the bipolar violation codes and replacing them at the receiving terminal with N zeros.

TABLE 4.1   B3ZS SUBSTITUTION RULES

Polarity of the Preceding Pulse	Number of Bipolar Pulses (ones) Since Last Substitution	
	Odd	Even
−	00−	+0+
+	00+	−0−

As an example, a three zero substitution algorithm (B3ZS) is described. This particular substitution algorithm is specified for the standard DS-3 signal interface in North America [13]. It is also used in the LD-4 coaxial transmission system in Canada [14].

In the B3ZS format, each string of three zeros in the source data is encoded with either 00V or B0V. A 00V line code consists of 2 bit intervals with no pulse (00) followed by a pulse representing a bipolar violation (V). A B0V line code consists of a single pulse in keeping with the bipolar alternation (B), followed by no pulse (0), and ending with a pulse with a violation (V). With either substitution, the bipolar violation occurs in the last bit position of the three zeros replaced by the special code. Thus the position of the substitution is easily identified.

The decision to substitute with 00V or B0V is made so that the number of B pulses (unviolated pulses) between violations (V) is odd. Hence if an odd number of ones has been transmitted since the last substitution, 00V is chosen to replace three zeros. If the intervening number of ones is even, B0V is chosen. In this manner all purposeful violations contain an odd number of intervening bipolar pulses. Also, bipolar violations alternate in polarity so that dc wander is prevented. An even number of bipolar pulses between violations occurs only as result of a channel error. Furthermore, every purposeful violation is immediately preceded by a 0. Hence considerable systematic redundancy remains in the line code to facilitate performance monitoring. Table 4.1 summarizes the substitution algorithm.

EXAMPLE 4.1

Determine the B3ZS line code for the following data sequence: 10100011000000010001. Use + to indicate a positive pulse, − to indicate a negative pulse, and 0 to indicate no pulse.

Solution. There are two possible sequences depending on whether an odd or even number of pulses have been transmitted following the previous violation.

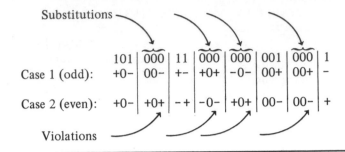

Example 4.1 indicates that the process of breaking up strings of zeros by substituting with bipolar violations greatly increases the minimum density of pulses in the line code. In fact, the minimum density is 33% while the average density is just over 60%. Hence the B3ZS format provides a continuously strong timing component. Notice that all BNZS coding algorithms guarantee continous timing information with no restrictions on source data. Hence BNZS coding supports any application in a completely transparent manner.

Another BNZS coding algorithm used by the Bell System is the B6ZS algorithm for T2 transmission lines [15]. Thus, unlike T1 lines, T2 lines do not require source code restriction to maintain timing. The B6ZS algorithm is defined in Table 4.2. This algorithm produces bipolar violations in the second and fifth bit positions of the substituted sequence.

CCITT recommends another BNZS coding format referred to as high density bipolar (HDB) coding [16]. A commonly used version of HDB replaces strings of four zeros with sequences containing, a bipolar violation in the last bit position. Since this coding format precludes strings of zeros greater than three, it is referred to as HDB3 coding. The encod-

TABLE 4.2 B6ZS SUBSTITUTION RULES

Polarity of Pulse Immediately Preceding Six Zeros to Be Substituted	Substitution
−	0−+0+−
+	0+−0−+
EXAMPLE	
	1000000101100000000000000001
+	−0−+0+−+0−+0+−0−+0+−0−+000−
−	+0+−0−+−0+−0−+0+−0−+0+−000+

TABLE 4.3 HDB3 SUBSTITUTION RULES

Polarity of Preceding Pulse	Number of Bipolar Pulses (ones) Since Last Substitution	
	Odd	Even
–	000–	+00+
+	000+	–00–

ing algorithm is basically the same as the B3ZS algorithm described earlier. Table 4.3 presents the basic algorithm. Notice that substitutions produce violations only in the fourth bit position, and successive substitutions produce alternating polarities for purposeful violations.

4.3.4 Pair Selected Ternary

The BNZS substitution algorithms described in the preceding section are examples of selecting codes in the ternary code space to increase the timing content of a binary signal. Pair selected ternary (PST) is another example [17].

The PST coding process begins by pairing the binary input data to produce sequences of 2 bit code words. These code words are then translated into two ternary digits for transmission. Since there are nine two-digit ternary codes but only four two-digit binary codes, there is considerable flexibility available in selecting the codes. The most useful of the possible coding formats is shown in Table 4.4. This particular format not only ensures a strong timing component but also it prevents dc wander by switching modes to maintain a balance between positive and negative pulses. The codes are selected from one column until a single pulse is transmitted. At that time the encoder switches modes and selects codes from the other column until another single pulse (of the opposite polarity) is transmitted.

TABLE 4.4 PAIR SELECTED TERNARY ENCODING

Binary Input	+ Mode	– Mode
00	–+	–+
01	0+	0–
10	+0	–0
11	+–	+–

EXAMPLE 4.2

Encode the following binary data stream into a PST line code:
01001110101100.

Solution. There are two possible solutions depending on whether the encoder is in the + or − mode at the beginning of the sequence.

	01	00	11	10	10	11	00
Case 1 (+ mode):	0+	− +	+ −	− 0	+0	+ −	− +
Case 2 (− mode):	0−	− +	+ −	+0	−0	+ −	− +

One potential drawback of the PST coding algorithm is that the binary data stream must be framed into pairs. Hence a PST decoder must recognize and maintain pair boundaries. Recognition of the boundaries is not difficult if random data is being transmitted since a pairwise misframe eventually produces unallowed codes (00, ++, − −). Furthermore, time division multiplex formats typically provide character and pairwise framing automatically.

The power spectrum of a PST line code with equal probabilities for 1's and 0's is obtained from Reference 17 and plotted in Figure 4.11. Also shown is the B6ZS power spectrum [15] and the conventional bipolar power spectrum.

An important point to notice in Figure 4.11 is that bipolar coding and its extensions require equal bandwidths. Their only significant difference is that B3ZS and PST have higher energy levels as a result of greater pulse densities. The higher energy levels have the undesirable ef-

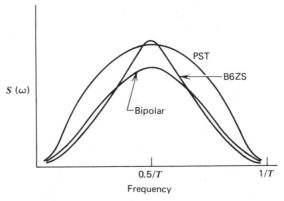

Figure 4.11 Spectrums of bipolar, B3ZS, and PST line codes for equally likely 1s and 0s.

fect of increasing crosstalk interference in multipair cables. However, the degradation from the increased crosstalk is somewhat offset by improved accuracy of the recovered sample clock.

4.3.5 Ternary Coding

Since bipolar and PST coding use a ternary code space to transmit binary data, they do not achieve as high an information rate as is possible with more efficient use of the code space. For example, an eight-element ternary code is capable of representing $3^8 = 6561$ different codes. In contrast, eight bits of binary data produce only $2^8 = 256$ different codes. The previously described line codes do not take advantage of the higher information content of ternary codes— they select codes for their timing content and spectral properties.

One ternary encoding procedure involves mapping successive groups of 4 bits into three ternary (4B3T) digits. Since binary words of 4 bits only require 16 of the 27 possible three-digit ternary code words, considerable flexibility exists in selecting the ternary codes. Table 4.5 presents one possible encoding procedure. Ternary words in the middle column are balanced in their dc content. Code words from the first and

TABLE 4.5 ENCODING TABLE FOR 4B3T LINE CODE

Binary Word	Ternary Word (Accumulated Disparity)		
	–	0	+
0 0 0 0	– – –		+ + +
0 0 0 1	– – 0		+ + 0
0 0 1 0	– 0 –		+ 0 +
0 0 1 1	0 – –		0 + +
0 1 0 0	– – +		+ + –
0 1 0 1	– + –		+ – +
0 1 1 0	+ – –		– + +
0 1 1 1	– 0 0		+ 0 0
1 0 0 0	0 – 0		0 + 0
1 0 0 1	0 0 –		0 0 +
1 0 1 0		0 + –	
1 0 1 1		0 – +	
1 1 0 0		+ 0 –	
1 1 0 1		– 0 +	
1 1 1 0		+ – 0	
1 1 1 1		– + 0	

third columns are selected alternately to maintain dc balance. If more positive pulses than negative pulses have been transmitted, column 1 is selected. When the disparity between positive and negative pulses changes, column 3 is chosen. Notice that the all-zeros code word is not used. Hence a strong timing content is maintained. Owing to the higher information efficiency, however, the ability to monitor performance is sacrificed, and framing is required on three-digit boundaries. 4B3T coding is used on the T148 span line developed by ITT Telecommunications [18]. This system provides T-carrier transmission for two DS-1 signals (48 channels) using a bandwidth that is only 50% greater than a T1 bandwidth (carrying 24 channels). A generalized discussion of 4B3T coding and other ternary coding techniques is contained in Reference 19.

4.3.6 Digital Biphase

Bipolar coding and its extensions BNZS and PST use extra encoding levels for flexibility in achieving desirable features such as timing transitions, no dc wander, and performance monitorability. These features are obtained by increasing the code space and not by increasing the bandwidth. (The first spectral null of all codes discussed so far, including an NRZ code, is located at the signaling rate $1/T$.)

Many varieties of line codes achieve strong timing and no dc wander by increasing the bandwidth of the signal while using only two levels for binary data. One of the most common of these codes providing both a strong timing component and no dc wander is the digital biphase code, also referred to as "diphase" or a "Manchester" code.

A digital biphase code uses one cycle of a square wave at a particular phase to encode a one and one cycle of an opposite phase to encode a zero. An example of a digital biphase coding sequence is shown in Figure 4.12. Notice that a transition exists at the center of every signaling interval. Hence strong timing components are present in the spectrum. Furthermore, logic 0 signals and logic 1 signals both contain equal amounts of positive and negative polarities. Thus dc wander is nonexistent. A digital biphase code, however, does not contain redundancy for performance monitoring. If in-service performance monitoring is desired, either parity bits must be inserted into the data stream or pulse

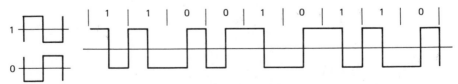

Figure 4.12 Digital biphase (Manchester) line code.

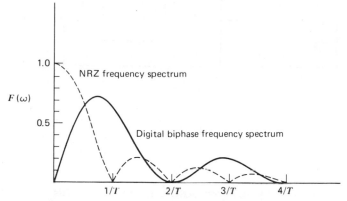

Figure 4.13 Power spectral density of digital biphase.

quality must be monitored. (A later section of this chapter discusses performance monitoring in more detail.)

The frequency spectrum of a digital biphase signal is derived in Appendix C and plotted in Figure 4.13, where it can be compared to the spectrum of an NRZ signal. Notice that a digital biphase signal has its first spectral null at $2/T$. Hence the extra timing transitions and elimination of dc wander come at the expense of a higher frequency signal. In comparison to three-level bipolar codes, however, the digital biphase code has a lower error rate for equal signal-to-noise ratios (see Appendix C).

Examination of the frequency spectra in Figure 4.13 shows that the diphase spectrum is similar to an NRZ spectrum but translated so it is centered about $1/T$ instead of direct current. Hence digital biphase actually represents digital modulation of a square wave carrier with one cycle per signal interval. Logic ones cause the square wave to be multiplied by $+1$ while logic zeros produce multiplication by -1. Diphase is primarily used on shorter links where terminal costs are more significant than bandwidth utilization. The "Ethernet," a local area data network being developed by Xerox, Digital Equipment and Intel Corporations uses digital biphase (Manchester) coding [20].

4.3.7 Differential Encoding

One limitation of NRZ and digital biphase signals, as presented up to this point, is that the signal for a 1 is exactly the negative of a signal for a 0. On many transmission media, it may be impossible to determine an absolute polarity or an absolute phase reference. Hence the decoder may decode all 1's as 0's and vice versa. A common remedy for this ambiguity is to use differential encoding which encodes a 1 as a change of state and encodes a 0 as no change in state. In this manner no absolute reference is necessary to

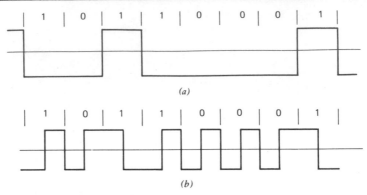

Figure 4.14 Differentially encoded NRZ and diphase signals. (*a*) Differentially encoded NRZ. (*b*) Differentially encoded diphase.

decode the signal. The decoder merely detects the state of each signal interval and compares it to the state of the previous interval. If a change occurred, a 1 is decoded. Otherwise, a 0 is determined.

Differential encoding and decoding does not change the encoded spectrum of purely random data (equally likely and uncorrelated ones and zeros), but it does double the error rate. If the detector makes an error in estimating the state of one interval, it also makes an error in the next interval. An example of a differentially encoded NRZ code and a differentially encoded diphase signal is shown in Figure 4.14. All signals of differentially encoded diphase retain a transition at the middle of an interval, but only the 0's have a transition at the beginning of an interval.

4.3.8 Coded Mark Inversion

A variety of line codes have evolved similar to the digital biphase code described previously. One of these is referred to as coded mark inversion (CMI) in CCITT recommendations [16]. CMI encodes ones (marks) as an NRZ level opposite to the level of the previous one. Zeros are encoded as a half-cycle square wave of one particular phase. Figure 4.15 shows a sample encoding for CMI. There is no dc energy in the signal and an abundance of signal transitions exist as in diphase. Furthermore, there is no ambiguity between 1's and 0's. Elimination of the ambiguity actually leads to a major drawback of CMI coding: its error performance is 3 dB worse than diphase.* CMI is the specified interface code for the fourth-level CCITT multiplex signal at 139.264 Mbps.

*The error performance of CMI is 3 dB worse than diphase when bit-by-bit detection is used. The inefficiency arises because for one-half of an interval a "1" looks like a "0." Because CMI has redundancy, some of the inefficiency can be recovered with maximum likelihood (Viterbi) detection.

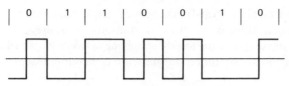

| 0 | 1 | 1 | 0 | 0 | 1 | 0 |

Figure 4.15. Coded mark inversion.

4.3.9 Multilevel Signaling

In the line codes discussed so far, two-level (binary) signaling has been assumed. In applications where the bandwidth is limited but higher data rates are desired, the number of levels can be increased while maintaining the same signaling rate. The data rate R achieved by a multilevel system is:

$$R = \log_2 (L) \left(\frac{1}{T}\right) \qquad (4.4)$$

where L = number of levels that can be freely chosen during each interval
T = signaling interval

The signaling rate $1/T$ is often referred to as the symbol rate and is measured in bauds. Within the data communications industry it is common practice to use "baud" as being synonymous with bit rate. Strictly speaking, however, the bit rate is only equal to the baud rate if binary signaling (1 bit per signal interval) is used. Figure 4.16 shows an example of an eight-level transmission format that achieves 3 bits per signal interval (i.e., 3 bits per baud).

Multilevel transmission systems achieve greater data rates within a given bandwidth but require much greater signal-to-noise ratios for a given error rate. If multilevel transmission were used on wire line media, the regenerative repeaters would have to be moved closer together to achieve a desired error rate. On the other hand, if the repeaters are closer together there is less attenuation, and a higher signaling rate with binary transmission could be used. Thus wire line transmission media are

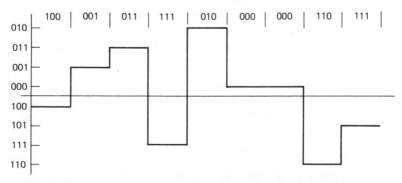

Figure 4.16 Multilevel transmission with 3 bits per signal interval.

essentially attenuation limited and not bandwidth limited and it follows that multilevel techniques are not as attractive for wire line transmission as they are in radio systems or for data transmission over the analog telephone network where adherence to strict bandwidths is required. Multilevel transmission is therefore discussed more fully in Chapter 6.

4.3.10 Partial Response Signaling

Conventional bandlimiting filters of a digital transmission system are designed to restrict the signal bandwidth as much as possible without spreading individual symbols so much that they interfere with sample values of adjacent intervals. One class of signaling techniques, variously referred to as duobinary [21], correlative level encoding [22], or partial response signaling [23], purposely introduces a prescribed amount of intersymbol interference that is accounted for in the detection cir-cuitry of the receivers. By overfiltering the encoded signal, the band-width is reduced for a given signaling rate, but the overlapping pulses produce multiple levels that complicate the detection process and in-crease the signal power requirements for a given error rate.

Figure 4.17 shows the pulse response of a typical partial response system. If the channel is excited by a pulse of duration T, channel filters (defined in Appendix C) limit the spectrum to the point that the main part of the pulse extends across three signal intervals and contrib-utes equally to two sample times. The reason for the term partial response is now apparent: the output only responds to one-half the amplitude of the input.

If the input pulse of Figure 4.17 is followed by another pulse of the same amplitude, the output will reach full amplitude by virtue of the overlap between pulses. However, if the next input pulse has a negative polarity, the overlap produces zero amplitude at the sample time. Thus, as shown in Figure 4.18 a partial response system with two-level inputs (+1, −1) produces an output with three levels (+1, 0, −1). In similar fashion, Figure 4.19 shows a system with four input levels (+3, +1, −1, −3) and seven output levels (+3, +2, +1, 0, −1, −2, −3).

Detection of a partial response signal is complicated by the addi-

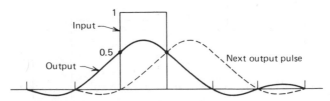

Figure 4.17 Output pulse of partial response channel.

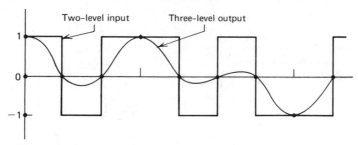

Figure 4.18   Three-level partial response inputs and outputs.

tional levels produced by the channel filters and the fact that sample values are dependent on two adjacent pulse amplitudes. One method of detecting a PRS signal involves subtracting the overlap of a previously detected pulse from the incoming signal to generate a difference signal representing the channel response to the new pulse. Only the overlap (intersymbol interference) at the sample times needs to be subtracted. The difference samples are nominally equal to one-half the amplitude of the unknown input pulse. This technique doubles the error rate in the same manner that differentially encoded error rates are doubled.

Another method of detecting partial response systems involves a technique called "precoding" at the source. Precoding transforms the input data in such a manner that the output level at the detector directly indicates the original data without comparison to the previous sample value. In a binary system, for example, a 1 is encoded with a pulse of the same polarity as the previous pulse. Hence logic 1's show up at the detector as either a +1 or a -1 sample. Transmission of a 0 is encoded with a pulse of opposite polarity to the last pulse. Hence logic 0's always show up at the detector as a zero-level signal. Similar precoding techniques exist for multilevel systems [24].

The partial response technique just described is actually a special case of a more general class of signaling techniques referred to as "correla-

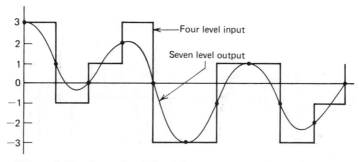

Figure 4.19   Seven level partial response inputs and outputs.

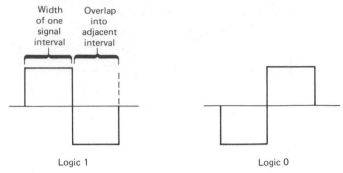

Figure 4.20 Individual signal element for 1 – D correlative level encoding.

tive level encoding" [22]. To describe the properties of generalized partial response or correlative level systems, it is convenient to introduce a delay operator D to denote delay equal to one signal interval T. Physically, D can be thought of as a delay line of length T. Two units of delay is implemented with two delay lines in series and is denoted as D^2.

Using this notation, the partial response system described is referred to as $1 + D$ PRS: the output represents the superposition of the input with a delayed version of the same input. Other forms of overlap are possible. These systems do not necessarily produce the overlapping pulses by overfiltering an input. An alternative approach is to overlap and add the pulses directly in the encoding process (hence the term: correlative level encoding). An interesting special case of correlative level encoding is the $1 - D$ system shown in Figure 4.20 and extended in Figure 4.21 to show the effect of overlapping pulses.

The $1 - D$ encoder uses a single cycle of a square wave across two signal intervals to encode each bit. Since neither of the two individual signals (+ –, or – +) produce dc energy, the encoded signal has no dc wander. Furthermore, the positive and negative levels of the composite signal alternate in a manner reminiscent of bipolar coding. In fact, if differential encoding is used (i.e. if a zero is encoded with the same

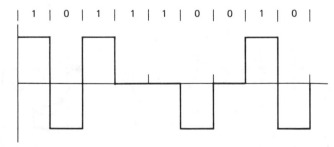

Figure 4.21 Representative waveform of 1 – D correlative level encoding.

phase as the previous interval and a one as the opposite phase) this form of encoding is identical to bipolar coding (assuming the levels are replaced by 50% duty cycle pulses). Thus $1 - D$ correlative level encoding is used to shape the bandwidth rather than to limit it. The spectra of unfiltered $1 + D$, $1 - D$, and $1 - D^2$ signals are obtained from Reference 23 and plotted in Figure 4.22.

The spectrum of the $1 - D^2$ signal is particularly interesting. It has no dc component and an upper limit equal to $1/2T$: the same upper limit as a maximally filtered NRZ code with no intersymbol interference. The $1 - D^2$ partial response systems have been used by GTE in a modified T-carrier system providing 48 voice channels [25] and for digital transmission over analog microwave radios using very low baseband frequencies for pilots and service channels (where no message spectrum exists) [26]. The 1A-RDS (data under voice) system of AT&T also uses $1 - D^2$ coding with four-level inputs to get 1.544 Mbps into 500 kHz of bandwidth.

Notice that the $1 - D^2$ spectrum is essentially the product of the $1 + D$ and the $1 - D$ spectra. Indeed, a $1 - D^2$ system can be implemented by concatenating a $1 - D$ encoder with a $1 + D$ channel response: $(1 - D)(1 + D) = 1 - D^2$. Thus correlative level system polynomials are very useful for simultaneously representing correlations and spectrum

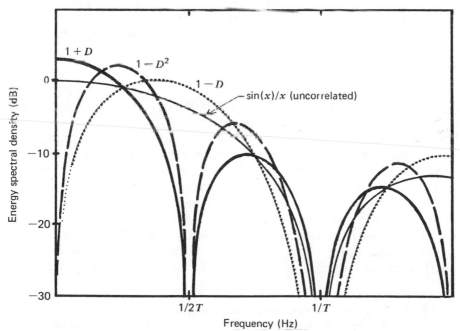

Figure 4.22 Spectra of unfiltered $1 + D$, $1 - D$, and $1 - D^2$ correlative encoded signals.

TABLE 4.6 DIGITAL TRANSMISSION SYSTEMS

Designation	Country or Administration	Bit Rate (Mbps)	Line Code	Media	Repeater Spacing
T1	AT&T	1.544	Bipolar	Twisted pair	6000 ft
Primary	CCITT	2.048	HDB3	Twisted pair	2000 m
T1C	AT&T	3.152	Bipolar	Twisted pair	6000 ft
T148	ITT	2.37, Ternary	4B3T	Twisted pair	6000 ft
9148A	GTE	3.152	$1 - D^2$, Duobinary	Twisted pair	6000 ft
T1D	AT&T	3.152	$1 + D$, Duobinary	Twisted pair	6000 ft
T2	AT&T	6.312	B6ZS	Low capacitance Twisted pair	14800 ft
LD – 4	Canada	274.176	B3ZS	Coax	1900 m
T4M	AT&T	274.176	Polar binary (NRZ)	Coax	5700 ft

shaping. Since all correlative level systems use more levels than necessary for encoding the data, they are inferior to uncorrelated or full response systems in terms of error performance.

A summary of digital transmission systems identifying various line codes is provided in Table 4.6.

4.4 ERROR PERFORMANCE

The preceding sections of this chapter emphasized the timing and power spectrum requirements of various transmission codes. Another fundamental consideration in choosing a line code is the error rate performance in the presence of noise. Except on relatively short lines, where noise may be insignificant, error performance requirements can impact system cost significantly. If a certain minimum error rate is specified, those coding schemes providing the desired error rate at lower signal-to-noise ratios allow regenerative repeaters to be spaced farther apart, thereby reducing installation and maintenance.* Repeater spac-

---

*On T1 lines the repeater spacing was predetermined by the locations where loading coils needed to be removed. Thus the error rate performance essentially determined the maximum data rate.

ing is an important economic factor in both wire line transmission and on point-to-point radio links.

The error rate results and comparisons presented in this section are based on white Gaussian noise. This is the most common form of noise and the best understood. In some applications, particularly within the existing public telephone network, impulse noise may actually be more prevalent. Thus these analyses do not provide a complete error rate analysis for some applications. The intent of this section is to present relative error performance comparisons of various line coding techniques. For this purpose a white noise analysis is most appropriate, since if impulses are large enough they cause errors independent of the coding scheme in use.

For the most part, the following sections present only the results of the error analyses in the form of graphs of error rate as a function of signal-to-noise ratios. Appendix C derives the basic equations used to produce the results.

4.4.1   Signal Detection

Invariably, the detection circuitry of a digital receiver processes incoming signal waveforms to measure each possible discrete signal. In most cases the measures are nothing more than samples of a filtered receive signal. Depending on the signal shape and the level of performance desired, we may use more sophisticated processing. In any case, the end measurement of a binary signal nominally produces one voltage level for a 0 and another voltage level for a 1. A decision of which signal was transmitted is made by comparing the measurement (at the appropriate time) to a threshold located halfway between these nominal voltages. Naturally, the error probability depends on the nominal distance between the voltages and the amount of fluctuation in the measurements caused by noise.

Since signal measurements are normally linear in nature, the error distance between 1's and 0's is proportional to the received signal voltage. Equivalently, the amount of noise power required to produce an error is a direct function of the signal-to-noise ratio. The optimum detector for a particular signal set maximizes the signal-to-noise ratio at the input to the decision circuit.

4.4.2   Noise Power

White Gaussian noise is characterized as having a uniform frequency spectrum across an arbitrarily large bandwidth and an amplitude distribution that varies according to a normal (Gaussian) probability distribution. A parameter N_0 conventionally represents the power spectral density of white noise. N_0 is the amount of power measured in a bandwidth of 1 Hz. Hence the rms power of white noise coming out of a filter with a bandwidth BW is $N_0 \cdot BW$.*

In order to determine the amount of noise power present at a decision circuit it is necessary to determine N_0 and the effective bandwidth of the detection circuitry. When the detection circuitry consists of nothing more than a filter, the effective bandwidth is usually very close to the 3 dB bandwidth of the filter. With other more sophisticated detectors, the effective bandwidth can be more difficult to determine. As derived in Appendix C, the effective bandwidth is usually referred to as the effective noise bandwidth (NBW) of the receiver. Hence the noise power at the decision circuit is $N_0 \cdot NBW$.

4.4.3 Error Probabilities

An error in detection occurs whenever noise causes the signal measurement to cross the threshold between the two nominal output levels of the detection circuitry. Appendix C derives this probability for white Gaussian noise as

$$\text{prob (error)} = \frac{1}{\sqrt{2\pi}\,\sigma} \int_V^\infty e^{-t^2/2\sigma^2}\,dt \qquad (4.5)$$

where V is the nominal distance (voltage) to a decision threshold and σ^2 is the noise power at the detector, $\sigma^2 = N_0 \cdot NBW$.

Equation 4.5 is nothing more than the area under the probability density function of a normal distribution. As shown in Figure 4.23, the equation represents the error probability as the probability of exceeding V/σ standard deviations in a normal distribution with zero mean and unit variance $N(1, 0)$.

The error rate is completely determined by the ratio of V to σ. Since V is the noise-free sample voltage and σ^2 is the rms noise power, V^2/σ^2 is a signal power to noise power ratio at the detector. This ratio is sometimes referred to as a post-detection SNR, since it is measured after the detection circuitry. It is usually more important to express error rates in terms of a signal-to-noise ratio at the input to the receiver. Figure 4.24 depicts a basic channel and detection model and the relation between a predetection SNR and a postdetection SNR. For reasons discussed in

---

*The power spectral density of white noise is also specified as $\frac{1}{2}N_0$ for a two-sided spectral density. As a practical matter, there is no difference in the specifications, since a real filter has the mathematical equivalent of identical positive and negative frequency bands. Thus the measured power coming through a filter with a one-sided (positive frequency) bandwidth BW is $N_0 BW$ in either case.

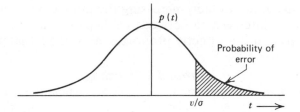

Figure 4.23   Probability of error for binary signaling.

Appendix C, the most appropriate predetection SNR for comparing line codes and digital modulation formats is an energy-per-bit to noise density ratio E_b/N_0. The relationship between E_b/N_0 and the signal power to noise power ratio is

$$\text{SNR} = \frac{\text{signal power}}{\text{noise power}}$$

$$= \frac{d \cdot E_s(1/T)}{N_0 \cdot NBW}$$

$$= \frac{d \cdot E_b \log_2 L \cdot (1/T)}{N_0 \cdot NBW} \tag{4.6}$$

where d is the pulse density, E_s is the energy per symbol, E_b is the energy per bit, $\log_2 L$ is the number of bits per symbol (i.e., L is the number of levels), $1/T$ is the signaling rate, and NBW is the effective noise bandwidth of the receiver.

In determining the signal power in Equation 4.6 notice the dependence on the pulse density d. In an NRZ line code the pulse density is 1, but in many of the other line codes the pulse density is dependent on the data and any substitution algorithms that may be in use. In these codes, increasing the pulse density increases the SNR but does not reduce the error rate. The error rate is determined by the energy-per-bit

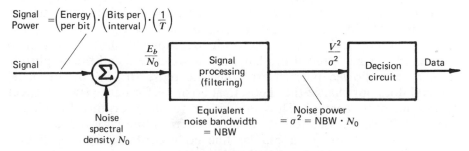

Figure 4.24   Signal detection model.

to noise density ratio. In fact, increasing the pulse density might worsen the error rate if interference between cable pairs is a significant consideration. (Interference is directly proportional to the signal power.)

Antipodal Signaling

The optimum signaling format for binary signaling maximizes the error distance for a given received signal power and simultaneously minimizes the noise bandwidth. This condition arises only when two signal levels are allowed, and only when one signal is the exact negative of the other. Since the signal for a 1 is the exact opposite (the antipode) of the signal for a 0, optimum signaling is often referred to as "antipodal signaling." Since no other bit-by-bit signaling scheme is any better, antipodal performance is often used as a basis for comparisons. Of the signaling schemes described previously, only balanced two-level (NRZ) encoding and digital biphase can provide antipodal performance. Figure 4.25 shows the optimum error performance provided by antipodal signaling as a function of E_b/N_0 and the signal-to-noise ratio SNR.

Error Rate of Level Encoded Signals

As already mentioned, a balanced, two-level line code is capable of providing optimum error rate performance. If an unsymmetric level code is used, such as unipolar, the same basic detector is used as in symmetric level encoding. The only difference is that the decision threshold must be moved from zero to half the amplitude of the "on" signal. In order to maintain the same error distance, the transmit power is increased by a factor of 2 (assuming 50% "on" pulses). Hence a unipolar code carries a performance penalty of 3 dB with respect to antipodal performance.

Figure 4.25 shows the ideal performance of a unipolar (on/off) code where it is compared to antipodal performance. Notice that for all signal-to-noise ratios the error rate of the on/off system is exactly equal to the error rate of an antipodal system with 3 dB less SNR.

Bipolar Signaling

With respect to error performance, bipolar signaling is basically identical to a unipolar code. During any particular signal interval, the receiver must decide between one of two possible levels: zero or a pulse with the appropriate polarity. Hence the decision threshold pertinent to a particular signal interval lies halfway between zero and the amplitude level of the allowed pulse. Narrowing the pulse to 50% of the signal interval does not change the theoretical error performance (with respect to average pulse energy). Thus the error rate curve for on/off keying in Figure 4.25 can be used to determine theoretical bipolar error rates.

One consideration in a bipolar line code contributes to a slightly higher error rate than in on/off keying. This increase occurs because

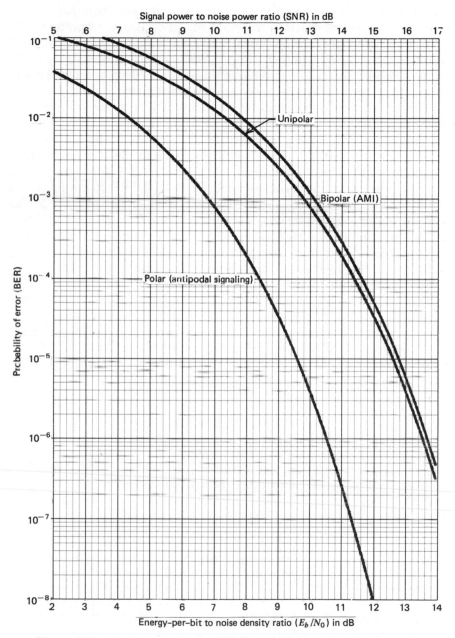

Figure 4.25 Error rates of polar (NRZ), unipolar, and bipolar line codes.

both positive and negative noise can cause an erroneous threshold crossing when a zero-level signal is transmitted. In contrast, a unipolar code
is affected only by positive noise when a lower-level signal is transmitted,
and by negative noise when the upper-level signal is transmitted. If the
bipolar detector treats the erroneous pulse as a 1 (despite a bipolar violation), the error probability when 0's are transmitted is doubled. Thus
the overall error probability is increased by 50% if zeros and ones are
equally likely.

Because of the steepness of the curve, an increase in the error rate of
50% does not represent much of a performance penalty. For example,
if the error rate is increased from 1×10^{-6} to 1.5×10^{-6}, the source
power of a line code needs to increase by only 0.2 dB to get back to a
1×10^{-6} error rate. At higher error rates, a larger penalty occurs because the curve is not as steep. These effects are demonstrated in Figure
4.25, where the ideal performance of a bipolar code (with 50% zeros)
can be compared to the performance of a unipolar code.

The fact that bipolar coding incurs a 3.2 dB penalty (at 10^{-6} error
rate) with respect to digital biphase is indicative that timing and dc
wander problems are solved by increasing the number of signal levels.
In contrast, digital biphase incurs a bandwidth penalty. Not all of the
3.2 dB penalty for bipolar coding, and its extensions BNZS and PST,
can be attributed to removing direct current and adding a consistent
timing pattern. Bipolar coding contains considerable redundancy for
performance monitoring or possible error correction.*

EXAMPLE 4.3

Assume each section of a T1 transmission system on 22 gauge cable is
near-end crosstalk limited and operating with a 10^{-6} error rate. What
design changes are needed to reduce the error rate to 10^{-8}?

Solution. Since the system is crosstalk limited, the error rate can not
be improved by increasing the power out of the regenerative repeaters.
The solution is to space the repeaters closer together (ignore the impractical aspects of this solution). From Figure 4.25 it can be determined that the signal power must be increased by 1.6 dB to improve
the error rate from 10^{-6} to 10^{-8}. In Figure 4.10 it is seen that the main
lobe of the bipolar spectrum extends up to 1.544 MHz. However, most
of the energy in the spectrum lies below 1 MHz. Using Figure 1.9, we

---

*In terms of logic-level decisions, a bipolar code can not provide error correction. If
a pulse is removed to eliminate a bipolar violation, half of the time the error is cor-
rected but half of the time another error is made. Error correction is possible only
with a Viterbi-like detector described previously. In this case a bipolar code provides better performance than a unipolar code.

determine that the attenuation of 22 gauge cable is approximately
20 dB/mile at 1 MHz. Hence $1.6/20 = 0.08$ miles or 422 ft is the reduction in repeater spacing required.

Example 4.3 demonstrates a number of important aspects of digital
transmission systems. First, arbitrarily good transmission quality can be
obtained with only a small penalty in transmit power or repeater spacing.* Thus, as mentioned in Chapter 2, a digital transmission and
switching network can be readily designed to impart no degradation to
voice quality except where analog-to-digital and digital-to-analog conversions take place.

Second, the dramatic improvement in error rate for a relatively small
increase in the signal-to-noise ratio implies extreme sensitivity in the
opposite direction also. A slight increase in noise power or signal
attenuation would cause a big increase in the error rate. Hence the
nominal design of a digital link often provides considerably better performance than normally necessary.

Third, the solution to Example 4.3 applies to any digital transmission
format using the same band of frequencies, and hence, the same attenuation per mile. Since all line codes have approximately the same steepness at a 10^{-6} error rate, the relative change in SNR is the same for all
systems. Thus a T1 system with suboptimum detection would exhibit
the same performance improvement if the repeaters were moved 422 ft
closer together. (T1 lines are designed to provide less than a 10^{-6} error
rate to begin with. See Reference 27 for a thorough report of error rate
surveys on T1 lines.)

Multilevel Error Rates

The multilevel transmission system shown in Figure 4.16 does not require bandwidth in excess of a binary system using the same signaling
rate—yet it achieves three times the data rate. The penalty for multilevel transmission comes in the form of greatly increased signal power
for a given error rate. For example, the average power of the eight-level
system in Figure 4.16 is 8.7 dB above the average power a symmetric
two-level system with the same error distance. To make matters worse,
some systems are peak power limited, in which case the eight-level system has a 12.4 dB disadvantage with respect to a two-level system. The
error rates of multilevel systems are derived in Appendix C and plotted
in Figure 4.26 as a function of E_b/N_0. For those cases where the peak
power is of interest, the relationship between the peak and average

---

*Some systems, such as the radio systems discussed in Chapter 6, require a greater
decrease in repeater spacing to achieve the same improvement in performance.

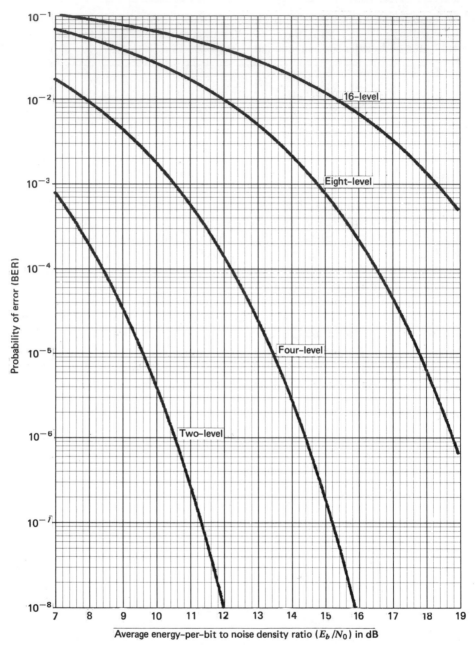

Figure 4.26 Error rates of balanced multilevel signals (all systems providing an identical data rate).

power of a multilevel system is derived in Appendix C as

$$\text{peak-to-average (dB)} = 10 \log_{10} \frac{(L-1)^2}{(2/L) \sum_{i=1}^{L/2} (2i-1)^2} \qquad (4.7)$$

where L is the number of equally spaced levels centered about zero (for example: $\pm 1, \pm 3, \pm 5, \ldots \pm L - 1$).

The error rates of $1 + D$ partial response systems are also derived in Appendix C and plotted in Figure 4.27. These error rates are derived under the assumption that bit-by-bit detection is used. Since partial response sytems exhibit redundancy (correlation) in adjacent samples, better performance can be achieved with Viterbi decoders [11].

4.5 PERFORMANCE MONITORING

Two basic techniques exist for directly monitoring the quality of a digital transmission link: redundancy checks and pulse quality measurements. Both techniques are designed to provide an indication of the bit error rate (BER) of the channel.

4.5.1 Redundancy Checks

Redundancy can be incorporated into a digital signal using one of two common methods. First, the line code itself may contain redundancy as in bipolar coding. In a random, independent error environment the frequency of bipolar violations is very nearly equal to the channel BER (except for extremely high BERs). Second, logic-level redundancy can be inserted into the data stream in the form of overhead bits. For example, parity bits are inserted into DS-3 and DS-4 signals for the express purpose of monitoring the channel error rate. (The frame structure of these and other higher-level multiplex signals are provided in Chapter 7). The main advantage of monitoring logic-level redundancy is its independence of the transmission link.

In contrast to line code redundancies, parity bits do not provide a one-to-one indication of the BER. The following equation relates the parity error rate (PER) to the channel bit error rate (BER):

$$\text{PER} = \sum_{i=1}^{N} \binom{N}{i} \rho^i (1-\rho)^{N-i} \qquad (i \text{ odd}) \qquad (4.8)$$

where N = the length of a parity field (number of bits over which parity is generated)

ρ = the bit error rate (BER) assuming random, independent errors

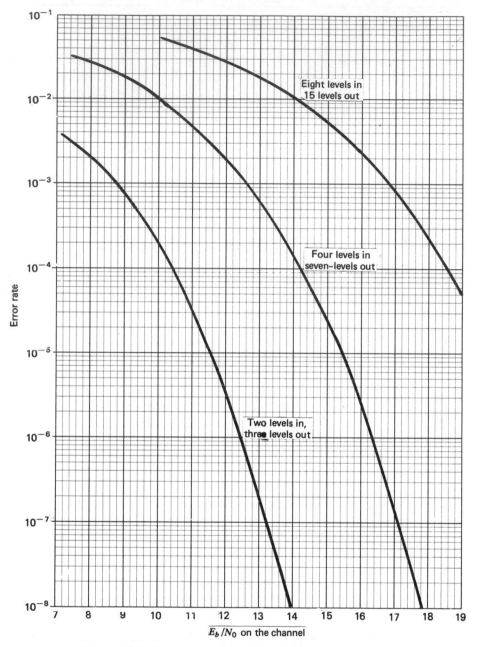

Figure 4.27 Error rates of $1 + D$ partial response systems.

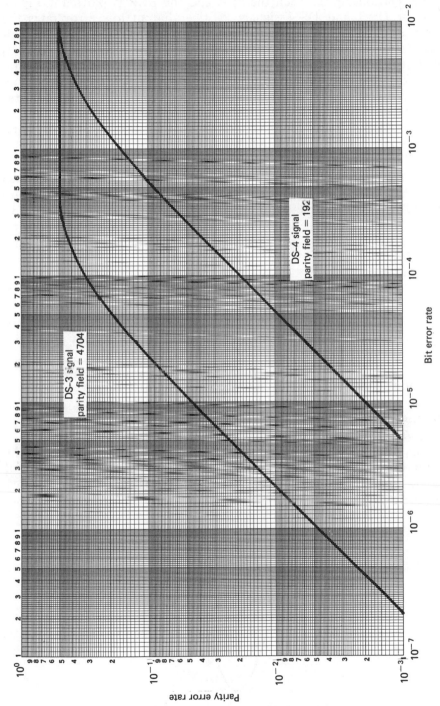

Figure 4.28 Parity error rates versus bit error rates for DS-3 and DS-4 signals.

DS-3 signal
parity field = 4704

DS-4 signal
parity field = 192

Bit error rate

Parity error rate

The relationship between the PER and the BER is plotted in Figure 4.28 for DS-3 and DS-4 signals. Notice that at low error rates the PER is essentially N times the BER. At high error rates, however, this relationship changes because any odd number of errors in a frame is indistinguishable from a single error. When the BER is high enough that more than one error in a parity field is likely, the PER is useless as an absolute estimate of the error rate. In these cases the PER indicates only that the BER is above a threshold approximately equal to $1/N$. Because a DS-4 frame format contains a higher density of parity bits, DS-4 signals can be monitored for error rates much lower than DS-3 signals.

To have confidence in error rate measurements, the sample sequence must be long enough to allow an average of about 10 errors in the sample size (e.g., the sample size must be 10/BER). Hence when trying to measure low BERs (e.g. 10^{-6} or 10^{-7}), the measurement time may be too long to respond to changing channel conditions such as radio channel fading.

4.5.2 Signal Quality Measurements

The second basic technique for monitoring digital transmission quality is to process the digital signal directly and measure certain properties related to the error rate. A simple approach involves merely measuring the received signal power—a common technique in analog systems. In a fixed noise environment this approach is adequate. However, on transmission links where the noise level can vary or where signal distortions can arise, the quality of the pulses themselves must be measured.

Figure 4.29 demonstrates the operation of a "pseudo" error detector designed to detect received pulses with abnormal amplitudes at the sample times. In the example shown binary data is detected by use of a single threshold located midway between the normal pulse amplitudes. Two additional thresholds are included to detect the presence of pulses

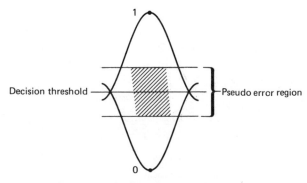

Figure 4.29 Pseudo error detection.

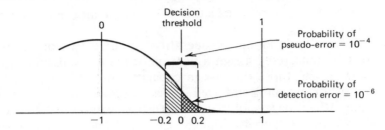

Figure 4.30 Pseudo error decision thresholds for error multiplication of 100.

with abnormal amplitudes. Sample values falling into the central decision region are not necessarily data errors, but a high pseudo error rate is a good indication that the channel is not performing properly.

In a random (Gaussian) noise environment the occurrence rate of pseudo errors is directly related to the actual error rate. Figure 4.30 shows a Gaussian noise distribution and decision thresholds chosen to produce pseudo errors at 100 times a 10^{-6} error rate. Hence an attractive feature of this error rate measurement is that it can measure very low error rates using comparatively short test intervals. Note, however, that the error multiplication factor is dependent on the error rate.

A major disadvantage of pseudo error detection is that pseudo error rates are not accurately related to actual error rates when the noise is non-Gaussian. Channel distortions, in particular, cause high pseudo-error rates when the actual error rates are small. Even though abnormal pulse amplitudes are present, errors do not occur unless noise adds to the distortion. On the other hand, impulse noise might produce high actual error rates but relatively low pseudo error rates.

The correspondence between measured and deduced error rates is improved when the decision threshold is narrowed. Unfortunately, narrower pseudo error decision regions are more difficult to implement and provide lower "error" rate multiplication factors.

4.6 TIME DIVISION MULTIPLEXING

Although frequency division multiplexing of digital transmission signals is possible and is used in some special situations, time division multiplexing is by far the most common and economical means of subdividing the capacity of a digital transmission facility. One application where FDM techniques have been used for digital signals is on multidrop data communications lines where the sources and destinations of the data are distributed along the line. Most telephone network applications, however, involve clusters of channels in the form of trunk groups between switching offices. Even for local distribution, where it may be tempting to add and drop channels on a distributed basis, clustered structures prevail because of maintenance and reliability considerations.

There are two basic modes of operation for time division multiplexing: those that repeatedly assign a portion of the transmission capacity to each source, and those that assign capacity only as it is needed. The first form of operation is referred to as *synchronous* time division multiplexing when necessary to distinguish it from the "as needed" mode of operation. Otherwise, time division multiplexing is understood to imply the synchronous variety. The "as needed" form of time division multiplexing is variously referred to as "asynchronous time division multiplexing" (ATDM); "statistical time division multiplexing" (Stat-Mux); or packet switching. Circuit-switched telephone networks predominantly use synchronous TDM whereas data networks typically use ATDM techniques. Discussion of these latter techniques is deferred to Chapter 8.

4.6.1   Bit Interleaving versus Word Interleaving

Two different structures of (synchronous) time division multiplex frames are shown in Figure 4.31. In the first instance each channel is assigned a time slot corresponding to a single bit—hence the term: bit interleaving. In the second instance each channel is assigned a longer time slot corresponding to some larger number of bits, referred to as a word—hence the term: word interleaving.

The decision as to which structure to use in a particular application is primarily dependent on the nature of the sources. In T-carrier systems each channel produces a complete 8 bit word at a time. Hence word interleaving is desirable so all bits can be transmitted as generated.

In contrast, delta modulation codecs interface more naturally to a bit interleaved system. If word interleaving were used, the periodic samples of a delta modulator would have to be accumulated until transmitted

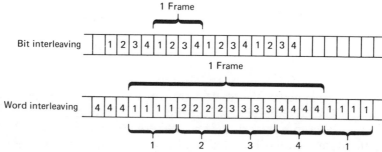

Figure 4.31 Bit interleaving and word interleaving of four-channel TDM multiplexers.

as a burst in an assigned time slot. Beyond the considerations of the nature of the sources, bit interleaved systems require less high-speed digital electronics and therefore are preferable as far as the multiplex equipment is concerned.

Higher-level digital TDM multiplexers use bit interleaving of the lower-level multiplex signals since the lower-level signals represent continuous, one bit at a time, data streams. The specific formats of some higher-level multiplex signals are described in Chapter 7 when synchronization of bit steams is considered.

4.6.2 Framing

In order to identify individual time slots within a TDM frame, a receiving terminal uses a counter synchronized to the frame format of the transmitter. Just as for synchronization of sample clocks, a certain amount of transmission overhead is required to establish and maintain frame synchronization. In fact, most of the techniques used for frame synchronization are directly analogous to clock synchronization techniques discussed previously. Specifically, the basic means of establishing frame synchronization are as follows:

1 Added digit framing.
2 Robbed digit framing
3 Added channel framing.
4 Statistical framing.
5 Unique line signal framing.

The main considerations in choosing a framing procedure are: time required to establish framing, effects of channel errors in maintaining frame synchronization, relationships between the line clock and sample clocks derived from the line clock, transmission overhead, and complexity of the framing circuitry.

The severity of a loss of framing and the time required to reestablish synchronization depends on the nature of the traffic. Since a loss of framing implies a loss of data on all channels, the mean time between misframes must be as long as possible. For voice traffic infrequent misframes can be tolerated if frame synchronization is reestablished rapidly enough to minimize the length of the "glitch" in the output speech. For data traffic the duration of reframe times is not as critical as the frequency of occurrence since most data communications protocols initiate recovery procedures and retransmit a message no matter how much data is lost.

A critical requirement for reframe time in the telephone network comes from the possibility that various in-channel control signals may be lost and interpreted as disconnects. Thus the maximum reframe time on a particular digital transmission link is often determined from analog network signaling conventions. A loss of framing is also used as a performance monitor for the transmission link and usually sets alarm conditions which in turn cause automatic switching to spare terminals or transmission facilities. Generally speaking, terminals serving 600 or more channels require automatic protection switching [28].

Out-of-frame conditions occur in two ways. First, the locally derived sample clock may lose synchronization with the line clock and produce a slip in the counter sequence. Normally, the timing information in the line clock is sufficiently strong to prevent misframes of this type. An exception occurs on radio links when deep fades reduce the signal power to the point that clock synchronization is impossible. Second, misframes can occur as a result of channel errors producing a false indication of an out-of-frame condition. Thus considerable redundancy in the framing pattern is required to minimize the probability of false misframes. A loss of framing is determined when the occurrence of framing pattern violations exceeds some short-term density threshold.

In all but one of the framing techniques (statistical framing) discussed in the following paragraphs, special framing bits or codes are inserted into the information stream. These insertions do not have to occur with each frame. Instead, they can be sent only once for a predetermined number of information frames. In this manner the transmission overhead for framing can be reduced—which is particularly important in bit-interleaved systems. A group of information frames contained between framing indicators is sometimes referred to as a "master frame" or "super frame."

Added Digit Framing

One common technique of framing a digital TDM information stream is to insert periodically a framing bit with an identifiable data sequence. Usually the framing bit is added once for every frame and alternates in value. This particular format, in fact, is the procedure used to establish framing in the original D1 channel banks. When the T1 line carries only voice traffic, this framing format is particularly useful since no information bits can sustain an alternating 1, 0 pattern. (An alternating pattern represents a 4 kHz signal component, which is rejected by the bandlimiting filter in the PCM coder.)

Framing is established in a receiving D1 channel bank by monitoring first one pulse position within a 193 bit frame and then another, until the alternating pattern is located. With this framing strategy, the expected framing time from a random starting point with random data is derived in Appendix A as

$$
\text{frame time} = \begin{pmatrix} \text{average number of} \\ \text{bit positions} \\ \text{searched before} \\ \text{framing bit is found} \end{pmatrix} \cdot \begin{pmatrix} \text{average number of bits} \\ \text{to determine that an} \\ \text{information position is} \\ \text{not a framing position} \end{pmatrix} \quad (4.9)
$$

$$
= (2N) \cdot \frac{N}{2}
$$

$$
= N^2 \text{ bit times}
$$

where N is the number of bits in a frame including the framing bit.

For T1 systems, $N = 193$ so that the framing time is 37,249 bits or 24.125 ms. Also of interest is the maximum framing time. Unfortunately, there is no absolute maximum framing time for a T1 system with random data. It is very unlikely, however, that the framing time would ever exceed the average search time for all bit positions, or 48.25 ms. This latter measure of framing time is referred to as the "maximum average frame time." It is the average time required to establish framing, but with the assumption that all bit positions must be tested for the framing sequence before the actual framing bit is found. Obviously, the maximum average frame time is twice the average value from a random starting point defined in Equation 4.9.

The framing time can be reduced by using more sophisticated frame search strategies. One approach examines one bit at a time, as before, but during a reframe the search begins a few clock positions in front of the present position under the assumption that short lapses of clock synchronization cause small counter offsets. A second approach uses a parallel search by monitoring all bit positions simultaneously for the framing pattern. With this framing procedure, framing is established when the last of the $N - 1$ information bit positions finally produces a framing pattern violation. The probability that all information bit positions produce a framing violation in n or less frames is derived in Appendix A as follows:

$$
\text{prob(frame time} < n) = (1 - (\tfrac{1}{2})^n)^{N-1} \quad (4.10)
$$

where N is the number of bits in a frame and the probability of ones $= \tfrac{1}{2}$

Using Equation 4.10, we determine the median framing time by setting prob(frame time $< n) = \tfrac{1}{2}$. Hence

$$
n = -\log_2 (1 - (\tfrac{1}{2})^{1/(N-1)}) \quad (4.11)
$$

Setting $N = 193$ for the frame length of D1 channel bank produces the result that $n = 8.1$ frames or approximately 1 ms.

An even more sophisticated framing strategy involves continually monitoring all bit positions, even while the system is synchronized. Then, when a misframe is detected, a new frame position is immediately established or the frame search time is significantly reduced. Furthermore, the continuous search for framing patterns provides additional information for declaring an out-of-frame condition. (There is little point in discarding the present frame position unless another bit position exhibits a more consistent framing pattern.)

The framing pattern of second generation channel banks (D2, D3, D4) in the Bell System was changed from the alternating 1, 0 pattern to establish a longer sequence for identification of signaling frames. As mentioned previously, the more recent channel banks provide 8 bits of voice for all time slots except in every sixth frame, which uses the least significant PCM bit for signaling. The signaling channels thus derived are divided into an A and a B subchannel—implying each subchannel sends a bit in every twelfth frame. Hence a 12 bit framing sequence is needed to identify the signaling bits and the two signaling subchannels. The selected framing sequence is provided in Figure 4.32.

When describing or processing the framing sequence shown in Figure 4.32, it is convenient to divide the framing bits into two separate sequences. During the odd frames the framing bit alternates, while during the even frames the framing bit sequence is 000111000111000. Figure 4.32 indicates that the A signaling frame can be identified by a

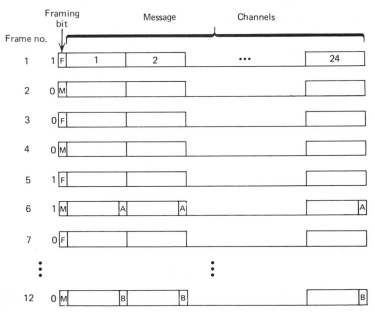

Figure 4.32 Framing sequence of primary digital signal of North America with in-channel signaling. F: Frame alignment signal-101010. M: Multiframe alignment signal = 001110.

0 to 1 transition in the even numbered frame sequence. Correspondingly, a 1 to 0 transition in the even numbered frame sequence signifies a B signaling frame. Frame acquisition begins by finding the alternating bit sequence (with 385 intervening bits). Then, the 000111 framing pattern is located.

Robbed-Digit Framing

The insertion of an extra digit into the transmitted data stream causes the line rate to be higher than the information rate. If the information digits are generated by a periodic sampling process, the inserted framing digits may create aperiodic sample or arrival times. In the case of T1 systems, the insertion rate is identical to the sample rate so no aperiodicities arise. However, in other applications when it is wasteful to insert framing bits at the sampling rate (delta mod systems or single channel PCM systems), aperiodicities occur and must be smoothed out to minimize sample time jitter.

Robbed-digit framing is a procedure that does not interrupt the transmission rate but periodically replaces information bits with a specific framing sequence. Hence robbed-digit framing is analogous to forced data errors to maintain clock synchronization. If the rate of replacement is suitably unrelated to the information frame rate, the replacements can be distributed across all channels.

Although robbed-digit framing is feasible for PCM voice traffic where the effects of the replacements can be minimized by making them occur in the least significant bit of a code word, this technique has limited general usage.

Added-Channel Framing

Added-channel framing is basically identical to added-digit framing except that framing digits are added in a group such that an extra channel is established. Hence the transmission rate of individual channels is integrally related to the line rate of the entire multiplex. This integral relationship greatly simplifies sample clock derivation and avoids any aperiodicities that may occur when a single digit is added.

The fact that the frame boundaries are identified by whole code words adds considerable performance and flexibility to the framing process. First, framing can be established more rapidly since random 8 bit code words are very unlikely to appear as framing codes. (See the problem set at the end of the chapter.) Second, the larger code space simplifies identification of auxiliary functions such as super frame boundaries, parity bits, or equipment status. In most systems the added channel contains more than framing bits.

The first-level digital multiplex signal of CCITT is an example of a system using added-channel framing. The CCITT multiplex standard establishes 32 channels per frame with one channel providing framing

and one more channel dedicated to signaling. Thus 30 out of 32 channels are available for message channels. Figure 4.33 shows the frame
structure of the primary multiplex signal of CCITT.

The average frame acquisition time of a multibit frame code is
derived in Appendix A as

$$\text{frame time} \atop \text{(in bits)} = \frac{N^2}{2(2^L - 1)} \tag{4.12}$$

where N is the length of a frame including the frame code, L is the
length of the frame code, and it is assumed that 1's and 0's are equally
likely.

From Figure 4.33 it can be seen that for the primary digital multiplex signal of CCITT, $N = 512$ and $L = 7$. Thus the average frame time
from a random starting point is determined from Equation 4.12 as
0.5 ms. Again, the "maximum average" frame time is twice the average
from a random starting point, or 1 ms.

Statistical Framing

The individual bits of PCM code words are not entirely random but
exhibit certain statistical patterns, dependent on the significance of the
bit in the code word. In particular, the sign bit of a D3 channel bank
format changes value less frequently than the other bits. Furthermore,
the most significant magnitude bit is normally 1 (lower magnitudes
prevail). Hence, in this case, framing of individual code words can be
established by monitoring statistical properties of the bits.

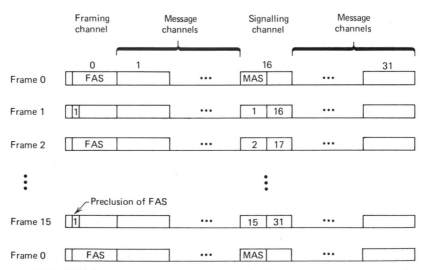

Figure 4.33 Framing sequence of primary digital signal of CCITT using channel-
associated signaling. FAS: 0011011 MAS: 0000.

Statistical framing has the obvious advantage of avoiding aperiodic timing sequences without forcing degradations into the information channels. The fact that statistical framing does not identify individual channels and is dependent on the nature of the traffic prohibits its use for most applications.

Unique Line Code Framing

Bipolar coding manages to shape the spectrum of the line code by adding extra signal levels to provide more flexibility in selecting signals. The same technique can be used to establish frame synchronization. In fact, with bipolar coding and added-digit framing, bipolar violations could be used to identify the framing boundaries uniquely and rapidly.

Even without added-digit framing, unique line codes can be used to carry information while simultaneously conveying frame positioning. If the number of signal levels is doubled for the framing bit only, the extra levels provide sufficient code space for the information but uniquely identify that bit position as a frame boundary. This procedure does not necessarily increase the error rate, since for any particular bit (information or framing), half of the levels can be disallowed.

The main advantage of using unique line codes for framing is that the information bit positions can not generate framing patterns. Hence framing is established as soon as a frame bit occurs, and misframes are detected almost immediately.

The main disadvantages of unique line codes are the added signal processing requirements (if new levels are established just for framing) and the dependence on the transmission terminals to locate framing. With the other framing techniques described in this section, the framing patterns are represented in the logic levels of the bit stream. Hence the transmission equipment can be changed independently of the multiplex equipment. Unique signal framing is used in Collins Radio Company's Time Division Exchange (TDX) loop [29]. For additional information on framing techniques and performance analyses see References 30 and 31.

4.7 TIME DIVISION MULTIPLEX LOOPS

In Chapter 2 it is mentioned that time division multiplexing is not as amenable to applications with distributed sources and sinks of traffic as is frequency division multiplexing. In this chapter a particular form of a TDM network is described that is quite useful in interconnecting distributed nodes. The basic structure of interest is referred to as a "time division multiplex loop" and is shown in Figure 4.34.

Basically, a time division multiplex loop is configured as a series of undirectional (two-wire) links arranged to form a closed circuit or loop. Each node of the network is implemented with two fundamental operational features. First, each node acts as a regenerative repeater merely to recover the incoming bit stream and retransmit it. Second, the network nodes recognize the TDM frame structure and communicate on the loop by removing and inserting data into specific time slots assigned to each node. As indicated in Figure 4.34, a full-duplex connection can be established between any two nodes by assigning a single time slot or channel for a connection. One node inserts information into the assigned time slot that propagates around the loop to the second node (all intervening nodes merely repeat the data in the particular time slot). The destination node removes data as the assigned time slot passes by and inserts return data in the process. The return data propagates around the loop to the original node where it is removed and replaced by new data, and so forth.

Since other time slots are not involved with the particular connection shown, they are free to be used for other connections involving arbitrary pairs of nodes. Hence a TDM loop with C time slots per frame can support C simultaneous full-duplex connections.

If, as channels become available, they are reassigned to different pairs of nodes, the transmission facilities can be highly utilized with high concentration factors and provide low blocking probabilities between all nodes. Thus a fundamental attraction of a loop network is that the transmission capacity can be assigned dynamically to meet changing

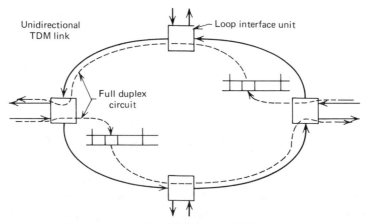

Figure 4.34 Time division multiplex loop.

traffic patterns. In contrast, if a star network with a centralized switching node is used to interconnect the nodes with four-wire links, many of the links to particular nodes would be underutilized since they can not be shared as in a loop configuration.

Another feature of the loop-connected network is the ease with which it can be reconfigured to accommodate new nodes in the network. A new access node is merely inserted into the nearest link of the network and the new node has complete connectivity to all other nodes by way of the TDM channels. In contrast, a star structured network requires transmission to the central node and expansion of the centralized switching facilities. In contrast, the addition of a node to a loop does not affect the operational aspects of any existing node.

The ability to reassign channels to arbitrary pairs of nodes in a TDM loop implies that the loop is much more than a multiplexer. It is, in fact, a distributed transmission and switching system. The switching capabilities come about almost as a by-product of TDM transmission. TDM loops represent the epitome of integrated transmission and switching.

TDM loops have not been used as integral parts of voice communications networks, although their usefulness for data has been recognized [32]. TDM loops have been used within computer complexes to provide high capacity and high interconnectivity between processors, memories, and peripherals [29]. The loop structure in this application is sometimes more attractive than more conventional bus structures since all transmission is unidirectional and therefore avoids timing problems on bidirectional buses that limit their physical length. Furthermore, as more nodes are added to a bus, the electrical loading increases—causing a limitation on the number of nodes that can be connected to a bus. Loops, on the other hand, have no inherent limits of transmission length or numbers of nodes.

One obvious limitation of a loop is its vulnerability to failures of any link or node. The effect of a node failure can be minimized by having bypass capabilities included in each node. When bypassed, a node becomes merely a regenerative repeater as on T-carrier transmission links. Link failures can be circumvented by providing alternate facilities. Figure 4.35 shows one particular structure using a second, reverse direction loop to provide backup capabilities in the case of failures. When fully operational, the network can use the reverse loop as a separate, independent network for traffic as needed. Whenever a failure occurs, the nodes adjacent to the break establish a new loop by connecting the forward path to the reverse path at both places. Hence all nodes continue to have full connectivity to any node on the new loop.

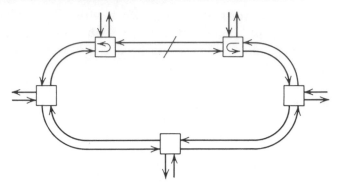

Figure 4.35 Use of reverse loop to circumvent link failures in TDM loops.

REFERENCES

1 F. deJager and M. Christiaens, "A Fast Automatic Equalizer for Data Links," *Philips Technical Review* Vol. 37, 1977.

2 R. K. Even, R. A. McDonald, and H. Seidel, "Digital Transmission Capability of the Loop Plant," *IEEE International Communications Conference*, 1979, pp 2.1.1–2.1.7.

3 N. Inoue, R. Komiya, and Y. Inoue, "Time-Shared Two-Wire Digital Transmission for Subscriber Loops," *IEEE International Communications Conference*, 1979, pp 2.4.1–2.4.5.

4 J. Meyer and T. Roste, "Field Trials of Two-Wire Digital Transmission in the Subscriber Loop Plant," *IEEE International Communications Conference*, 1979, pp 2.5.1–2.5.5.

5 F. D. Waldhauer, "A 2-Level, 274 Mb/s Regenerative Repeater for T4M," *IEEE International Communications Conference*, 1975, pp 48–13 to 48-17.

6 N. Karlovac and T. V. Blalock, "An Investigation of the Count Rate Performance of Baseline Restorers," *Nuclear Sciences Symposium*, Washington, D.C., 1974.

7 F. D. Waldhauer, "Quantized Feedback in an Experimental 280-Mb/s Digital repeater for coaxial transmission," *IEEE Transactions on Communications*, January 1974, pp 1–5.

8 M. R. Aaron, "PCM Transmission in the Exchange Plant," *Bell System Technical Journal*, January 1962, pp 99–141.

9 Members of Technical Staff, Bell Telephone Laboratories, *Transmission Systems for Communications*, Western Electric Co., 1971, p 668.

10 "1.544 Mbps Digital Service," *Bell System Technical Reference Publication*, No. 41451, May 1977.

11 G. D. Forney, "The Viterbi Algorithm," *Proceedings of IEEE*, March 1973, pp 268–278.

12 V. I. Johannes, A. G. Kaim, and T. Walzman, "Bipolar Pulse Transmission with Zero Extraction," *IEEE Transactions on Communications*, April 1969, pp 303–310.

13 "The D3 Channel Bank Compatibility Specification—Issue 3," *Technical Advisory No. 32*, American Telephone and Telegraph Company, October 1977.

14 B. Johnston and W. Johnston, "LD-4 A Digital Pioneer in Action," *Telesis*, Vol. 5, No. 3, June 1977, pp 66–72.

15 J. H. Davis, "T2: A 6.3 Mb/s Digital Repeatered Line," *IEEE International Communications Conference*, 1969, pp 34–9 to 34–16.

16 Recommendation G. 703, *CCITT Orange Book*, Vol. 3, No. 2.

17 J. M. Sipress, "A New Class of Selected Ternary Pulse Transmission Plans For Digital Transmission Lines," *IEEE Transactions on Communication Technology*, September 1965, pp 366–372.

18 E. E. Schnegelberger and P. T. Griffiths, "48 PCM Channels on T1 Facilities," *National Electronics Conference*, 1975, pp 201–205.

19 P. A. Franaszek, "Sequence-State Coding for Digital Transmission," *Bell System Technical Journal*, December 1967, pp 143–157.

20 The Ethernet—Data Link Layer and Physical Layer Specifications, Version 1.0, Xerox Corporation Document, September 30, 1980.

21 A. Lender, "The Duobinary Technique for High Speed Data Transmission," *IEEE Transactions on Communication Electronics*, May 1963, pp 214–218.

22 Adam Lender, "Correlative Level Coding for Binary Data Transmission," *IEEE Spectrum*, February 1960, pp 104–110.

23 P. Kabal and S. Pasupathy, "Partial Response Signaling," *IEEE Transactions on Communications*, September 1975, pp 921–934.

24 A. M. Gerrish and R. D. Howson, "Multilevel Partial Response Signaling," *International Communications Conference*, 1967, pp 186.

25 D. W. Jurling and A. L. Pachynski, "Duobinary PCM System Doubles Capacity of T1 Facilities," *International Communications Conference*, 1977, pp 32.2-297 to 32.2-301.

26 T. Seaver, "An Efficient 96 PCM Channel Modem for 2 GHz FM Radio," *National Telecommunications Conference*, 1978, pp 38.4.1–38.4.5.

27 M. B. Brilliant, "Observations of Errors and Error Rates on T1 Digital Repeatered Lines," *Bell System Technical Journal*, March 1978, pp 711–746.

28 Members of Technical Staff, Bell Telephone Laboratories, *Engineering and Operations in the Bell System*, Western Electric Co., 1978

29 *C-System General Description*, Collins Radio Company Report No. 523-0561697-20173R, Collins Radio Company, Dallas, Texas, May 1, 1970.

30 R. A. Scholtz, "Frame Synchronization Techniques," *IEEE Transactions on Communications*, August 1980, pp 1204–1212.

31 D. T. R. Munhoz, J. R. B. deMarca, and D. S. Arantes, "On Frame Synchronization of PCM Systems," *IEEE Transactions on Communications*, August 1980, pp 1213–1218.

32 J. R. Pierce, "Network for Block Switching of Data," *Bell System Technical Journal*, July–August 1972, pp 1133–1145.

PROBLEMS

4.1 Using the symbols +, 0, – to represent a positive pulse, no pulse, and a negative pulse respectively, determine the following line code sequences of the binary data sequence:

0110 1000 0100 0110 0000 0010

(a) Bipolar with most recent pulse being positive.
(b) Bipolar with most recent pulse being negative.
(c) Pair selected ternary beginning in "–" mode.
(d) B3ZS with a +0+ substitution having just been made.
(e) B6ZS with the most recent pulse being positive.

4.2 Assume that two identical cable systems are used for digital transmission with equal pulse amplitudes. One system uses conventional bipolar signaling and the other uses pair selected ternary. Compare the crosstalk levels of the two systems. (Assume that ones and zeros are equally probable.)

4.3 A digital transmission system operating at an error rate of 10^{-6} is to have its data rate increased by 50%. Assuming that the same transmitted power is to be used in the second system, what is the new error rate?

4.4 What is the average reframe time of a D3 channel bank if we assume a random starting point? What is the "maximum average" reframe time of a D3 channel bank? (Assume 1's and 0's in the message traffic are equally likely.)

4.5 Repeat Problem 4.4 for the primary TDM multiplex signal specified by CCITT.

4.6 A TDM system operating at 2 Mbps is to have a "maximum average" reframe time of 20 ms. What is the maximum possible frame length if framing is established with a bit-by-bit frame search? (Assume that 1's and 0's in message channels are equally likely.)

4.7 A T1 transmission system using a D1 frame format is to have an average reframe time (from a random starting point) of 10 ms. How large a block of bit positions must be examined in parallel to achieve the desired result?

4.8 What is the expected framing time for a T1 line (D3 frame format) if the framing strategy is bit-by-bit and the data stream has 60% ones and 40% zeros?

4.9 What is the average pulse density of 4B3T coding? (Assume ones and zeros are equally likely.)

4.10 A TDM transmission link using 4B3T coding can transmit thirty-two 64 kbps voice channels using the same symbol rate as a T1 line (1544 kbps). Assuming a fixed Gaussian noise environment, how much must the average transmit power of the 4B3T system be increased to provide the equivalent error rate of a bipolar code?

4.11 Assume that crosstalk interference in a multipair cable system produces an effect equivalent to Gaussian noise at an equal power level. Using an error rate of 10^{-6} as a design objective, determine the effective degradation of the crosstalk on binary (polar) NRZ coding under each of the following conditions. (The effective degradation is determined as the increased transmit power, in decibels, required to achieve the desired error rate.)

(a) The crosstalk level is 16 dB below the average signal level, but the crosstalk is to be overcome on only one pair. (i.e. all other pairs stay at a power level for 10^{-6} BER with no crosstalk.)

(b) The crosstalk level is 16 dB below the average signal level, but the effects of the crosstalk are to be overcome on all pairs. (Hint: Use signal-power to noise-power ratios, not E_b/N_0.)

4.12 Repeat Problem 4.11 for bipolar coding.

4.13 How much does -18 dB of crosstalk degrade the error rate performance of a polar binary NRZ signal for a BER of 10^{-7}? (Assume that all transmitters are at equal power levels.) If crosstalk increases 15 dB per decade increase in frequency, what is the relative crosstalk level of a four-level NRZ code carrying the same data rate? What is the overall performance penalty of the four-level system compared to the two-level system?

4.14 How many distinct code words of length four can be constructed from ternary symbols? How many of these code words contain an equal number of positive and negative pulses? How many code words can be used to ensure a minimum of one timing pulse occurs in each code word?

4.15 Can a set of ternary code words of length eight be constructed to encode binary data using exactly four pulses per 8-bit word and containing equal numbers of positive and negative pulses?

4.16 Given the following sequence of signal levels determine the sequence of output signal levels for each of the following correlative encodings.

Input sequence: $+1, -3, +1, -1, +3, +3, -3$

(a) $1 + D$ Encoder:
(b) $1 - D$ Encoder:
(c) $1 - D^2$ Encoder:

CHAPTER FIVE

DIGITAL SWITCHING

Of the three basic elements in a communications network (terminals, transmission media, and switches), switches are the most invisible to the users yet represent the most important element in terms of available service offerings. As mentioned in Chapter 1, a major milestone was established in 1965 when stored program control switching was first introduced into the U.S. public telephone network [1]. Stored program control provides the means for implementing many innovative user services and for greatly simplifying switch administration and maintenance. Most of the metropolitan areas of the United States are now serviced by stored program switches.

The use of computers to control the switching functions of a central office led to the designation "electronic" switching (e.g., electronic switching system: ESS or Electronic Automatic Exchange: EAX). However, the switching matrices of these first generation electronic switches are actually electromechanical in nature. The first use of electronic switching matrices occurred in France in 1971 when digital switching was applied to an end-office environment. Curiously, these first digital switches did not use stored program control. Electronic switching matrices were first introduced into the U.S. public network in 1976 with the Bell System's No. 4ESS digital switch.

Beginning in 1978, numerous U.S. independent telephone companies began installing digital switches in end offices, mostly to replace older step-by-step switches. Virtually all U.S. and international switch manufacturers have now introduced or announced digital switching systems for end-office applications.

This new generation of switching equipment will initially provide only those services that are available in the earlier stored program machines. The digital equipment is necessarily introduced into analog transmission environments and is therefore invisible to local distribution facilities. As time passes, however, special service requests will enable users to communicate over switched, digital connections to accommodate high-rate, high-quality data communications. Unfortunately, for data communications users, new digital end-office switches will not replace recently installed stored program control machines like the No. 1ESS. Thus the advent of digital switching in many metropolitan areas of the United States will be deferred until depreciation schedules of existing equipment elapse. As an alternative, Bell System engineers are developing the capability of providing switched digital connections through No. 1ESS switching machines [2]. Digitized voice or data signals will pass through the analog switch to digital trunk circuits and on to digital switching machines like the No. 4ESS. In this manner new digital services can be offered in the absence of a local digital switch.

Since neither digital end-office switching nor digital toll switching provide direct benefits to network customers, the motivation for the digital equipment is reduced costs for the operating companies. These cost reductions are derived from reduced maintenance, reduced floor space, simplified expansion, reduced TDM interface costs, and, as time passes, continually reduced manufacturing costs [3].

In contrast to the public network, private voice networks are beginning to evolve using digital transmission and switching exclusively. In some PBX systems the digitization takes place in the telephone itself. Thus these systems can be used for efficient data communications as well as voice. The motivation for digitizing the voice in the telephone is not necessarily derived from a desire to provide a digital link for data transmission. The installed wiring for many PBX facilities is limited to a few wires per telephone. This severely restricts installation of multiwire mechanical key sets. Analog or digital electronic key sets, on the other hand, have greatly reduced wiring requirements and can therefore be installed without prohibitive rewiring costs.

This chapter describes the basic operation and implementation of digital time division switching as applied to PBXs, end offices, and toll switches. Before digital switching is discussed, certain basic switching concepts and terminology are introduced.

5.1 SWITCHING FUNCTIONS

Obviously, the basic function of any switch is to set up and release connections between transmission channels on an "as-needed basis." The structure and operation of a switch varies significantly, however, depending on particular applications. Three main switching categories for voice circuits are local (line-to-line) switching, transit (tandem) switching, and call distribution.

The most common switching function involves direct connections between subscriber loops at an end office or between station loops at a PBX. These connections inherently require setting up a path through the switch from the originating loop to a specific terminating loop. Each loop must be accessible to every other loop. This level of switching is sometimes referred to as line switching.

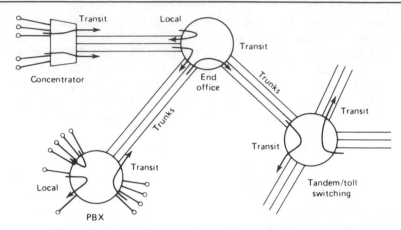

Figure 5.1   Local and transit traffic switching examples.

Transit connections require setting up a path from a specific incoming (originating) line to an outgoing line or trunk group. Normally, more than one outgoing circuit is acceptable. For example, a connection to an interoffice trunk group can use any one of the channels in the group. Hence transit switching structures can be simplified because alternatives exist as to which outgoing line is selected. Furthermore, it is not even necessary that every outgoing line be accessible from every incoming line. Transit switching functions are required by all switching machines in the telephone network. Some machines such as remote concentrators and toll or tandem switches service only transit traffic (e.g. do not provide local connections). These concepts are illustrated in Figure 5.1.

Call distributors are often implemented with the same basic equipment as PBXs. The mode of operation (software) is significantly different, however, in that incoming calls can be routed to any available attendant. Normally, the software of an automatic call distributor (ACD) is designed to evenly distribute the arriving calls among the attendants. Although it is not an inherent requirement that every incoming line (trunk) be connectable to every attendant, call distributors are normally designed to provide accessibility to all attendants. Furthermore, it is often desirable that nonblocking operations be provided. (No matter what switch paths are in use, a new request can be serviced if an attendant is available.)

5.2   SPACE DIVISION SWITCHING

Conceptually, the simplest switching structure is a rectangular array of crosspoints as shown in Figure 5.2. This switching matrix can be used

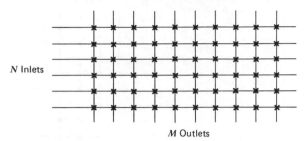

Figure 5.2 Rectangular crosspoint array.

to connect any one of N inlets to any one of M outlets. If the inlets and outlets are connected to two-wire circuits, only one crosspoint per connection is required.*

Rectangular crosspoint arrays are designed to provide intergroup (transit) connections only, that is, from an inlet group to an outlet group. Applications for this type of an operation occur in the following:

1 Remote concentrators.
2 Call distributors.
3 The portion of a PBX or end-office switch that provides transit switching.
4 Single stages in multiple stage switches.

In most of the foregoing applications, it is not necessary that the inlets be connectable to every outlet. In situations involving large groups of outlets, considerable savings in total crosspoints can be achieved if each inlet can access only a limited number of outlets. When such a situation occurs "limited availability" is said to exist. By overlapping the available outlet groups for various inlet groups, a technique called "grading" is established. An example of a graded switching matrix is shown in Figure 5.3. Notice that if outlet connections are judiciously chosen, the adverse effect of limited availability is minimized. For example, if inlets 1 and 8 in Figure 5.3 request a connection to the outlet group, outlets 1 and 3 should be chosen instead of outlets 1 and 4 to avoid future blocking for inlet 2.

Graded switching structures are often used for access to large trunk groups in electromechanical switches where crosspoints are expensive and individual switching modules are limited in size. Gradings are also used in individual switching stages of large multiple stage switches where more than one path to any particular outlet exists.

Intragroup switching, as for loop-to-loop switching, requires each

*In actual fact, two (and sometimes three) switching contacts are associated with each crosspoint of a two-wire switch. Since these contacts are part of a single unit and operate in unison, they are considered a single crosspoint.

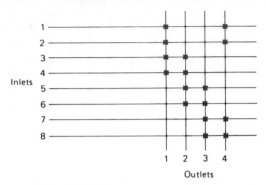

Figure 5.3 Graded rectangular switching matrix.

loop to be connectable to every other loop. Thus full availability from all inlets to all outlets of the switching matrix is required. Figure 5.4 shows two matrix structures that can be used to fully interconnect two-wire lines. The dashed lines indicate that corresponding inlets and outlets of two-wire switching matrices are actually connected together to provide bidirectional transmission on two-wire circuits. For purposes of describing switching matrices, however, it is convenient to consider the inlets and outlets of two-wire switching matrices as being distinct.

Both structures in Figure 5.4 allow any connection to be established by selecting a single crosspoint. However, the square matrix, which is also called a two-sided matrix, allows any particular connection to be established in two ways. For example, if input link i is to be connected to input link j, the selected crosspoint can be at the intersection of inlet i and outlet j—or at the intersection of inlet j and outlet i. For simplicity these crosspoints are referred to as (i, j) and (j, i) respectively. In a typical implementation crosspoint (i, j) is used when input i requests service, and crosspoint (j, i) is used when input j requests service.

In the triangular matrix of Figure 5.4, the redundant crosspoints are eliminated. The crosspoint reduction does not come without complica-

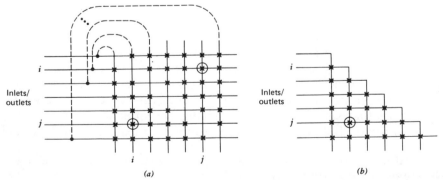

Figure 5.4 Two-wire switching matrices. (*a*) Square matrix. (*b*) Triangular (folded).

tions, however. Before setting up a connection between switch input i and switch input j, the switch control element must determine which is larger: i or j. If i is larger, crosspoint (i, j) is selected. If i is smaller, crosspoint (j, i) must be selected. With computer controlled switching, the line number comparison does not represent a significant imposition. In the older, electromechanically controlled switches, however, the added complexity of the switch control is more significant.

Switching machines for four-wire circuits require separate connections for the go and return branches of a circuit. Thus two separate connections must be established for each service request. Figure 5.5 depicts a square matrix structure used to provide both connections. The structure is identical to the square matrix shown in Figure 5.4 for two-wire switching. The difference, however, is that corresponding inlets and outlets are not connected to a common two-wire input. All of the inlets of the four-wire switch are connected to the wire pair carrying the incoming direction of transmission, and all of the outlets are connected to the outgoing pairs. When setting up a connection between four-wire circuits i and j, the matrix in Figure 5.5 must select both crosspoints (i, j) and (j, i). In actual operation these two crosspoints may be selected in unison and implemented as a common module.

5.2.1 Multiple Stage Switching

In the switching structures described to this point, an inlet is connected directly to an outlet through a single crosspoint. (Four-wire switches use two crosspoints per connection, but only one for an inlet to outlet connection.) For this reason, these switching structures are referred to as "single stage" switches. Single stage switches have the property that each individual crosspoint can only be used to interconnect one particular inlet/outlet pair. Since the number of inlet/outlet pairs is equal to

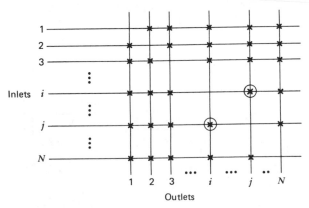

Figure 5.5 Four-wire switching matrix.

$N(N - 1)/2$ for a triangular array, or $N(N - 1)$ for a square array, the number of crosspoints required for a large switch is prohibitive. Furthermore, the large number of crosspoints on each inlet and outlet line imply a large amount of capacitive loading on the message paths. Another fundamental deficiency of single stage switches is that one specific crosspoint is needed for each specific connection. If that crosspoint fails, the associated connection can not be established. (An exception is the square, two-wire switch that has a redundant crosspoint for each potential connection. Before the redundant crosspoint could be used as an alternate path, however, the inlet oriented selection algorithm would have to be modified to admit outlet oriented selection.)

Analysis of a large single stage switch reveals that the crosspoints are very inefficiently utilized. Only one crosspoint in each row or column of a square switch is ever in use, even if all lines are active. In order to increase the utilization efficiency of the crosspoints and thereby reduce the total number, it is necessary that any particular crosspoint be usable for more than one potential connection. If crosspoints are to be shared, however, it is also necessary that more than one path be available for any potential connection so that blocking does not occur. The alternate paths serve to eliminate or reduce blocking and also to provide protection against failures. The sharing of crosspoints for potential paths through the switch is accomplished by multiple stage switching. A block diagram of one particular form of a multiple stage switch is shown in Figure 5.6.

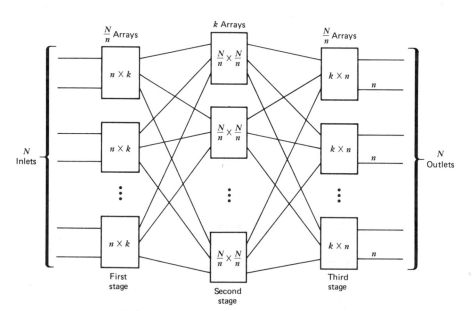

Figure 5.6   Three-stage switching matrix.

The switch of Figure 5.6 is a three-stage switch in which the inlets and outlets are partitioned into subgroups of n inlets and n outlets each. The inlets of each subgroup are serviced by a rectangular array of crosspoints. The inlet arrays (first stage) are $n \times k$ arrays where each one of the k outputs is connected to one of the k center stage arrays. The interstage connections are often called junctors. The third stage consists of $k \times n$ rectangular arrays that provide connections from each center stage array to the groups of n outlets. All center stage arrays are N/n by N/n arrays that provide connections from any first-stage array to any third-stage array. Notice that if all arrays provide full availability, there are k possible paths through the switch for any particular connection between inlets and outlets. Each of the k paths utilizes a separate center stage array. Thus the multiple stage structure provides alternate paths through the switch to circumvent failures. Furthermore, since each switching link is connected to a limited number of crosspoints, capacitive loading is minimized.

The total number of crosspoints N_x required by a three-stage switch, as shown in Figure 5.6 is:

$$N_x = 2Nk + k\left(\frac{N}{n}\right)^2 \qquad (5.1)$$

where N = the number of inlets/outlets
n = the size of each inlet/outlet group
k = the number of center stage arrays

As is demonstrated shortly, the number of crosspoints defined in Equation 5.1 can be significantly lower than the number of crosspoints required for single stage switches. First, however, we must determine how many center stage arrays are needed to provide satisfactory service.

Nonblocking Switches

One attractive feature of a single stage switch is that it is strictly nonblocking. If the called party is idle, the desired connection can always be established by selecting the particular crosspoint dedicated to the particular input/output pair. When crosspoints are shared, however, the possibility of blocking arises. In 1953 Charles Clos [4] of Bell Laboratories published an analysis of three-stage switching networks showing how many center stage arrays are required to provide a strictly nonblocking operation. His result demonstrated that if each individual array is nonblocking, and if the number of center stages k is equal to $2n - 1$, the switch is strictly nonblocking.

The condition for a nonblocking operation can be derived by first observing that a connection through the three-stage switch requires locating a center stage array with an idle link from the appropriate first stage and an idle link to the appropriate third stage. Since the individual arrays themselves are nonblocking, the desired path can be set up any time a center stage with the appropriate idle links can be located. A key point in the derivation is to observe that since each first-stage array has n inlets, only $n - 1$ of these inlets can be busy when the inlet corresponding to the desired connection is idle. If k is greater than $n - 1$, it follows that, at most, $n - 1$ links to center stage arrays can be busy. Similarly, at most $n - 1$ links to the appropriate third-stage array can be busy if the outlet of the desired connection is idle.

The worst case situation for blocking occurs (as shown in Figure 5.7) if all $n - 1$ busy links from the first-stage array lead to one set of center stage arrays, and if all $n - 1$ busy links to the desired third-stage array come from a separate set of center stage arrays. Thus these two sets of center stage arrays are unavailable for the desired connection. However, if one more center stage array exists, the appropriate input and output links must be idle, and that center stage can be used to set up the connection. Hence if $k = (n - 1) + (n - 1) + 1 = 2n - 1$ the switch is strictly nonblocking. Substituting this value of k into Equation 5.1 reveals that for a strictly nonblocking operation of a three stage switch:

$$N_x = 2N(2n - 1) + (2n - 1)\left(\frac{N}{n}\right)^2 \tag{5.2}$$

As expressed in Equation 5.2, the number of crosspoints in a nonblocking three-stage switch is dependent on how the inlets and outlets are partioned into subgroups of size n. Differentiating Equation 5.2 with respect to n and setting the resulting expression equal to 0 to determine the minimum reveals that (for large N) the optimum value of

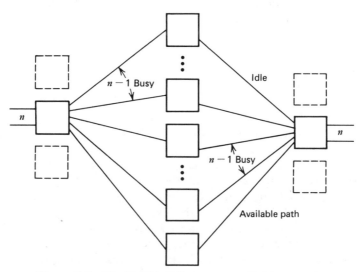

Figure 5.7 Nonblocking three-stage switching matrix.

TABLE 5.1 CROSSPOINT REQUIREMENTS
OF NONBLOCKING SWITCHES

Number of Lines	Number of Crosspoints for Three-Stage Switch	Number of Crosspoints for Single-Stage Switch
128	7,680	16,256
512	63,488	261,632
2,048	516,096	4.2 million
8,192	4.2 million	67 million
32,768	33 million	1 billion
131,072	268 million	17 billion

n is $(N/2)^{1/2}$. Substituting this value of n into Equation 5.2 then provides an expression for the minimum number of crosspoints of a nonblocking three-stage switch.

$$N_x(\text{min}) = 4N(\sqrt{2N} - 1) \qquad (5.3)$$

where N = total number of inlets/outlets.

Table 5.1 provides a tabulation of $N_x(\text{min})$ for various sized nonblocking three-stage switches and compares the values to the number of crosspoints in a single stage square matrix. Both switching structures inherently provide four-wire capabilities, which is of most interest to us since voice digitization implies four-wire circuits.

As indicated in Table 5.1, a three-stage switching matrix provides significant reductions in crosspoints, particularly for large switches. However, the number of crosspoints for large three-stage switches is still quite prohibitive. Large switches typically use more than three stages to provide greater reductions in crosspoints. For example, the No. 1ESS uses an eight-stage switching matrix that can service up to 65,000 lines. However, the most significant reductions in crosspoint numbers are achieved not so much from additional stages but by allowing the switch to introduce acceptably low probabilities of blocking.

5.2.2 Blocking Probabilities: Lee Graphs

Strictly nonblocking switches are rarely needed in most voice telephone networks. Both the switching systems and the number of circuits in interoffice trunk groups are sized to service most requests as they occur, but economics dictates that network implementations have limited capacities that are occasionally exceeded during peak traffic situations. Equipment for the public telephone network is designed to provide a certain maximum probability of blocking for the busiest hour of the day. The value of this blocking probability is one aspect of the telephone company's grade of service. (Other aspects of grades of service are: availability, transmission quality, and delay in setting up a call.)

A typical residential telephone is busy 5 to 10% of the time during the busy hour. Business telephones are often busy for a larger percentage of their busy hour (which may not coincide with a residential busy hour). In either case, network blocking occurrences on the order of 1%* during the busy hour do not represent a significant reduction in the ability to communicate since the called party is much more likely to have been busy anyway. Under these circumstances, end-office switches and, to a lesser degree, PBXs can be designed with significant reductions in crosspoints by allowing acceptable blocking probabilities.

There are a variety of techniques that can be used to evaluate the blocking probability of a switching matrix. These techniques vary according to complexity, accuracy, and applicability to different network structures. One of the most versatile and conceptually straightforward approaches of calculating blocking probabilities involves the use of probability graphs as proposed by C. Y. Lee [5]. Although this technique requires several simplifying approximations, it can provide reasonably accurate results, particularly when comparisons of alternate structures are more important than absolute numbers. The greatest value of this approach lies in the ease of formulation and the fact that the formulas directly relate to the underlying network structures. Thus the formulations help provide insight into the network structures and how these structures might be modified to change the performance.

In the following analyses we are determining the blocking probabilities of various switching structures using utilization percentages or "loadings" of individual links. The notation p will be used, in general, to represent the fraction of time that a particular link is in use (i.e., p is the probability that a link is busy). In addition to a utilization percentage or loading, p is also sometimes referred to as an occupancy. The probability that a link is idle is denoted by $q = 1 - p$.

When any one of n parallel links can be used to complete a connection, the composite blocking probability B is the probability that all links are busy:[†]

$$B = p^n \tag{5.4}$$

When a series of n links are all needed to complete a connection, the blocking probability is most easily determined as 1 minus the probability that they are all available:[†]

$$B = 1 - q^n \tag{5.5}$$

---

*Transmission and switching equipment in the public network is normally designed for even lower blocking probabilities to provide for growth in the traffic volume.
[†] These formulas assume each link is busy or idle independently of other links.

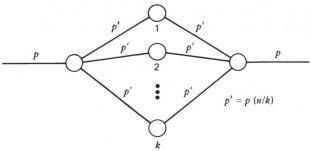

Figure 5.8 Probability graph of three-stage network.

A probability graph of a three-stage network is shown in Figure 5.8. This graph relates the fact that any particular connection can be established with k different paths: one through each center stage array. The probability that any particular interstage link is busy is denoted by p'. The probability of blocking for a three-stage network can be determined as:

B = probability that all paths are busy
  = (probability that an arbitrary path is busy)^k
  = (probability that at least one link in a path is busy)^k (5.6)
  = $(1 - q'^2)^k$

where k is the number of center stage arrays, and $q' = 1 - p'$ is the probability that an interstage link is idle.

If the probability p that an inlet is busy is known, the probability p' that an interstage link is busy can be determined as:

$$p' = p/\beta \qquad (5.7)$$

where $\beta = k/n$. Equation 5.7 presents the fact that when some number of inlets (or outlets) are busy, the same number of first-stage outputs (or third-stage inputs) are also busy. However, there are $\beta = k/n$ times as many interstage links as there are inlets or outlets. Hence the percentage of interstage links that are busy is reduced by the factor β.

The factor β is defined as though k is greater than n, which implies that the first stage of the switch is providing space expansion (i.e., switching some number of input links to a larger number of output links). Actually, β may be less than 1, implying that the first stage is concentrating the incoming traffic. First-stage concentration is usually employed in end-office or PBX switches where the inlets are lightly used (5 to 10%). In tandem or toll offices, however, the incoming trunks are heavily utilized, and expansion is usually needed to provide adequately low-blocking probabilities.

Substituting Equation 5.7 into Equation 5.6 provides a complete expression for the blocking probability of a three-stage switch in terms

TABLE 5.2 THREE-STAGE SWITCH DESIGNS FOR BLOCKING PROBABILITIES OF .002 AND INLET UTILIZATIONS OF 0.1

Switch Size N	n	k	β	Number of Crosspoints	Number of Crosspoints in Nonblocking Design	
128	8	5	0.625	2,560	7,680	($k = 15$)
512	16	7	0.438	14,336	63,488	($k = 31$)
2,048	32	10	0.313	81,920	516,096	($k = 63$)
8,192	64	15	0.234	491,520	4.2 million	($k = 127$)
32,768	128	24	0.188	3.1 million	33 million	($k = 255$)
131,072	256	41	0.160	21.5 million	268 million	($k = 511$)

of the inlet utilization p.

$$B = (1 - (1 - p/\beta)^2)^k \tag{5.8}$$

Table 5.2 tabulates numbers of crosspoints obtained from Equation 5.1 for the same switch sizes presented in Table 5.1. The number of center arrays was chosen in each case to provide a blocking probability on the order of .002. The inlet utilization in each example was assumed to be 10%. Notice that the designs with small but finite blocking probabilities are significantly more cost effective than nonblocking designs.

The switch designs in Table 5.2 assume that the inlets are only 10% busy, as might be the case for an end-office switch or a PBX. The dramatic savings in crosspoints for large switches is brought about by introducing significant concentration factors $(1/\beta)$ into the middle stage. When the inlet utilization is higher (as typically occurs in tandem switches), high concentration factors are not acceptable, and the crosspoint requirements therefore increase. Table 5.3 lists corresponding crosspoint requirements and implementation parameters for inlet loadings of 70%.

TABLE 5.3 THREE-STAGE SWITCH DESIGNS FOR BLOCKING PROBABILITIES OF .002 AND INLET UTILIZATIONS OF 0.7

Switch Size N	n	k	β	Number of Crosspoints	Number of Crosspoints in Nonblocking Design	
128	8	14	1.75	7,168	7,680	($k = 15$)
512	16	22	1.38	45,056	63,488	($k = 31$)
2,048	32	37	1.16	303,104	516,096	($k = 63$)
8,192	64	64	1.0	2.1 million	4.2 million	($k = 127$)
32,768	128	116	0.91	15.2 million	33 million	($k = 255$)
131,072	256	215	0.84	113 million	268 million	($k = 511$)

The results presented in Tables 5.2 and 5.3 indicate that very large switches still require prohibitively large numbers of crosspoints, even when blocking is allowed. As mentioned previously, very large switches use more than three stages to provide further reductions in crosspoints. Figure 5.9 shows a block diagram of a five-stage switch obtained by replacing every center stage array in Figure 5.6 with a three-stage array. This particular structure is not optimum in terms of providing a given level of performance with the fewest crosspoints, but it is a useful design because of its modularity. (Furthermore, it is a lot easier to analyze than some other five-stage structures.)

If the middle three stages of a five-stage switch as shown in Figure 5.9 are strictly nonblocking $(k_2 = 2n_2 - 1)$, the design provides a savings of 8704 crosspoints in each center stage array of the 32,768 line, three-stage switch designs presented earlier. Hence a little over 1 million crosspoints are saved in the 32,768 line tandem switch design of Table 5.3. Since the middle stages do not introduce blocking, the performance of this five-stage switch is identical to the performance of the three-stage design. Naturally, a more cost effective design could be obtained by allowing small amounts of blocking in the middle stages. The probability graph of the five-stage switch is shown in Figure 5.10. From this graph, the blocking probability is determined as follows:

$$B = \{1 - (q_1)^2[1 - (1 - q_2^2)^{k_2}]\}^{k_1} \qquad (5.9)$$

where $q_1 = 1 - p_1$ and $q_2 = 1 - p_2$

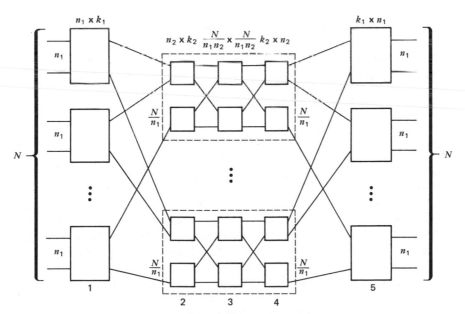

Figure 5.9 Five-stage switching network.

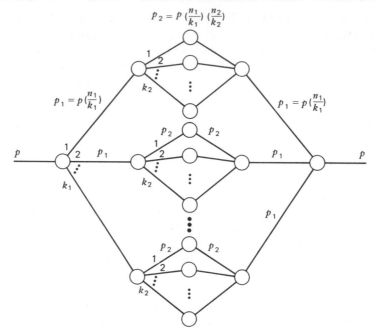

Figure 5.10 Probability graph of five-stage network.

Even greater crosspoint reductions can, of course, be achieved by using more stages to replace the rather large first- and third-stage arrays. The problem set at the end of the chapter demonstrates how the total number of crosspoints in the 32,000 line switch can be reduced to less than 3 million. The 130,000 line switch is not practical with electromechanical switching matrices, but, is well within the capabilities of a digital time division switch.

5.2.3 Blocking Probabilities: Jacobaeus

The formulations of blocking probability obtained from probability graphs entail several simplifying assumptions. One of these assumptions involves expressing the composite blocking probability of the alternate paths as the product of the blocking probabilities of each individual path. This step assumes that the individual probabilities are independent. In actual fact, the probabilities are not independent, particularly when significant amounts of expansion are present. Consider a switching matrix with $k = 2n - 1$. Equation 5.8 produces a finite blocking probability even though the switch is known to be strictly nonblocking. The inaccuracy results because when $2n - 2$ paths are busy, the remaining path is assumed to be busy with a probability of $1 - (q')^2$. In fact, the remaining path is necessarily idle.

In general, when space expansion exists, the assumption of independent individual probabilities produces higher than actual blocking probabilities. The inaccuracy results because as more and more paths in a switch are found to be busy, the remaining paths are less and less likely to be in use (only a subset of n of the interstage links can ever be busy at any one time).

A more accurate but not exact analysis of multistage switching matrices was presented in 1950 by C. Jacobaeus [6]. Although the analysis is conceptually straightforward, it does involve a considerable amount of manipulation that is not presented here. The resulting equation for a three-stage switch is obtained from Reference 7 as:

$$B = \frac{(n!)^2}{k!\,(2n-k)!}\, p^k (2-p)^{2n-k} \tag{5.10}$$

where n = number of inlets (outlets) per first (third) stage array
$\quad k$ = number of second-stage arrays
$\quad p$ = inlet utilization

In the interest of comparing the two methods, Equations 5.6 and 5.10 have been evaluated for three-stage switches with varying amounts of space expansion. The results were obtained for inlet utilizations of .7 and are presented in Table 5.4.

Table 5.4 reveals that the two analyses are in close agreement for near-unity expansion factors. In fact, if β is exactly equal to 1, the two formulations produce identical results. As expected, the Lee graph analysis (Equation 5.8) produces overly pessimistic values for the blocking probability when β is greater than 1.

TABLE 5.4 COMPARISON OF BLOCKING
PROBABILITY ANALYSES ($p = .7$)[a]

Number of Center Stages k	Space Expansion β	Lee Equation 5.8	Jacobaeus Equation 5.10
14	0.875	.548	.598
16	1.0	.221	.221
20	1.25	.014	.007
24	1.50	$3.2(10)^{-4}$	$2.7(10)^{-5}$
28	1.75	$3.7(10)^{-6}$	$7.7(10)^{-9}$
31[b]	1.94	$8.5(10)^{-8}$	$1(10)^{-12}$

[a] Switch size $N = 512$; inlet group size $n = 16$; inlet utilization $p = .7$.
[b] Nonblocking.

TABLE 5.5 COMPARISON OF BLOCKING PROBABILITY ANALYSES $(p = .1)^a$

Number of Center Stages k	Space Expansion β	Lee Equation 5.8	Jacobaeus Equation 5.10
6	0.375	.0097	.027
8	0.5	$2.8(10)^{-4}$	$8.6(10)^{-4}$
10	0.625	$4.9(10)^{-6}$	$1.5(10)^{-5}$
12	0.75	$5.7(10)^{-8}$	$1.4(10)^{-7}$
14	0.875	$4(10)^{-10}$	$7.8(10)^{-10}$
16	1.0	$2.9(10)^{-12}$	$2.9(10)^{-12}$

^{a}Switch size $N = 512$; inlet group size $n = 16$; inlet utilization $p = .1$.

As another comparison between the two approaches, Table 5.5 is included to demonstrate the use of Equations 5.8 and 5.10 for switches with significant amounts of concentration made possible by a relatively low inlet utilization of .1.

Table 5.5 reveals that a Lee graph analysis (Equation 5.8) consistently underestimates the blocking probability when concentration exists. Actually, the Jacobaeus analysis presented in Equation 5.10 also underestimates the blocking probability if large concentration factors and high blocking probabilities are used. When necessary, more accurate techniques can be used for systems with high concentration and high blocking. However, switches with high blocking probabilities normally have no practical interest so they are not considered here.

Users of corporate PBXs sometimes experience high blocking probabilities, but blocking in these cases usually arises from too few tie lines to other corporate locations or too few trunk circuits to the public network. The subject of blocking in trunk groups is treated in Chapter 9.

Up to this point, the blocking probability analyses have assumed that a specific inlet is to be connected to a specific outlet. Also, it has been assumed that the requests for service on the individual lines are independent. These assumptions are generally valid for switching one subscriber line to another in an end-office switch, or for connecting one station to another in a PBX. When connecting to or from trunks, however, these assumptions are not valid.

When connecting to a trunk circuit, any one circuit in a trunk group is acceptable. Thus the blocking probability to a specific circuit is only as important as its significance in the overall blocking to the trunk group. The blocking probability to any particular circuit in a trunk group can be relatively large and still achieve a low composite blocking probability to the trunk group as a whole. If the blocking probabilities to the individual trunks are independent, the composite blocking probability is the product of the individual probabilities. However the paths to the individual trunk circuits normally involve some common links (e.g., the junctors from a first-stage array to all second-stage arrays). For this reason the individual blocking probabilities are normally dependent, which must be considered in an accurate blocking probability analysis.

As an extreme example, consider a case where all trunks in a trunk group are assigned to a single outlet array in a three-stage switch. Since the paths from any particular inlet to all trunks in the group are identical, the ability to select any idle trunk is useless. In actual practice the individual circuits of a trunk group should be assigned to separate outlet arrays.

Another aspect of trunk groups that must be considered when designing a switch or analyzing the blocking probabilities involves the interdependence of activity on the individual circuits within a trunk group. In contrast to individual subscriber lines or PBX stations, individual circuits in a trunk group are not independent in terms of their probabilities of being busy or idle. If some number of circuits in a trunk group are tested and found to be busy, the probability that the remaining circuits are busy is increased. The nature of these dependencies is discussed more fully in Chapter 9. At this point it is only necessary to point out that these dependencies cause increased blocking probabilities if the individual trunks are competing for common paths in the switch. Again, the effect of these dependencies is minimized by assigning the individual trunks to separate inlet arrays so that independent paths are involved in connections to and from the trunk group. This process is sometimes referred to as decorrelating the trunk circuits.

One last aspect of the blocking probability as a grade of service parameter that must be mentioned involves variations in the loading of the network by individual users. In the design examples for an end office presented earlier, it was tacitly assumed that all subscribers are busy 10% of the time during a busy hour. In actual fact, some subscribers are active much more than 10% of the time, and other subscribers are active less than 10% of the time. In terms of traffic theory, some subscribers present more than 0.1 erlangs* of traffic to the network, but others present less.

When a switch is partitioned into subgroups (as all large switches must be) and the traffic is concentrated by first-stage switching arrays, a few overactive subscribers in one subgroup can significantly degrade service for the other subscribers in the subgroup. It does not matter

---

*An erlang is a measure of traffic intensity specifying the proportion of time that a device is busy. A circuit is said to carry 0.1 erlangs if it is busy 10% of the time.

that the subscribers in some other subgroup may be experiencing lower than average blocking. Their essentially nonblocking service is no compensation for those subscribers experiencing a relatively poor grade of service.

Operating companies have traditionally solved the problem of overactive subscribers by specifically assigning the most active lines (businesses) to separate inlet groups of the switch. Sometimes this procedure requires making traffic measurements to determine which lines are most active and reassigning these lines to different parts of the switch. These procedures fall into the general category of line administration. If the subgroups are large enough, or if the designs provide adequate margin for overactive users, this aspect of line administration can be minimized. One feature of a digital switch that can be utilized in this regard is the ability to design economical switches with very low nominal blocking probabilities so that wide variations in traffic intensities can be accommodated.

Call Packing

A number of techniques are possible for reducing blocking probabilities in a given switching network by judiciously choosing the paths through the switch when setting up new connections. In a manner similar to that mentioned earlier for graded matrices the path that imposes the fewest restrictions on other potential connections should be chosen. For example, if a choice exists in a three-stage switch for selecting a path through a center stage array that is already in use versus a path through an array that is not in use, the former path should be chosen. If a whole center stage array is kept available, any subsequent connect request can be serviced. This procedure is referred to as call packing.

Generally speaking, call packing implies new connections should use the most congested portions of a switch if possible. An unfortunate implication of testing the most congested paths first is that the average pathfinding times are increased since initial attempts at finding a path are less likely to be successful than if paths are tested at random.

Another technique that can be used to reduce the blocking probability of a given matrix is to rearrange existing connections to make room for a new connection that would otherwise be blocked. This technique has greater potential for reducing the blocking probability since it essentially allows the "packing" to be done with the knowledge of what the next call is. A major disadvantage, however, is that existing connections may experience temporary breaks in continuity when they become rearranged. Temporary breaks produce noise impulses that may be unnoticeable in a voice circuit. High-speed modem signals, however, are less tolerant of the noise impulses and must be considered carefully in such a design. In the CBX 2000 PBX provided by Danray, rearrangement is used for voice circuits, but lines for data must be excluded from the rearrangement operations. For a further discussion of call rearrangement see Reference 8. Although these more sophisticated pathfinding procedures can improve the performance of a switching system, their usefulness in digital time division switches is somewhat limited because of the ease with which these switching networks can be designed for very low blocking probabilities.

5.2.4 Folded Four-Wire Switches

Multiple stage switches can be used for either two-wire or four-wire switching operations. Figure 5.11 depicts a four-wire connection through a four-stage switch. Notice that two paths must be established for the complete connection. A two-wire connection requires only one path since each outlet is externally connected to its corresponding inlet.

The two paths shown in Figure 5.11 demonstrate a particularly useful relationship: one path is a mirror image of the other path. If the switch diagram is folded about the vertical center line, the paths coincide. Hence this method of setting up connections is sometimes referred to as a folded operation. When all connections in the switch are set up with a folded relationship, several benefits result.

First of all, only one pathfinding operation is needed since the reverse path is automatically available as a mirror image of the forward path. In essence, every crosspoint on one side is paired with another crosspoint in a corresponding array on the opposite side of the switch. Whenever one crosspoint of a pair is used in a connection, the other crosspoint in the pair is also used. For example, the third inlet array in the first stage uses crosspoint 6, 4 to connect its sixth inlet to its fourth

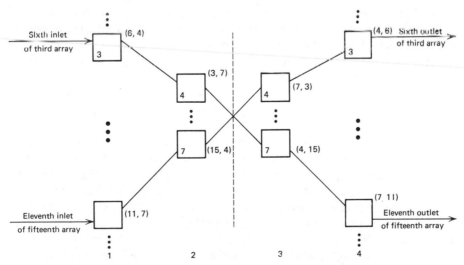

Figure 5.11 Four-wire connection through four-stage switch.

outlet (leading to the fourth array of the second stage). The corresponding crosspoint in the third outlet array of the last stage connects its fourth inlet (coming from the fourth array in the fourth stage) to its sixth outlet. In general, crosspoint i, j in one array is paired with crosspoint j, i in the corresponding array on the opposite side of a switch. Since the availability of one crosspoint in a pair ensures the availability of the other, the reverse path is automatically specified and available.

A second advantage of the folded four-wire operation results because the amount of information specifying the status of the switch can be cut in half. Only the status of each pair of crosspoints or associated junctor is needed to find an available path through the switch.

A third benefit of the folded structure occurs because the blocking probability is one-half of the probability of finding two paths independently. It might seem that pairing the crosspoints in the described manner would restrict the paths available for a particular connection. On the contrary, the crosspoint pairings merely guarantee that a reverse path is automatically available for any selected path in the forward direction.

A fourth potential advantage of the folded operation can be obtained by sharing the crosspoint selection logic for corresponding arrays in the switch. For example, the same control signal that selects crosspoint 6, 4 in the first stage can be used to select crosspoint 4, 6 in the last stage. In an actual implementation, the paired crosspoints may even be in a common module.

The folded operation in the preceding paragraphs referred to a switch with an even number of switching stages. An even number was chosen because the concept is easiest to demonstrate when no center stage is present. The basic approach can be extended to odd numbers of switching stages if the center stage contains an even number of arrays and is folded about a horizontal line at the center of the stage. In this manner, crosspoint i, j in the top center-stage array is paired with crosspoint j, i in the bottom center-stage array, and so on.

5.2.5 Pathfinding

Determining a path through a single stage switch is virtually automatic since the necessary crosspoint is uniquely specified by the inlet/outlet pair to be connected. In contrast, availability of more than one path in a multiple stage switch complicates the path selection process in several ways. First, the control element of a switch must keep track of (or be able to sense) which potential paths for a particular connection are available. With computer controlled switching, the information as to which switching paths are available is retained in memory and referred to as a state store. A programmed pathfinding routine is then used to process the state store information and select an available path. Whenever a new connection is established or an old one released, the state store is updated with the appropriate information.

In older, electromechanically controlled switches, status of the switching paths is retained in the switching matrix itself. A path is set up by sequentially testing links until an appropriate set of idle links is found. In crossbar switches a control element, called a marker, simultaneously tests a number of paths and selects one of the available ones. In many electromechanical switching arrays the path selection process is sequenced or randomized in some manner to evenly distribute wear on the mechanical crosspoints. This procedure not only extends the life of what would otherwise be a heavily used portion of the switch, but also allows repeated connection attempts to use different paths through the switch. Thus if a bad connection occurs because of faulty switching equipment, a caller can hang up and repeat the request to obtain a new connection with satisfactory performance.

Pathfinding Times

Pathfinding operations, whether implemented in a stored program computer or in specialized equipment such as a marker, involves the use of common equipment and must therefore be analyzed to determine the rate at which connect requests can be processed. The time required to find an available path is directly dependent on how many potential paths are tested before an idle one is found. Some systems can test a number of paths in parallel and thereby shorten the processing time. Since the expected number of potential paths that must be tested to find an idle path is a function of link utilizations, pathfinding times unfortunately increase when the common control equipment is busiest.

Assume that the probability of a complete path through the switch being busy is denoted by p. If each of k possible paths through the switch has an equal and independent probability of being busy, the expected number of paths N_p that must be tested before an idle path is found is determined in Appendix A as follows:

$$N_p = \frac{1 - p^k}{1 - p}$$

(5.11)

EXAMPLE 5.1

What is the expected number of potential paths that must be tested to find an idle path in the three-stage, 8 192 line switch defined in Table 5.2?

Solution. As indicated in the table, a space expansion factor of 0.234 is used to provide a blocking probability of .002. Hence the utilization of each interstage link is .1/0.234 = .427. The blocking probability of each path through the switch is merely the probability that one of two links in series is busy. Hence the probability $p = 1 - (1 - .427)^2 = .672$, and the expected number of paths to be tested is:

$$N_p = \frac{1 - (0.672)^{15}}{1 - 0.672}$$

$$= 3.04$$

Example 5.1 demonstrates that, on average, only three of the 15 potential paths need to be tested before an idle path is found. However, when the switch is experiencing greater than normal traffic loads, the average number of paths tested increases. For example, if the input line utilization increases from 10% to 15%, the blocking probability increases from .002 to .126, and the expected number of paths to be tested in the pathfinding operation increases from 3 to 4.9.

5.2.6   Switch Matrix Control

When an available path through a common control switching network is determined, the control element of the switch transfers the necessary information to the network to select the appropriate crosspoints. Crosspoint selection within a matrix is accomplished in one of two ways. The control may be associated with the output lines and therefore specify which inputs are to be connected to the associated outputs, or the control information may be associated with each input, and subsequently specify to which outputs the respective inputs are to be connected. The first approach is referred to as "output-associated control," while the second is accordingly referred to as "input-associated control." These two control implementations are presented in Figure 5.12.

Input-associated control is inherently required in step-by-step switches where the information (dial pulses) arrive on the input link and is used to directly select the output links to each successive stage. In common control systems, however, the address information of both the originating line and the terminating line is simultaneously available. Hence the connection can be established by beginning at the desired outlet and proceeding backward through the switch while selecting inputs to each stage.

The implementation of both types of digital crosspoint arrays using standard components is shown in Figure 5.13. Output-associated control uses a conventional data selector/multiplexer for each matrix output. The number of bits required to control each data selector is $\log_2 N$ where N is the number of inlets. Thus the total number of bits required to completely specify a connection configuration is $M \log_2 N$.

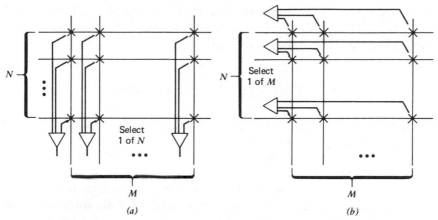

Figure 5.12 Switch matrix control. (*a*) Output associated control. (*b*) Input associated control.

Input-associated control can be implemented using conventional line decoders/demultiplexers. The outputs are commoned using a "wired or" logic function. Thus the output gates of each decoder circuit must be open collector or tristate devices if TTL logic is used. The total number of bits required to specify a connection configuration in this case is $N \log_2 M$.

A significant drawback of input-associated control arises from the need to disable unused inputs to prevent crossconnects when another input selects the same output. With output-associated control, unused outputs can remain connected to an input without preventing that input from being selected by another output. For this reason and for generally greater speeds of operation, digital switching networks typically use output-associated control. Notice, however, that the total amount of information needed to specify a connection configuration with input-associated control is less than that with output control if the

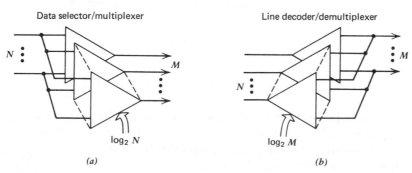

Figure 5.13 Standard component implementations of digital crosspoint array. (*a*) Output associated control. (*b*) Input-associated control.

number of inputs N is much smaller than the number of outputs M ($N \log_2 M < M \log_2 N$). Furthermore, input-associated control is more flexible in terms of "wired-or" expansion.

5.3   TIME DIVISION SWITCHING

As evidenced by multiple stage switching, sharing of individual crosspoints for more than one potential connection provides significant savings in implementation costs of space division switches. In the cases demonstrated, the crosspoints of multistage space switches are shared from one connection to the next, but a crosspoint assigned to a particular connection is dedicated to that connection for its duration. Time division switching involves the sharing of crosspoints for shorter periods of time so that individual crosspoints and their associated interstage links are continually reassigned to existing connections. When the crosspoints are shared in this manner, much greater savings in crosspoints can be achieved. In essence, the savings are accomplished by time division multiplexing the crosspoints and interstage links in the same manner that transmission links are time division multiplexed to share interoffice wire pairs.

Time division switching is equally applicable to either analog or digital signals. Analog time division switching is attractive when interfacing to analog transmission facilities, since the signals are only sampled and not digitally encoded. However, large analog time division switches experience the same limitations as do analog time division transmission links: the PAM samples are particularly vulnerable to noise, distortion, and crosstalk. In digital switches, the voice signals are regenerated every time they pass through a logic gate. Thus at some point in switching sizes, the cost of digitizing PAM samples is mandated by the need to maintain end-to-end signal quality. As the cost of digitization continues to decline, digital switching of analog signals is more and more competitive for small switch sizes. If a sufficient number of the signals to be switched are already digital or will be converted to digital at a future date, digital switching becomes more attractive than analog switching, even for the smallest PBXs.

5.3.1   Analog Time Division Switching

A vast majority of PBXs presently available in the United States use time division technology for the switching matrix. Of these, approximately one-half use analog time division switching while the other half use digital time division switching. Figure 5.14 depicts a particularly simple analog time division switching structure. A single switching bus supports a multiple number of connections by interleaving PAM

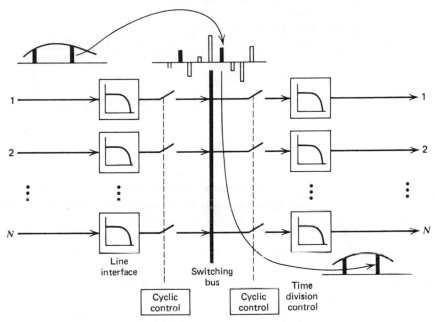

Figure 5.14   Analog time division switching.

samples from receive line interfaces to transmit line interfaces. The operation is depicted as though the receive interfaces are separate from the respective transmit interfaces. When connecting two-wire analog lines, the two interfaces are necessarily implemented in a common module. Furthermore, in some PAM-PBX systems, analog samples are simultaneously transferred in both directions between the interfaces [9].

Included in Figure 5.14 are two cyclic control stores. The first control store controls gating of inputs onto the bus one sample at a time. The second control store operates in synchronism with the first and selects the appropriate output line for each input sample. A complete set of pulses, one from each active input line, is referred to as a frame. The frame rate is equal to the sample rate of each line. For voice systems the sampling rate ranges from 8 to 12 kHz. The higher sampling rates are used to simplify the band-limiting filter and reconstructive filters in the line interfaces.

5.3.2   Digital Time Division Switching

The analog switching matrix described in the preceding section is essentially a space division switching matrix. By continually changing the connections for short periods of time in a cyclic manner, the configuration of the space division switch is replicated once for each time slot. This mode of operation is referred to as time multiplexed switching. While this mode of operation can be quite useful for both analog and

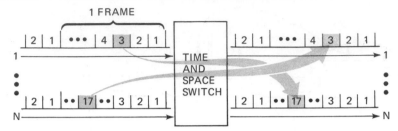

Figure 5.15   Time and space division switching.

digital signals, digital time division multiplexed signals usually require switching between time slots, as well as between physical lines. This second mode of switching represents a second dimension of switching and is referred to as time switching.

In the following discussion of digital time division switching it is assumed, unless otherwise stated, that the switching network is interfaced directly to digital time division multiplex links. This assumption is generally justified since, even when operating in an analog environment, the most cost-effective switch designs multiplex groups of digital signals into TDM formats before any switching operations take place. Thus most of the following discussion is concerned with the internal structures of time division switching networks, and possibly not with the structure of an entire switching complex.

The basic requirement of a time division switching network is shown in Figure 5.15. As an example connection, channel 3 of the first TDM link is connected to channel 17 of last TDM link. The indicated connection implies that information arriving in time slot 3 of the first link is transferred to time slot 17 of the last link. Since the voice digitization process inherently implies a four-wire operation, the return connection is required and realized by transferring information from time slot 17 of the last input link to time slot 3 of the first link. Thus each connection requires two transfers of information: each involving translations in both time and space.

A variety of switching structures are possible to accomplish the transfers indicated in Figure 5.15. All of these structures inherently require at least two stages: a space division switching stage and a time division switching stage. As discussed later, larger switches use multiple stages of both types. Before discussing switching in both dimensions, however, we discuss the characteristics and capabilities of time switching alone.

A Digital Memory Switch

Primarily owing to the low cost of digital memory, time switching implementations provide digital switching functions more economically than space division implementations. Basically, a time switch operates

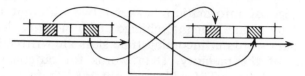

Figure 5.16 Time slot interchange operation.

by writing data into and reading data out of a single memory. In the process the information in selected time slots is interchanged as shown in Figure 5.16. When digital signals can be multiplexed into a single TDM format, very economical switches can be implemented with time switching alone. However, practical limitations of memory speed limit the size of a time switch so that some amount of space division switching is necessary in large switches. As demonstrated in later sections, the most economical multistage designs usually perform as much switching as possible in the time stages.

The basic functional operation of a memory switch is shown in Figure 5.17. Individual digital message circuits are multiplexed and demultiplexed in a fixed manner to establish a single TDM link for each direction of travel. The multiplexing and demultiplexing functions can be considered as part of the switch itself, or they may be implemented in remote transmission terminals. If the multiplexing functions are implemented locally, the multiplexer and demultiplexer may be connected in parallel, directly to the memory. Otherwise, a serial-to-parallel converter is used to accumulate the information in a time slot before it is written into the memory. In either case, a memory write access is required for each incoming time slot, and a memory read access is required for each outgoing time slot.

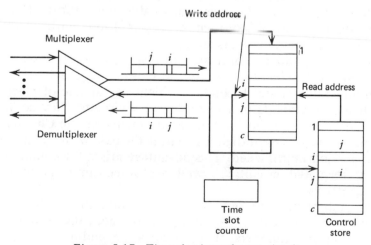

Figure 5.17 Time slot interchange circuit.

The exchange of information between two different time slots is accomplished by a time slot interchange (TSI) memory. In the TSI of Figure 5.17 data words in incoming time slots are written into sequential locations of the memory. Data words for outgoing time slots, however, are read from TSI addresses obtained from a control store. As indicated in the associated control store, a full-duplex connection between TDM channel i and TDM channel j implies that TSI address i is read during outgoing time slot j and vice versa. The TSI memory is accessed twice during each link time slot. First, some control circuitry (not shown) selects the time slot number as a write address. Second, the content of the control store for that particular time slot is selected as a read address.

Since a write and a read is required for each channel entering (and leaving) the TSI memory, the maximum number of channels c that can be supported by the simple memory switch is

$$c = \frac{125}{2t_c} \tag{5.12}$$

where 125 is the frame time in microseconds for 8 kHz sampled voice, and t_c is the memory cycle time in microseconds.

As a specific example, consider the use of a 500 ns memory. Equation 5.12 indicates that the memory switch can support 125 full-duplex channels (62 connections) in a strictly nonblocking mode of operation. The complexity of the switch (assuming digitization occurs elsewhere) is quite modest: the TSI memory stores one frame of data organized as c words by 8 bits each. The control store also requires c words, but each word has a length equal to $\log_2(c)$ (which is 7 in the example). Thus both memory functions can be supplied by 128×8 bit random access memories (RAMs). The addition of a time slot counter and some gating logic to select addresses and enable new information to be written into the control store can be accomplished with a handful of conventional integrated circuits (ICs).

This switch should be contrasted to a space division design that requires 7,680 crosspoints for a nonblocking three-stage switch. Although modern integrated circuit technology might be capable of placing that many digital crosspoints in a few integrated circuits, they could never be reached because of pin limitations. As mentioned in Chapter 2, one of the main advantages of digital signals is the ease with which they can be time division multiplexed. This advantage arises for communication between integrated circuits as well as for communication between switching offices.

If the multiplexer and demultiplexer combination in Figure 5.17 is enhanced to provide concentration and expansion, the system can service much greater numbers of input lines, depending on the average circuit utilization. For example, if the average line is busy 10% of the time, a concentrator/memory switch/expandor operation could support 1000 circuits with a blocking probability of less than .002. The concentration and expansion operations, however, imply a significant increase in the complexity of the system. In fact, these equipments essentially represent time multiplexed space division switches that must be controlled accordingly. The concentrator/memory switch/expandor structure essentially becomes a simple form of a space-time-space (STS) switch discussed later.

Time Stages in General

Time switching stages inherently require some form of delay element to provide the desired time slot interchanges. Delays are most easily implemented using random access memories that are written into as data arrives and read from when data is to be transferred out. If one memory location is allocated for each time slot in the TDM frame format, the information from each TDM channel can be stored for up to one full frame time without being overwritten.

There are two basic ways in which the time stage memories can be controlled: written sequentially and read randomly, or written randomly and read sequentially. Figure 5.18 depicts both modes of operation and indicates how the memories are accessed to translate information from time slot 3 to time slot 17. Notice that both modes of operation use a cyclic control store that is accessed in synchronism with the time slot counter.

The first mode of operation in Figure 5.18 implies that specific memory locations are dedicated to respective channels of the incoming TDM link. Data for each incoming time slot is stored in sequential locations within the memory by incrementing a modulo-*c* counter with

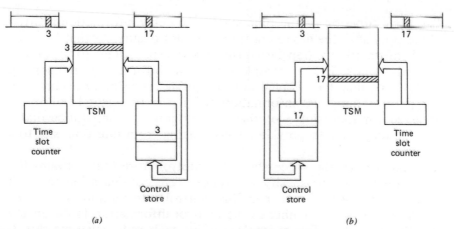

(a)   (b)

Figure 5.18 Time stage modes of operation. (*a*) Sequential writes/random reads. (*b*) Random writes/sequential reads.

every time slot. As indicated, the data received during time slot 3 is automatically stored in the third location within the memory. On output, information retrieved from the control store specifies which address is to be accessed for that particular time slot. As indicated, the seventeenth word of the control store contains the number 3, implying that the contents of time stage memory (TSM) address 3 is transferred to the output link during outgoing time slot 17.

The second mode of operation depicted in Figure 5.18 is exactly the opposite of the first one. Incoming data is written into the memory locations as specified by the control store, but outgoing data is retrieved sequentially under control of an outgoing time slot counter. As indicated in the example, information received during time slot 3 is written directly into TSM address 17 where it is automatically retrieved during outgoing TDM channel number 17. Notice that the two modes of time stage operation depicted in Figure 5.18 are forms of output-associated control and input-associated control, respectively. In a multiple stage design example presented later, it is convenient to use one mode of operation in one time stage and the other mode of operation in another time stage.

5.4 TWO-DIMENSIONAL SWITCHING

Larger digital switches require switching operations in both a space dimension and a time dimension. There are a large variety of network configurations that can be used to accomplish these requirements. To begin with, consider a simple switching structure as shown in Figure 5.19. This switch consists of only two stages: a time stage T followed by a space stage S. Thus this structure is referred to a time-space TS switch.

The basic function of the time stage is to delay information in arriving time slots until the desired output time slot occurs. At that time the delayed information is transferred through the space stage to the appropriate output link. In the example shown the information in incoming time slot 3 of link 1 is delayed until outgoing time slot 17 occurs. The return path requires that information arriving in time slot 17 of link N be delayed for time slot 3 of the next outgoing frame. Notice that a time stage may have to provide delays ranging from one time slot to a full frame.

Associated with the space stage is a control store that contains the information needed to specify the space stage configuration for each individual time slot of a frame. This control information is accessed cyclically in the same manner as the control information in the analog time division switch. For example, during each outgoing time slot 3, control information is accessed that specifies interstage link number 1

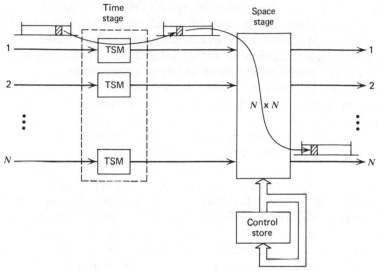

Figure 5.19 Time-space (TS) switching matrix.

is connected to output link N. During other time slots, the space switch is completely reconfigured to support other connections.

As indicated, a convenient means of representing a control store is a parallel end-around-shift register. The width of the shift register is equal to the number of bits required to specify the entire space switch configuration during a single time slot. The length of the shift register conforms to the number of time slots in a frame. Naturally, some means of changing the information in the control store is needed so that new connections can be established. In actual practice, the control stores may be implemented as random access memories with counters used to generate addresses in a cyclic fashion.

Implementation Complexity of Time Division Switches

In previous sections, alternative space division switching structures were compared in terms of the total number of crosspoints required to provide a given grade of service. Other factors that should be considered in a comprehensive analysis are: modularity, pathfinding requirements, effects of failures, serviceability, wiring or interconnection requirements, electrical loadings, and others. Despite the need to assess these other considerations, a crosspoint count is a useful, single measure of a space division switch cost, particularly with electromechanical crosspoints.

In the case of solid state electronic switching matrices, in general, and time division switching, in particular, the number of crosspoints alone is a less meaningful measure of implementation cost. Switching structures that utilize integrated circuits with relatively large numbers of internal crosspoints are generally more cost effective than other structures that may have fewer crosspoints but more packages. Hence a more relevant design parameter for solid state switches would be the total number of integrated circuit packages. If alternate designs are implemented from a common set of integrated circuits, the number of packages may closely reflect the number of crosspoints.

Another useful cost parameter is the total number of IC pin-outs required in a particular implementation. Although this parameter is obviously related closely to the total number of packages, it is generally more useful since it more accurately reflects package cost and circuit board area requirements. Pin-out measurements also provide a direct indication of implementation reliability, since external interconnections are generally less reliable than internal connections of an IC.

Medium scale integrated (MSI) circuits typically provide the equivalent of one crosspoint (AND gate) for $1\frac{1}{2}$ external pins to access the crosspoint. Thus when MSI technology is used, the total number of crosspoints is a useful indication of the total number of pins. We therefore continue to use crosspoints as an implementation cost measurement with the understanding that medium scale integration would be used in all comparative designs. To do so we must be sure that all systems operate at about the same speed, since higher speeds require lower levels of integration.

In addition to the number of crosspoints in space division stages, a digital time division switch uses significant amounts of memory that must be included in an estimate of the overall cost.* The memory count includes the time stage memory arrays and the control stores for both the time stages and the space stages. In the following analyses we assume that 100 bits of memory corresponds to $1\frac{1}{2}$ IC interconnections. (A 1024 bit random access memory typically requires 14 pins.) With this assumption, we can relate memory costs to crosspoint costs by a 100 bits per crosspoint factor. Hence the following analyses of implementation complexity for digital time division switching matrices include the total number of crosspoints and the total number of bits of memory divided by 100.

The implementation complexity is expressed as follows:

$$\text{complexity} = N_X + \frac{N_B}{100} \tag{5.13}$$

where N_X = number of space stage crosspoints
N_B = number of bits of memory

---

*It is worth noting that digital memories are inherently implemented with at least two crosspoints per bit. In this case the crosspoints are gates used to provide write and read access to the bits. These crosspoints, however, are much less expensive than message crosspoints that are accessed from external circuits.

EXAMPLE 5.2

Determine the implementation complexity of the TS switch shown in Figure 5.19 where the number of TDM input lines $N = 80$. Assume each input line contains a single DS-1 signal (24 channels). Furthermore, assume a one-stage matrix is used for the space stage.

Solution. The number of crosspoints in the space stage is determined as

$$N_X = 80^2 = 6400$$

(The crosspoints on the main diagonal are included since two channels within a single TDM input may have to be connected to each other.) The total number of memory bits for the space stage control store is determined as

N_{BX} = (number of links) (number of control words)
(number of bits/control word)

$= (80) (24) (7)$

$= 13440$

The number of memory bits in the time stage is determined as the sum of the time slot interchange and the control store bits:

N_{BT} = (number of links) (number of channels)
(number of bits/channel) + (number of links)
(number of control words) (number of bits/control word)

$= (80) (24) (8) + (80) (24) (5)$

$= 24,960$

Thus the implementation complexity is determined as

complexity $= N_X + (N_{BX} + N_{BT})/100 = 6784$ equivalent crosspoints

The implementation complexity determined in Example 5.2 is obviously dominated by the number of crosspoints in the space stage. A significantly lower complexity (and generally lower cost) can be achieved if groups of input links are combined into higher-level multiplex signals before being switched. The cost of the front end multiplexers is relatively small if the individual DS-1 signals have already been synchronized for switching. In this manner, the complexity of the space stage is decreased appreciably while the overall complexity of the time stage increases only slightly. (See the problem set at the end of this chapter.) The implementation costs are reduced proportionately—

up to the point that higher speeds dictate the use of a more expensive technology.

Multiple Stage Time and Space Switching

As discussed in the preceding section, an effective means of reducing the cost of a time division switch is to multiplex as many channels together as practical and perform as much switching in the time stages as possible. Time stage switching is generally less expensive than space stage switching—primarily because digital memory is much cheaper than digital crosspoints (AND gates). To repeat, the crosspoints themselves are not so expensive, it is the cost of accessing and selecting them from external pins that makes their use relatively costly.

Naturally, there are practical limits as to how many channels can be multiplexed into a common TDM link for time stage switching. When these practical limits are reached, further reductions in the implementation complexity can be achieved only by using multiple stages. Obviously, some cost savings result when a single space matrix of a TS or ST switch can be replaced by multiple stages.

A generally more effective approach involves separating the space stages by a time stage, or, conversely, separating two time stages by a space stage. The next two sections describe these two basic structures. The first structure, consisting of a time stage between two space stages, is referred to as a space-time-space (STS) switch. The second structure is referred to as a time-space-time (TST) switch.

5.4.1 STS Switching

A functional block diagram of an STS switch is shown in Figure 5.20. Each of the space switches is assumed to be a single stage (nonblocking) switch. For large switches, it may be desirable to implement the space switches with multiple stages. Establishing a path through an STS switch requires finding a time switch array with an available write access during the incoming time slot and an available read access during the desired outgoing time slot. When each individual stage (S, T, S) is nonblocking, the operation is functionally equivalent to the operation of a three-stage space switch. Hence a probability graph in Figure 5.21 of an STS switch is identical to the probability graph of Figure 5.8 for three-stage space switches. Correspondingly, the blocking probability of an STS switch is

$$B = (1 - q'^2)^k \tag{5.14}$$

where $q' = 1 - p' = 1 - p/\beta$ $(\beta = k/N)$
     k = number of center stage time switch arrays

Assuming the space switches are single stage arrays and that each TDM link has c message channels, we may determine the implementa-

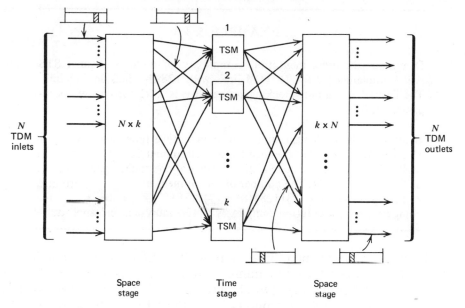

Figure 5.20   Space-time-space (STS) switching structure.

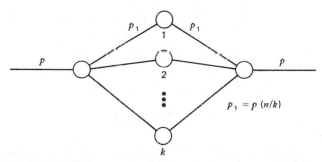

Figure 5.21   Probability graph of STS switch with nonblocking stages.

tion complexity of an STS switch as*

complexity = (number of space stage crosspoints)
        + [(number of space stage control bits)
        + (number of time stage memory bits)
        + (number of time stage control bits)]/100

$$= 2kN + \frac{[2 \cdot k \cdot c \cdot \log_2 N + k \cdot c \cdot 8 + k \cdot c \cdot \log_2 c]}{100}$$

$$(5.15)$$

*This derivation assumes output-associated control is used in the first stage and input-associated control is used in the third stage. A slightly different result occurs if the space stages are controlled in different manners.

EXAMPLE 5.3

Determine the implementation complexity of a 2048 channel STS switch implemented for 16 TDM links with 128 channels on each link. The desired maximum blocking probability is .002 for channel occupancies of 0.1.

Solution. The minimum number of center stage time switches to provide the desired grade of service can be determined from Equation 5.14 as $k = 7$. Using this value of k, the number of crosspoints is determined as $(2)(7)(16) = 224$. The number of bits of memory can be determined as $(2)(7)(128)(4) + (7)(128)(8) + (7)(128)(7) = 20,608$. Hence the composite implementation complexity is 430 equivalent crosspoints.

The value of implementation complexity obtained in Example 5.3 should be compared to the number of crosspoints obtained for an equivalent sized three-stage switch listed in Table 5.2. The space switch design requires 41,000 crosspoints while the STS design requires only 430 equivalent crosspoints. The dramatic savings comes about as a result of the voice signals having already been digitized and multiplexed (presumably for transmission purposes). When the STS switch is inserted into an analog environment, the dominant cost of the switch occurs in the line interface. Modern digital switches are not unlike modern digital computers. The cost of a central processing unit has become relatively small compared to the costs of peripheral devices.

5.4.2 TST Switching

A second form of multiple stage time and space switching is shown in Figure 5.22. This switch is usually referred to as a time-space-time (TST) switch. Information arriving in a TDM channel of an incoming link is delayed in the inlet time stage until an appropriate path through the space stage is available. At that time the information is transferred through the space stage to the appropriate outlet time stage where it is held until the desired outgoing time slot occurs. Assuming the time stages provide full availability (i.e. all incoming channels can be connected to all outgoing channels), any space stage time slot can be used to establish a connection. In a functional sense the space stage is replicated once for every internal time slot. This concept is reinforced by the TST probability graph of Figure 5.23.

An important feature to notice about a TST switch is that the space stage operates in a time divided fashion, independently of the external TDM links. In fact, the number of space stage time slots l does not have to coincide with the number of external TDM time slots c.

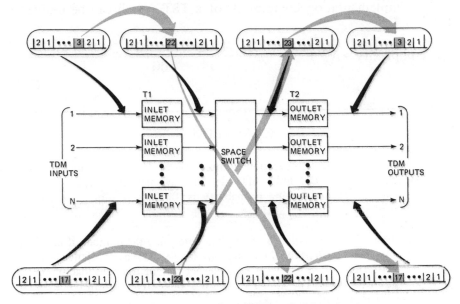

Figure 5.22 Time-space-time (TST) switching structure.

If the space stage is nonblocking, blocking in a TST switch occurs only if there is no internal space stage time slot during which the link from the inlet time stage and the link to the outlet time stage are both idle. Obviously, the blocking probability is minimized if the number of space stage time slots l is made to be large. In fact, as a direct analogy of three-stage space switches, the TST switch is strictly nonblocking if $l = 2c - 1$. The general expression of blocking probability for a TST switch with nonblocking individual stages (T, S, T) is:

$$B = [1 - q_1^2]^l \qquad (5.16)$$

where $q_1 = 1 - p_1 = 1 - p/\alpha$
α = time expansion (l/c)
l = number of space stage time slots

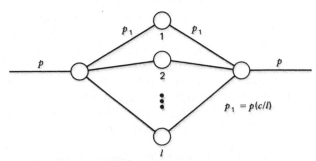

Figure 5.23 Probability graph of TST switch with nonblocking stages.

The implementation complexity of a TST switch can be derived as follows:*

$$\text{complexity} = N^2 + \frac{[N \cdot l \cdot \log_2 N + 2 \cdot N \cdot c \cdot 8 + 2 \cdot N \cdot l \log_2 c]}{100}$$

(5.17)

EXAMPLE 5.4

Determine the implementation complexity of a 2048 channel TST switch with 16 TDM links and 128 channels per link. Assume the desired maximum blocking probability is .002 for incoming channel occupancies of 0.1.

Solution. Using Equation 5.16, we can determine the number of internal time slots required for the desired grade of service as 25. Hence time concentration of $1/\alpha = 5.12$ is possible because of the light loading on the input channels. The implementation complexity can now be determined from Equation 5.17 as 656.

The results obtained in Examples 5.3 and 5.4 indicate that the TST architecture is more complex than the STS architecture. Notice, however, that the TST switch operates with time concentration whereas the STS switch operates with space concentration. As the utilization of the input links increase, less and less concentration is acceptable. If the input channel loading is high enough, time expansion in the TST switch and space expansion in the STS switch are required to maintain low blocking probabilities. Since time expansion can be achieved at less cost than space expansion, a TST switch becomes more cost effective than an STS switch for high channel utilizations. The implementation complexities of these two systems are compared in Figure 5.24 as a function of the input utilization.

As can be seen in Figure 5.24, TST switches have a distinct implementation advantage over STS switches when large amounts of traffic are present. For small switches, the implementation complexities favor STS architectures. The choice of a particular architecture may be more dependent on such other factors as modularity, testability, expandability. One consideration that generally favors an STS structure is its relatively simpler control requirements [10]. For very large switches with heavy traffic loads, the implementation advantage of a TST switch is dominant. Evidence of this fact is provided by the No. 4ESS, a TST structure that is the largest capacity switch built to date.

---

*This derivation assumes that the inlet time stage uses output-associated control (random reads) and the outlet time stage uses input-associated control (random writes). A slightly different result occurs if the time stages operate differently.

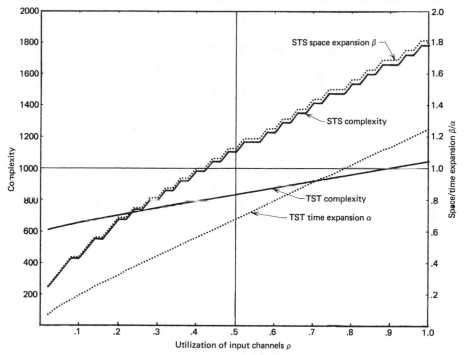

Figure 5.24 Complexity comparison of STS and TST switching structures for a blocking probability of .002.

TSSST Switches

When the space stage of a TST switch is large enough to justify additional control complexity, multiple space stages can be used to reduce the total crosspoint count. Figure 5.25 depicts a TST architecture with a three-stage space switch. Because the three middle stages are all space stages, this structure is sometimes referred to as a TSSST switch. The

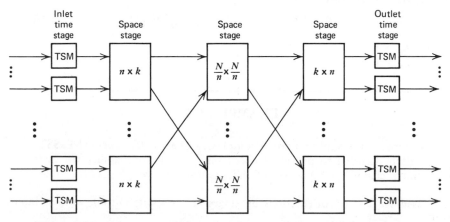

Figure 5.25 Time-space-space-space-time (TSSST) switching structure.

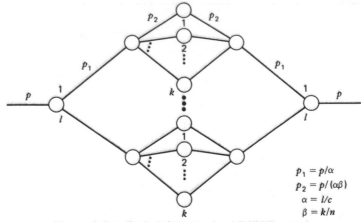

$$p_1 = p/\alpha$$
$$p_2 = p/(\alpha\beta)$$
$$\alpha = 1/c$$
$$\beta = k/n$$

Figure 5.26 Probability graph of TSSST switch.

implementation complexity of a TSSST switch can be determined as follows:*

$$\text{complexity} = \frac{N_X + (N_{BX} + N_{BT} + N_{BTC})}{100} \tag{5.18}$$

where N_X = number of crosspoints = $2 \cdot N \cdot k + k \cdot (N/n)^2$
$\quad N_{BX}$ = number of space stage control store bits = $2 \cdot k \cdot (N/n) \cdot l \cdot \log_2 (n) + k \cdot (N/n) \cdot l \cdot \log_2 (N/n)$
$\quad N_{BT}$ = number of bits in time stages = $2 \cdot N \cdot c \cdot 8$
$\quad N_{BTC}$ = number of time stage control store bits = $2 \cdot N \cdot l \cdot \log_2 (c)$

The probability graph of a TSSST switch is shown in Figure 5.26. Notice that this diagram is functionally identical to the probability graph of a five-stage space switch shown in Figure 5.10. Using the probability graph of Figure 5.26, we can determine the blocking probability of a TSSST switch as

$$B = \{1 - (q_1)^2 \, [1 - (1 - q_2^2)^k]\}^l \tag{5.19}$$

where $q_1 = 1 - p_1 = 1 - p/\alpha$
$\quad q_2 = 1 - p_2 = 1 - p/\alpha\beta$

EXAMPLE 5.5

Determine the implementation complexity of a 131,072 channel TSSST switch designed to provide a maximum blocking probability of .002 under channel occupancies of 0.7. Assume the switch services 1024

---

*The assumed control orientations by stages are: output, output, output, input, input.

TDM input links with 128 channels on each link. Also assume that unity time expansion is used on the space stages.

Solution. The space switch can be designed in a variety of ways depending on how many links are assigned to each array in the first (and third) space stages. A value of 32 is chosen as a convenient binary number near the theoretical optimum presented earlier $(N/2)^{1/2}$. With this value for n, the only unknown in Equation 5.19 is the number of center stage arrays k. Thus k is determined to be 24 for a blocking probability of .0015. The complexity is determined from Equation 5.18 as

$$N_X = (2)(1024)(24) + (24)(32)^2 = 73,728$$

$$N_{BX} = 2(24)(32)(128)(5) + (24)(32)(128)(5) = 1,474,560$$

$$N_{BT} - 2(1024)(128)(8) = 2,097,152$$

$$N_{BTC} = 2(1024)(128)(7) = 1,835,008$$

$$73,728 + 3,571,712/100 = 127,795 \text{ equivalent crosspoints}$$

The result of Example 5.5 demonstrates that very large capacity switches can be implemented with digital time division techniques at practical levels of complexity. In the middle 1960s it became apparent that switches of this size would be needed in the U.S. telephone network. Since a comparably sized eight-stage space division switch requires on the order of 10 million crosspoints, traditional space division technology was rejected, and Bell Laboratories began development of the No. 4ESS. This switch was the first digital switch in the U.S. telephone network when it became operational in 1976. The No. 4ESS (a TSSSST switch) has the capability of terminating 107,520 trunks with a blocking probability of less than .005 at a channel occupancy of 0.7 [11].

Naturally, there are not many places in the United Stages where a switch of this size is needed. In some 200 metropolitan areas, however, the No. 4ESS is capable of providing significant cost savings because it can service all of the toll traffic in the area with one switch instead of several of the largest space division switches previously available (No. 4A crossbars).

5.4.3 Modular Switch Structures

The preceding sections describe the two most basic switching structures for digital time division switches: STS and TST. Of the two, TST has become the most popular. However, some manufacturers have departed from these basic structures in the interest of using one type of module for all switching stages. A single module contains both time and space switching. For a description of a modular design see Reference 12.

Obviously, a main attraction of the modular approach is the ability to implement a wide range of switch sizes. Other advantages are simplified manufacturing, maintenance and test; reduced redundancy costs and spare parts inventories; and typically more graceful expansion capabilities. A principle disadvantage of the modular approach is the introduction of potentially long delays through the switch. On average, a time stage introduces one-half of a frame-time delay into each message circuit. A multiplicity of time stages increases the average delay through a switch accordingly. Delays of as much as a full millisecond (eight frames) are of no direct consequence to voice traffic. However, as discussed more fully later, propagation delays corresponding to just a few frames can lead to singing conditions on local connections that were formerly only possible with long-distance circuits.

Propagation delays through a multiplicity of time stages can be minimized in the path and time slot selection process, or the time stages can be implemented with less than a full frame's worth of memory [13]. In the latter case, incoming time slots of a TSI do not have full access to outgoing time slots. Hence restricting the delays will impact the blocking probability.

5.4.4 Custom Switch Designs

The digital time division switch examples presented in the preceding sections demonstrated that very cost effective switching networks can be implemented with standard memory and logic components. This section carries the design process one step farther to show the potentials of modern technology if customized integrated circuits are designed specifically for switching functions. Since the cost of digital switching networks implemented with readily available components is already a small percentage of a typical switch cost, it is reasonable to ask why customized circuits are of interest. The answers are summarized as follows:

1 Fewer components and, in particular, fewer interconnections increase the reliability of the implementation.

2 The cost of one or more levels of redundancy becomes less important so that standby modules or entire standby networks are feasible.

3 Blocking probability performance can be reduced to even lower levels than those economically realizable with standard components.

4 As time passes, more and more of the cost intensive interface functions will be implemented with integrated circuits so that a

standardized switch matrix will represent larger percentages of the total switch cost.

The following customized circuits do not contain all of the features that would be desirable to make them adaptable to a variety of applications. They do, however, demonstrate the most important functional features that are desired in a particular application. Specifically, the control store is integrated with the switching elements so that control information for each time slot is transferred internally to the integrated circuit package. When the control memories are implemented separately, the control bandwidth entering the switching elements can be almost as high as the message bandwidth passing through. A second feature of the following implementation is that only a few different types of circuits are required to implement a complete time and space switching network. In fact, an entire TSSST switching matrix for 32,000 channels is implemented with two types of custom circuits and one type of standard circuit.

A customized space switch element [14] is shown in Figure 5.27. This device consists of nothing more than a 16 X 1 input selector that is controlled in a time divided manner using information coming out of an end-around shift register. The shift register is clocked at the time slot

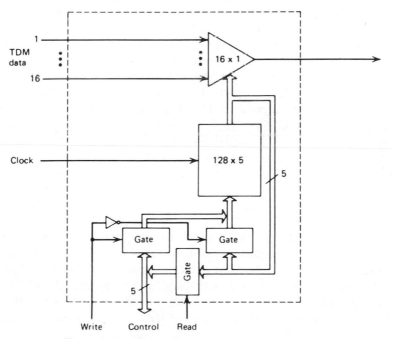

Figure 5.27 Time division space switch element.

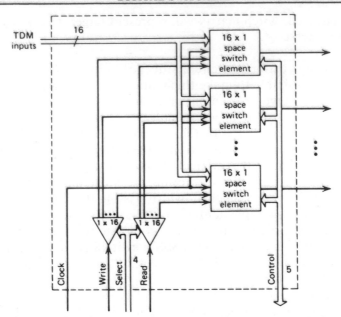

Figure 5.28   Sixteen by sixteen time division space switch.

rate while data is shifted through the selector at the bit rate. As indicated, the shift register length is set for a 128 channel frame.* Each
control word is shown to contain 5 bits: 4 bits for crosspoint selection
and 1 bit to indicate whether the output link is in use during the particular time slot. The remaining circuitry provides the ability to change
or merely read control information as it is shifted back into the bottom
of the shift register. As indicated, the entire element requires 26 pins.

By commoning the inputs to 16 space switch elements, a 16 × 16
time divided switch matrix is established as shown in Figure 5.28. The
only additional circuitry required are two 1 × 16 selectors used to gate
control store read or write commands to individual elements.

A diagram of a basic time switch element that can be implemented in
a single LSI circuit is shown in Figure 5.29. The control store of this
circuit operates in a manner similar to the control store of the space
switch element. The main difference is that an external control signal is
included to select between the time slot counter and the control information. The time slot counter is selected for sequential accesses, and
the control store is selected for random accesses. Data entering and
leaving the TSI memory must pass through serial-to-parallel and parallel-
to-serial converters to minimize the pin count. As indicated, the TSI
memory is designed for 120 channels of information while the control

---

*A more flexible control store design would use a resetable counter to address a
random access memory so that the number of channels per frame could be varied.

controlled by designing a prescribed amount of net attenuation (by wa
of net loss) into the transmission path of shorter toll network circuits
On the longer circuits, the echos and singing are eliminated by ech
suppressors. Class 5 digital switches could prevent instability in th
same manner: by designing a certain amount of loss into the encoding
decoding process. While the necessary amount of signal attenuation
(approximately 2 to 3 dB in each path [17]) could be tolerated on
local connections, the added loss to some toll connections would be
unacceptably large.

From the foregoing discussion it can be seen that the use of selectable attenuations is one solution to the instability problems. The necessary loss is inserted into the talking path for local connections but not
used on long-distance calls, which already have loss designed into them.
A second solution involves matching the impedances at the hybrids
more closely. Although nearly complete elimination of the unwanted
coupling is prohibitively expensive, adequate isolation of the transmission paths can be accomplished with just two different matching networks: one for loaded loops and one for unloaded loops [18]. This
solution also has the advantage of providing better matching, and consequently greater stability, for long-distance connections. Another pos-
ibility for zero loss digital switching is to use echo cancellors [19]. For
more complete discussions of the zero loss and other transmission prob-
ems associated with local digital switching see References 17 and 18.

Notice that an all-digital network (with four-wire telephones) avoids
he instability problems because there are no two-wire analog lines. The
oice is digitized at the telephone and uses a separate path from the re-
eive signal all the way to the destination telephone. Thus an end-to-
nd four-wire circuit completely eliminates echos and singing—allowing
ll connections to operate on a zero loss basis.

5.5.2 BORSCHT

Chapter 1 the basic functional requirements of the subscriber loop
terface are described. These requirements are repeated here with two
ditional requirements for a digital switch interface: coding and hy-
id. The complete list of interface requirements is unaffectionately
own as BORSCHT [20]:

B: Battery feed
O: Overvoltage protection
R: Ringing
S: Supervision
C: Coding
H: Hybrid
T: Test

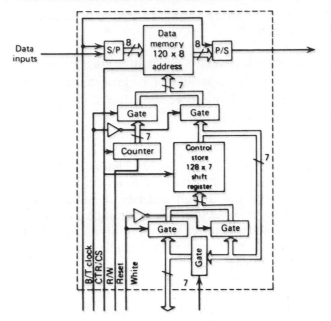

Figure 5.29 Time switch element.

store has 128 control words. Thus a time expansion of 128/120 is implied.

Complete TST switching matrices can be implemented using the two
custom circuits just described and a small assortment of standard components for element selection and signal fan-out. As a specific example,

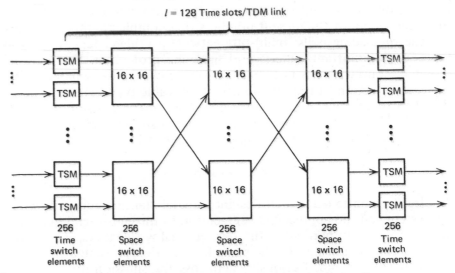

Figure 5.30 TSSST switch design with custom circuits.

the TST switching structure shown in Figure 5.30 contains the following set of components:

1   256 Time switch elements (inlet time stage).
2   768 Space switch elements (three-stage space switch).
3   256 Time switch elements (outlet time stage).
4   170 Selectors—1 × 16.

The switching matrix of Figure 5.30 services 256 TDM inputs with 120 channels per input for a total of 30,720 channels. At input occupancies of 0.8 erlangs, this switch provides a blocking probability of .005 (Equation 5.19). The LSI design consists of some 1500 integrated circuits with a little over 30,000 interconnections. A space division switch of comparable capacity requires several million crosspoints. The custom switch design can be packaged into a single equipment frame (no redundancy included) [15].

5.5   DIGITAL SWITCHING IN AN ANALOG ENVIRONMENT

When digital end-office switches are installed in an analog environment, the analog interfaces to the local loops are necessarily unchanged. Local digital transmission will evolve as digital subscriber carrier systems are integrated into the switch and as customers desiring the digital facilities pay special rates to have their telephones and service lines converted. Although some local loops will pass higher bandwidth digital signals adequately, conversion difficulties will vary depending on various analog implementation and maintenance practices in effect. Some of the present practices that complicate this conversion are use of loading coils, build-out networks, bridged taps, high resistance or intermittant splices, and lightning protection. Furthermore, crosstalk considerations in a digital system may require separate cables for each direction of transmission or newer cable types with an internal shield to isolate the go and return directions of transmission. For a discussion of a conversion strategy to provide a smooth transition to all-digital local loops in Great Britain see Reference 16.

5.5.1   Zero Loss Switching

As already mentioned, a well designed digital transmission and switching system adds no appreciable degradation to the end-to-end quality of digitized voice. In particular, the derived analog signal coming out of a decoder can be adjusted to the same level as that presented to the encoder at the far end. Curiously, zero loss transmission presents some

significant problems when digital switching is used in the an mission environment of a class 5 central office.

Analog end-office switches are two-wire switches designe connect bidirectional two-wire customer loops. The voice process, however, inherently requires separation of the go signal paths involved in a connection. Thus, a class 5 digita must be a four-wire switch. When inserted into a two-wir vironment, hybrids are required to separate the two d transmission. As shown in Figure 5.31, hybrids at each en ternal digital subnetwork produce a four-wire circuit with t for echos and singing. (Amplifiers are shown in conjuncti encoders to offset forward path loss inherent in the hybrids arises as a result of impedance mismatches at the hybrid wanted coupling from the receive portion to the transm the four-wire connection. Impedance mismatches occur b variability in the lengths and wire sizes of the subscriber ticular, loaded and unloaded wire pairs have markedly dif ance characteristics. Completely eliminating the unwante matching impedances on a line-to-line basis would be expensive.

The instability problems are compounded by certai artificial delay that are required in a digital time divisi though the delay through a digital switch (several hun onds, typically) is basically unnoticeable to a user, it equivalent of as much as 30 to 40 miles of wire. This has the effect of lowering the oscillation frequencies th wise be outside the voiceband and effectively remov decoder filters.

As mentioned in Chapter 1, hybrids are used exten terface to the toll network where the transmission m four-wire. In these instances, the instability of the fo

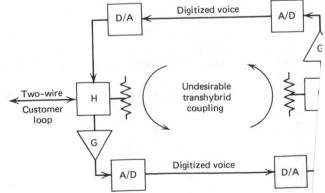

Figure 5.31   Four-wire digital switch circuit with two-wire

As mentioned in Chapter 1, the high voltage, low resistance, and current requirements of many of these functions is particularly burdensome to integrated circuit implementations. First generation digital switches reduced the termination costs by using analog switching (concentrators) to common codecs [21]. Integrated circuit manufacturers are working diligently to reduce the cost of this interface as a subscriber loop interface circuit (SLIC). For a description of an SLIC that took five years to develop see Reference 22. Because the costs of these functions are dropping dramatically, the current trend is per-line SLICs [23]. For an extensive overview of the application of integrated circuit technology in telephony see Reference 24.

5.5.3 Conferencing

In an analog network conference calls are established by merely adding individual signals together using a conference bridge. If two people talk at once, their speech is superposed. Furthermore, an active talker can hear if another conferee begins talking. Naturally, the same technique can be used in a digital switch if the signals are first converted to analog, added, and then converted back to digital.

As described in Chapter 3, μ255 and the A-law code recommended by CCITT were designed with the specific property of being easily digitally linearizable (EDL). With this property, the addition function can be performed digitally by first converting all codes to linear formats, adding them, and then converting back to compressed formats. To the user, the operation is identical to the customary analog summation. For a conference involving N conferees, N separate summations must be performed—one for each conferee and containing all signals but his own.

Another conferencing technique, particularly suited to adaptive delta modulated speech, involves monitoring the activity of all conferees and switching the digital signal of the loudest talker to all others. Hence at least a partial decoder for each conferee must be used to measure the activity of each conferee. Although this technique is functionally different from a customary analog conference circuit, the operation is not different from a long-distance circuit with echo suppressors.

These same basic conferencing considerations also arise when a digital telephone has digital extensions. Either the individual signals must be added or the channel is given to the loudest talker. For a discussion of conferencing fundamentals for digital PBXs see Reference 25.

5.5.4 Transmultiplexing

A transmultiplexer (transmux) is designed to directly convert TDM signals in the digital hierarchy to FDM signals in the analog hierarchy and vice versa. A common example is shown in Figure 5.32 where five

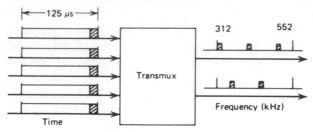

Figure 5.32 Transmultiplexer for converting five DS-1 signals into two supergroups.

DS-1 signals with 24 channels each are translated to two supergroups with 60 channels each. The primary need for transmultiplexing occurs at the interface between digital toll switches and long haul analog transmission systems such as TD-2, TH-3 and AR-6A analog radios.

Functionally, the transmultiplexer in Figure 5.32 is equivalent to a back-to-back interconnection of a digital demultiplexer (digital channel bank) and an analog FDM multiplexer. A transmux is much less expensive than back-to-back demultiplexing/multiplexing because it avoids the costly signaling interfaces provided with voice channel interfaces in conventional mutliplex equipment.

Transmultiplexers have been developed using two basic techniques. The first technique merely implements back-to-back demultiplexers and multiplexers, but without the costly signaling interfaces. Examples of these systems are Bell's LT-1 and GTE's 4691A/B FDM/PCM converter [26].

The second and more recent approach involves digitizing the FDM signals directly and using digital signal processing to perform traditional analog domain operations such as filtering, mixing, and amplification. Thus, digital equivalents of single sideband signals are processed to produce digitized baseband (PCM) signals. In the reverse direction (TDM to FDM) the digital TDM signals are digitally processed to produce digitized representations of single sideband multiplex groups. Then the FDM signal is produced by a conventional digital-to-analog converter. Examples of the latter type of transmux can be obtained in References 27 and 28.

REFERENCES

1 A. Feiner and W. S. Hayward, "No. 1ESS Switching Network Plan," *Bell System Technical Journal*, September 1964, pp 2193–2220.

2 Irwin Dorros, "ISDN," *IEEE Communications Magazine*, March 1981, pp 16–19.

3 M. R. Aaron, "Digital Communications—The Silent (R)evolution?" *IEEE Communications Magazine*, January 1979, pp 16–26.

4   C. Clos, "A Study of Non-Blocking Switching Networks," *Bell System Technical Journal*, March 1953, pp 406–424.

5   C. Y. Lee, "Analysis of Switching Networks," *Bell System Technical Journal*, November 1955, pp 1287–1315.

6   C. Jacobaeus, "A Study of Congestion in Link Systems," *Ericsson Techniques*, No. 48, Stockholm, 1950, pp 1–70.

7   A. A. Collins and R. D. Pedersen, *Telecommunications, A Time For Innovation*, Merle Collins Foundation Dallas, Texas, 1973.

8   V. E. Benes, "Traffic in Connecting Networks When Existing Calls are Rearranged," *Bell System Technical Journal*, Vol. 49, 1970, pp 1471–1482.

9   T. E. Browne, D. J. Wadsworth, and R. K. York, "New Time Division Switching Units for No. 101 ESS," *Bell System Technical Journal*, February 1969, pp 443–476.

10 S. G. Pitroda, "Telephones Go Digital," *IEEE Spectrum*, October 1979, pp 51–60.

11 A. E. Ritchie and L. S. Tuomenoksa "No. 4ESS: System Objectives and Organization," *Bell System Technical Journal*, September 1977, pp 1017–1027.

12 P. Charransol, J. Hauri, C. Athenes, and D. Hardy, "Development of a Time Division Switching Network Usable in a Very Large Range of Capacities," *IEEE Transactions on Communications*, July 1979, pp 982–988.

13 H. Inose, T. Saito, and K. Ichijo, "Time Division Networks with Partial-Access Pulse Shifters Performing Serial-Parallel Conversion, *IEEE Transactions on Communications*, August 1972, pp 762–768.

14 A. A. Collins, J. C. Bellamy, and R. L. Christensen, "Time Space Time (TST) Control Store," *U.S. Patent #3,956,593;* May 11, 1976.

15 Arthur A. Collins, "Are We Committed to Change in Telecommunication Systems," *Signal*, August 1974, pp 62–65.

16 D. Sibbald, H. W. Silcock, and E. S. Usher, "Digital Transmission Over Existing Subscriber Cable," *International Symposium on Subscriber Loops and Services*, Atlanta, Georgia, 1978.

17 James L. Neigh, "Transmission Planning for an Evolving Local Switched Digital Network," *IEEE Transactions on Communications*, July 1979, pp 1019–1024.

18 R. Bunker, F. Scida, and R. McCabe, "Zero Loss Considerations in Digital Class 5 Office," *IEEE Transactions on Communications*, July 1979, pp 1013–1018.

19 D. L. Duttweiler, "A Twelve Channel Echo Cancellor," *IEEE Transactions on Communications*, May 1978, pp 647–653.

20 F. D. Reese, "Memo to Management—You Must Appraise How New Technology Fits Customers," *Telephone Engineering and Management*, October 1, 1975, pp 116–121.

21 J. B. Terry, D. R. Younge, and R. T. Matsunaga, "A Subscriber Line Interface for the DMS-100 Digital Switch," *IEEE National Telecommunications Conference*, 1979, pp 28.3.1–28.3.6.

22 J. R. Sergo, Jr., "DSS Quad Line Circuit, *International Symposium of Subscriber Loops and Services*, 1978, pp 182–184.

23 S. E. Pucini and R. W. Wolff, "Architecture of GTD-5 EAX Digital Family," *IEEE International Conference on Communications*, 1980, pp 18.2.1–18.2.8.

24 P. R. Gray and D. G. Messerschmitt, "Integrated Circuits for Telephony," *Proceedings of IEEE*, August 1980, pp 991–1009.

25 R. J. D'Ortenzio, "Conferencing Fundamentals for Digital PABX Equipments," *IEEE International Conference on Communications*, 1979, pp 2.5-29 to 2.5-36.

26 D. W. Surling and G. McGibbon, "The 4691B FDM/PCM Converter," *International Conference on Communications*, 1980.

27 M. G. Bellanger, G. Bonnerot, and P. Caniquit, "Specification of A/D and D/A Converters for FDM Telephone Signals," *IEEE Transactions on Circuits and Systems*, July 1978, pp 461–467.

28 T. Aoyama, F. Mano, K. Wakabayashi, R. Maruta, and A. Tomozawa, "120-channel Transmultiplexer Design and Performance," *IEEE Transactions on Communications*, September 1980, pp 1709–1716.

PROBLEMS

5.1 How many four-wire voice circuit connections can be provided by a bidirectional PAM switching bus if the minimum achievable pulse width is 250 ns?

5.2 The TS switch of Figure 5-19 uses DS-1 signals on each TDM link. What is the implementation complexity if groups of 5 DS-1 inputs are first multiplexed to form 16 input links with 120 channels on each link? What is the implementation complexity of the multiplexers (i.e., how many crosspoints and how many bits of memory are in the multiplexers)?

5.3 Determine the Lee graph and Jacobaeus blocking probabilities of the first switch in Table 5.5 ($k = 6$) if two inlets out of 16 become connected to 1 erlang subscribers. (Possibly these lines could be dial-up ports to a computer) (Hint: two inlets are permanently busy but the remaining inlets continue to be .1 erlang ports.)

5.4 Repeat Problem 5.3 but instead of two inlets being busy assume that two of the output links of the first stage module have failed.

5.5 How many crosspoints are needed in a 1024 line, three-stage space switch if the input loading is 6 CCS per line and the maximum acceptable blocking probability (using a Lee graph analysis) is .005?

    (a) $N = 16$, $N/n = 64$
    (b) $N = 32$, $N/n = 32$
    (c) $N = 64$, $N/n = 16$

5.6 What is the (Lee graph) blocking probability of the TS switch in Figure 5-19 for channel loadings of 0.2 erlangs?

(a) Assume each TDM input is a 24 channel interoffice trunk group.

(b) Assume the TDM inputs are derived from 24-channel channel banks with each analog interface connected to a dedicated 0.2 erlang line.

5.7 Design an STS switch for 128 primary TDM signals of the CCITT hierarchy (30 voice channels per input). Blocking should be less than .002 and the loading is 0.2 erlangs per channel. How many time slot interchange modules are needed? What is the complexity of the switch?

5.8 Repeat Problem 5.7 for a TST design.

5.9 Determine the number of crosspoints N_X and the total number of memory bits N_B required for a TST switch defined as follows:

Number of lines = 32
Single stage space switch
Number of channels per frame = 30
Time expansion = 2

5.10 What is the blocking probability of the switch in Problem 5.9 if the channel loading is 0.9 erlangs per channel?

5.11 How many bits of memory are needed in a time slot interchange circuit for a sixty channel signal with nine bits per time slot?

CHAPTER SIX

DIGITAL RADIO

Chapter 4 discusses various techniques of encoding digital information for transmission over wire-line facilities. Generally speaking, the bandwidth of the signal is not a critical consideration since (unloaded) wire pairs have no rigid bandwidth limitations. If the data rate of a wire-line transmission system is to be increased, the signaling rate can be increased and the regenerative repeaters moved closer together to compensate for increased attenuation and crosstalk. For this reason, Chapter 4 emphasizes various forms of binary transmission. In radio systems, however, rigid transmission bandwidths are normally specified in the interest of establishing well defined, noninterfering channels. Maximum use of these channels often implies that multilevel signaling techniques are needed to achieve high data rates despite bandwidth restrictions. In fact, in the interest of ensuring a certain minimum utilization efficiency of the radio frequency spectrum, the FCC in the United States has established certain data rates that must be achieved by digital microwave radios [1]. Basically, these minimum rates specify approximately the same number of voice circuits (at 64 kbps) as are available in existing analog FM radios.

Recall that existing analog microwave radios use frequency modulation of FDM single-sideband signals. Since the frequency modulation process typically increases the effective bandwidth of each voice channel by a factor of 2 to 4 (see Tables 1.6 and 1.7), digital systems compete with 8 to 16 kHz per voice channel instead of 4 kHz. By using advanced modulation techniques, digital systems can provide the same number of 64 kbps voice circuits as the analog FM counterparts. However, the introduction of the AR-6A single sideband analog radio currently being field tested by the Bell System implies that digital radios will be unable to compete in terms of bandwidth efficiency unless lower rate digitization algorithms are used.

Over and above the minimum bit rates established by the FCC, competition among digital microwave manufacturers and the economics of maximizing the number of voice circuits per radio have stimulated the development of advanced digital modulation techniques, achieving even greater transmission rates. The first part of this chapter describes these basic modulation techniques and the transmission efficiencies they provide. The latter sections describe basic system design considerations for point-to-point microwave relay systems.

Information Density

A useful parameter for characterizing the bandwidth efficiency of a digital modulation system is the information density, defined as

$$\delta = \frac{R}{BW} \tag{6.1}$$

where R = data rate in bits per second
BW = bandwidth of the digital signal in hertz.

The units of information density are sometimes referred to loosely as bits per hertz. However, as defined in Equation 6.1, the units should be bits per second per hertz or simply bits per cycle. Since bits per second per hertz conveys the nature of information density more completely, it is the preferred unit.

The bandwidth factor in Equation 6.1 can be defined in a variety of ways. In theoretical studies the bandwidth of a signal is usually determined as the width of the ideal filter used to strictly bandlimit the signal (i.e. the Nyquist bandwidth). In practical systems, where the spectrum can never be strictly bandlimited, the bandwidth is more difficult to define. In radio systems a channel is usually specified with a certain minimum signal attenuation at the band edges. In this case, the information density is easily determined as the bit rate of a signal meeting these requirements divided by the alloted bandwidth.

When no particular emission specifications are provided, however, the bandwidth required by a digital signal must be defined in another more or less arbitrary manner. Using a 3 dB bandwidth is usually inappropriate since the signal spectrum may drop rather slowly and thereby allow significant amounts of energy to spill over into adjacent channels. Using a greater attenuation level to define the band edges would be more appropriate, but is still somewhat arbitrary. A generally more relevant criterion for practical systems defines the bandwidth of a signal as the channel spacing required to achieve a specified maximum level of interference into identical adjacent channels [2]. Yet another definition of bandwidth, commonly used in Europe, is the 99% power bandwidth.

The theoretical maximum information density for binary signaling is 2 bps/Hz for a two-level line code or 1 bps/Hz for a modulated double-sideband signal. If a four-level line code is used to achieve 2 bits per signal interval, the theoretical information density is 4 bps/Hz for the line code, or 2 bps/Hz for a double-sideband carrier signal. The information density of amplitude modulated signals can be doubled by using single-sideband transmission to effectively achieve the same efficiency as the line codes (baseband signals).

TABLE 6.1 INFORMATION DENSITIES REQUIRED BY FCC FOR
COMMON CARRIER MICROWAVE CHANNELS USING
64 kbps PER VOICE CIRCUIT

Frequency Band (gHz)	Channel BW (MHz)	Required Bit Rate (Mbps)	Information Density (bps/Hz)
2	3.5	6.144	1.8
4	20	73.7	3.7
6	30	73.7	2.5
11	40	73.7	1.8

As a practical example, the maximum guaranteed data rate on dial-up analog telephone lines is 4800 bps (usually with 2 bits per signal interval and a 2400 Hz signaling rate). Since the usable bandwidth of a telephone channel is approximately 3 kHz, dial-up lines provide an information density of 1.6 bps/Hz. Correspondingly, the maximum guaranteed data rate on leased analog lines is 9600 bps, implying an information density of 3.2 bps/Hz.*

As a starting point for digital microwave radios, Table 6.1 lists the minimum bit rates and corresponding information densities required in common carrier bands of the United States. Competition and compatibility with the digital hierarchy have led to realizations of even greater transmission efficiencies.

6.1 DIGITAL MODULATION

Digital transmission on a radio link differs from wire-line transmission in two important regards. First, the digital information must modulate a carrier in some manner to produce the radio frequency (RF) signal. In many cases the modulation process can be viewed as a special form of amplitude modulation by a nonreturn to zero (NRZ) line code signal. Thus the line code represents a baseband signal that amplitude modulates the carrier in the transmitter and is reproduced by demodulation in the receiver. Representing the modulation process in this manner has the advantage that the RF spectra can be determined by merely translating the line code (baseband) spectra to the selected carrier frequency.

Second, a radio link differs from wire-line transmission due to the necessity of strictly bandlimiting the transmitted signals to prevent interference into other channels. Although wire-line transmission links automatically filter the line signal to some extent, explicit filter require-

---

*Recently, 14.4 kbps voiceband modems have been introduced. These modems use very sophisticated digital signal processing and equalization techniques to achieve the higher data rate.

ments sometimes occur only in the receivers to reject as much noise as possible. Since radio links are *bandlimited* in the transmitter and *noise-filtered* in the receiver, the end-to-end filter function must be partitioned between the two ends. Figure 6.1 shows a block diagram of a radio link showing representative baseband and RF waveforms along with corresponding frequency spectrums. For modulation, Figure 6.1 shows multiplication of the carrier by the baseband waveform.

In the actual design of a digital radio, the modulation process must be designed in conjunction with the filter functions. For ease of description, however, modulation techniques are considered first. Later on, the filtering requirements of each type of modulation are discussed.

6.1.1 Amplitude Modulation

Historically, the simplest form of modulation to generate and detect is amplitude modulation (AM). A conceptual illustration of amplitude modulation is shown in Figure 6.2. The mathematical definition is

$$x(t) = [1 + a \cdot m_n(t)] \cos \omega_c t \qquad (6.2)$$

where a = modulation index $(0 < a \leqslant 1)$,
   $m_n(t)$ = an n-level, symmetric NRZ baseband signal normalized to a maximum amplitude of 1
   $\omega_c = 2\pi f_c$, the radian carrier frequency

Amplitude modulation is an example of a class of special modulation techniques referred to as "linear modulation." Linear modulation implies that the spectrum of the modulated signal is obtained by translating the baseband spectrum to the selected carrier frequency band. As shown in Figure 6.2, amplitude modulation by a two-level digital baseband signal essentially translates the $\sin(x)/x$ baseband spectrum up to the carrier frequency f_o. Other linear modulation techniques are: double-sideband modulation, single-sideband modulation, and vestigial-sideband modulation.

Inspection of Equation 6.2 or Figure 6.2 indicates that if 100% modulation is used $(a = 1)$, no carrier is produced for a logic 0. For obvious reasons, this form of amplitude modulation is often referred to as on/off keying, or "amplitude shift keying" (ASK). As shown in Figure 6.3, on/off keying can be obtained by direct multiplication of a carrier with a two-level (unipolar) line code described in Chapter 4.

Amplitude modulated signals are usually demodulated with a simple envelope detector. The cost-effectiveness of this detector is the basic reason that commercial analog broadcasting uses amplitude modulation. Unfortunately, the error performance of digital amplitude modulation in general, and envelope detection in particular, is inferior to other forms of digital modulation and detection. For this reason, amplitude

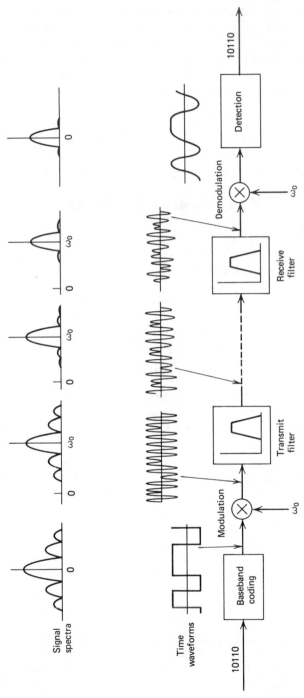

Figure 6.1 Simplified block diagram of end-to-end radio channel.

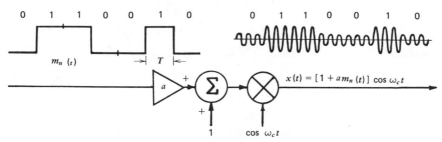

Figure 6.2 Digital binary amplitude modulation.

modulation is used only where the cost of the receiver is a significant consideration. Digital microwave links use other forms of modulation and demodulation to minimize the error rate for a given signal-to-noise ratio.

Conventional amplitude modulation provides suboptimum error performance for two basic reasons. First, if $a < 1$, a discrete (information-less) spectral line occurs at the carrier frequency. Although the existence of this spectral line simplifies carrier recovery, it increases the transmitter power without adding to discrimination between information signals.

With 100% modulation (on/off keying) no line spectra are produced, but the system is still inefficient in its use of transmitted power. As discussed in Chapter 4 for two-level line coding, the maximum use of transmitted power is achieved when one signal is the negative of the other. Thus, a second deficiency of amplitude modulation arises because a 0 signal is not the exact negative of a 1 signal. To achieve optimum performance, a symmetric two-level baseband signal should directly modulate (multiply) the carrier. As shown in Figure 6.4, this form of modulation produces two identical signals except for a 180° phase reversal. Hence this form of modulation is sometimes referred to as phase reversal keying (PRK), or more often, two-level phase shift keying (2-PSK). In classical modulation terminology, PRK signaling belongs to the class of double-sideband modulation systems. However, in

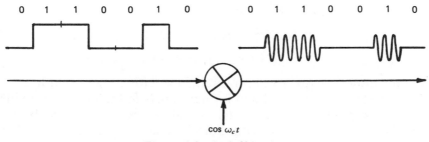

Figure 6.3 On/off keying.

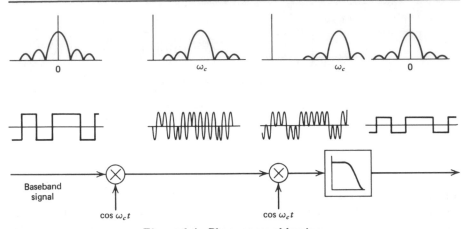

Figure 6.4   Phase reversal keying.

some digital modulation systems closely related to PRK (or 2-PSK) direct multiplication of the carrier by a digital baseband signal is referred to as amplitude modulation. Thus in keeping with this terminology, PRK can be considered as a special form of amplitude modulation. Notice that PRK can not, however, be detected by envelope detection. Instead, a PRK signal must be detected by comparing it to a coherent carrier reference.

Coherent detection involves comparing the incoming signal to a local carrier synchronized in phase to that used at the transmitter. With conventional amplitude modulation there is no information in the phase. With PRK signaling, however, all of the information is in the phase. The use of a coherent reference for PRK signals allows optimum (antipodal) error rate performance. The transmitter must modulate the carrier in a phase-coherent fashion to allow coherent detection in the receiver.

The basic equation defining a PRK signal is

$$x(t) = m_2(t) \cdot \cos \omega_c t \qquad (6.3)$$

where $m_2(t)$ is the binary digital baseband signal [$m_2(t) = +1$ for a one and $m_2(t) = -1$ for a zero]. The demodulation equation is

$$y(t) = x(t) \cdot [2 \cos \omega_c t]$$

$$= [m_2(t) \cdot \cos \omega_c t] \cdot [2 \cos \omega_c t] \qquad (6.4)$$

$$= m_2(t) + m_2(t) \cdot \cos 2\omega_c t$$

where the double frequency term is removed with a low-pass filter.

As indicated in Figure 6.4 and Equation 6.4, the coherent demodulation process produces a symmetric two-level signal that can be processed further to detect the data. Since the demodulated baseband signal is, in essence, equivalent to a line code, it must include timing considerations as discussed in Chapter 4. In particular, there must be enough signal transitions in the baseband signal to allow recovery of a sample clock. However, the baseband signal need not preclude dc wander since dc levels are translated to the carrier frequency and passed adequately by the double-sideband system.*

6.1.2 Frequency Shift Keying

In addition to an inefficient use of signal power, conventional amplitude modulated signals have one other undesirable characteristic. By definition, an amplitude modulated signal uses multiple signal levels. Hence AM is quite vulnerable to signal saturations that narrow the distance between amplitude levels. A common source of saturation in a radio system occurs in the output power amplifier of the transmitter. In some cases output amplifiers are operated at less than maximum power to eliminate saturation and other nonlinearities—so they can accomodate amplitude modulated signals [3].

Angle modulated systems: frequency modulation (FM) or phase modulation (PM) use constant amplitude signals not adversely affected by signal saturations. Hence FM and PM can be transmitted at higher power levels than AM systems. The ability to use saturating power amplifiers is one of the reasons why FM was originally chosen for analog microwave radios. This section discusses digital frequency modulation, commonly referred to as frequency shift keying (FSK). The next section discusses digital phase modulation, commonly referred to as phase shift keying (PSK). Both systems provide a constant amplitude signal. Systems using constant amplitude carriers are also referred to as constant envelope systems.

The general expression for an n-ary FSK signal is

$$x(t) = \cos\left[\left(\omega_c + \frac{m_n(t) \cdot \Delta\omega}{2}\right) t\right] \tag{6.5}$$

where ω_c is the radian center frequency, m_n is a symmetric n-level NRZ digital baseband signal, and $\Delta\omega$ is the radian difference frequency between signals.

A typical binary FSK signal is shown in Figure 6.5 along with a simple (but low performance) means of implementing the modulator and demodulator. The modulator is a voltage controlled oscillator (VCO) that is biased to produce the center frequency when no modulation is

*A single-sideband system does not pass dc energy. Thus if single-sideband modulation is used, the baseband signal must exclude dc energy from its spectrum. Some double-sideband systems might also require the elimination of baseband dc energy so that a carrier tone can be inserted into the center of the passband without affecting the signal.

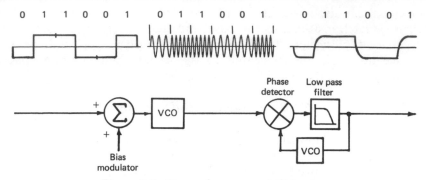

Figure 6.5   Binary frequency shift keying.

applied. The amplitude a of the symmetric two-level baseband signal produces a frequency deviation of $+ \Delta\omega/2$ for a 1 and $- \Delta\omega/2$ for a 0.

The demodulator is implemented as a phase-locked-loop: a VCO, a phase detector, and a loop filter. The phase detector measures the difference in phase between the FSK signal and the VCO output. A positive voltage is produced when the receive signal leads the VCO, and a negative voltage is produced otherwise. After being filtered to minimize the effects of noise, the phase detector output drives the VCO in such a way as to reduce the phase difference. Ideally, the input control voltage of the demodulator VCO will be identical to the input of the modulator VCO. The loop filter, however, necessarily slows the demodulator response to minimize the effects of noise.

Frequency shift keying generally provides poorer error performance than phase shift keying, particularly for multilevel signaling in a confined bandwidth. For this reason, FSK has not been used in digital microwave applications where high performance and high information densities are desired. The most prevalent usage of FSK is in low rate, low cost, asynchronous modems for data transmission over the analog telephone network.

Minimum Shift Keying

One particular form of frequency shift keying receiving considerable attention recently is minimum shift keying (MSK) [4]. Basically, MSK is binary FSK with the two signaling frequencies selected so that exactly 180° difference in phase shift exists between the two frequencies in one signal interval. In this manner MSK produces a maximum phase differential at the end of an interval using a *minimum* difference in signaling frequencies. Furthermore, an MSK signal maintains continuous phase at signaling transitions. For this reason MSK belongs to a class of FSK signals referred to as continuous phase frequency shift keying (CPFSK). Figure 6.6 depicts a representative MSK waveform. Notice

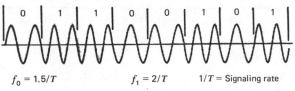

$f_0 = 1.5/T$ $f_1 = 2/T$ $1/T$ = Signaling rate

Figure 6.6   Minimum shift keying.

that there is exactly one-half cycle difference between a 1 signal and a 0 signal.

A mathematical expression for MSK signaling can be derived from Equation 6.5 where $m_n(t) = m_2(t)$ is a symmetric binary NRZ signal, and $\Delta\omega = \pi/T$. The result for any particular signal interval is

$$x(t) = \cos\left(\omega_c t + \frac{\pi t}{2T} + \phi_0\right) \qquad \text{(logic 1)}$$

$$= \cos\left(\omega_c t - \frac{\pi t}{2T} + \phi_0\right) \qquad \text{(logic 0)}$$

(6.6)

where $1/T$ is the signaling rate, and ϕ_0 is the phase at the beginning of the signal interval ($\pm\pi$ in Figure 6.6).

The main attraction of minimum shift keying is its comparatively compact spectrum. Furthermore, with an appropriate means of detection, MSK can provide optimum error performance in terms of the energy-per-bit to noise-density ratio (E_b/N_0). The expression for the power spectral density of the MSK signal defined in Equation 6.6 is:

$$S(\omega) = 32\pi^2 T \left(\frac{\cos z}{\pi^2 - 4z^2}\right)^2$$

(6.7)

where $z = |\omega - \omega_c|T$. The frequency spectrum of an MSK signal is plotted in Figure 6.7 where it is compared to the frequency spectrum of PRK (2-PSK) signaling with the same data rate. Notice that the MSK spectrum is more compact and has its first spectral null at $3/4T$ instead of $1/T$ for PRK.*

6.1.3   Phase Shift Keying

The second category of angle modulated, constant envelope signals is referred to as phase shift keying (PSK). Actually, we already discussed one form of PSK when phase reversal keying (PRK) was introduced. PRK is more commonly referred to as 2-PSK, indicating that each signal

---

*MSK is actually more closely related to 4-PSK and therefore is compared to it in a later section of this chapter.

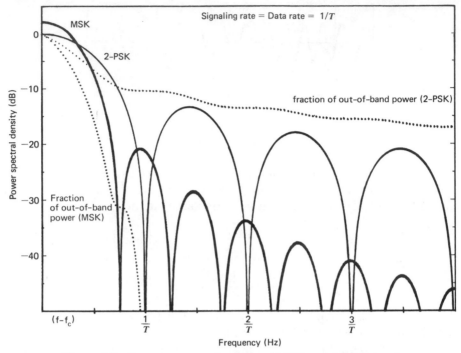

Figure 6.7 Power spectra of unfiltered MSK and 2-PSK signals.

interval uses one of two phases that are 180° apart to encode binary data. Multiple-phase phase shift keying is also possible. 4-PSK (also called QPSK) and 8-PSK are the most common examples of multiple phase PSK.

Phase shift keying has become the most popular modulation technique for intermediate information density, high performance applications. The popularity is primarily due to its: constant envelope, insensitivity to level variations, and good error performance. Both 2-PSK and 4-PSK theoretically provide optimum error performance in terms of a signal-to-noise ratio (E_b/N_0).

A general expression for n-ary phase shift keying is provided in Equation 6.8. This expression assumes 100% modulation is employed. That is, the phase shift from one interval to the next can range anywhere from $-180°$ to $+180°$. It is possible to devise PSK systems with lower modulation indices that allow only transitions to neighboring phases.

$$x(t) = \cos\left(\omega_c t + \frac{m_n(t) \cdot \Delta\phi}{2}\right) \tag{6.8}$$

where $\Delta\phi = 2\pi/n$ is the separation between adjacent signal phases, and $m_n(t)$ is a symmetric n-level NRZ baseband signal with levels: ± 1, $\pm 3, \ldots$.

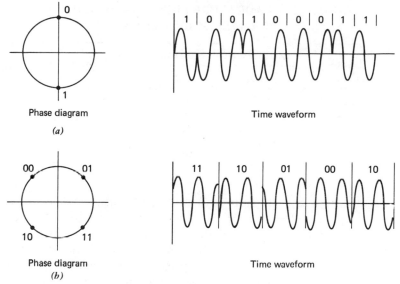

Phase diagram

(a)

Time waveform

Phase diagram

(b)

Time waveform

Figure 6.8 Phase shift keying. (a) 2-PSK. (b) 4-PSK.

Examples of typical 2-PSK and 4-PSK waveforms are shown in Figure 6.8. The signaling rate for the 4-PSK system is shown to be exactly one-half the 2-PSK signaling rate so that equal data rates are provided. The same figure also shows corresponding phasor diagrams of the signaling phases of a cosine wave as defined in Equation 6.8. Other phase orientations are possible. The particular phases shown, however, are convenient for later discussions relating 4-PSK systems to other types of digital modulation.

Quadrature Signal Representations

Despite the somewhat exotic sounding name, quadrature signal representations are a very convenient and powerful means of describing PSK signals, and many other digitally modulated signals. Quadrature signal representations involve expressing an arbitrary phase sinusoidal waveform as a linear combination of a cosine wave and a sine wave with zero starting phases. The derivation of this representation is provided by the trigonometric identity:

$$\cos (\omega_c t + \phi) = \cos \phi \cdot \cos \omega_c t - \sin \phi \cdot \sin \omega_c t \qquad (6.9)$$

Notice that $\cos \phi$ and $\sin \phi$ are constants over a signaling interval, and hence represent coefficients for expressing $\cos (\omega_c t + \phi)$ as a linear combination of the signals $\cos \omega_c t$ and $\sin \omega_c t$. Since $\cos \omega_c t$ and $\sin \omega_c t$ are 90° out of phase with respect to each other, they are orthogonal in a phasor diagram and hence are said to be "in quadrature."

**TABLE 6.2 QUADRATURE SIGNAL COEFFICIENTS
FOR 4-PSK MODULATION**

Data	Quadrature Coefficients		Composite
Values	$\cos \omega_c t$	$\sin \omega_c t$	Signal
01	0.707	-0.707	$\cos (\omega_c t + \pi/4)$
00	-0.707	-0.707	$\cos (\omega_c t + 3\pi/4)$
10	-0.707	0.707	$\cos (\omega_c t - 3\pi/4)$
11	0.707	0.707	$\cos (\omega_c t - \pi/4)$

In essence, $\cos \omega_c t$ and $\sin \omega_c t$ represent basis vectors in a two-dimensional phasor diagram. The cosine signal is usually referred to as the in-phase or I signal, and the sine signal is referred to as the out-of-phase or Q signal. Table 6.2 provides an example of quadrature signal representations for the 4-PSK signals presented in Figure 6.8. Table 6.3 provides a corresponding specification for an 8-PSK system using signal phases as provided in the phasor diagram of Figure 6.9. The phasor diagram assumes counterclockwise rotation, and hence the sine function lags the cosine function by 90°.

Most of the rest of this chapter relies heavily on quadrature signal representations to describe various modulation concepts, analyses, and implementations.

Modulator Implementations

A variety of techniques are possible for implementing PSK modulators. As mentioned when discussing PRK modulation, a 2-PSK modulator can be implemented by merely inverting the carrier (multiplying by -1)

**TABLE 6.3 QUADRATURE SIGNAL COEFFICIENTS
FOR 8-PSK MODULATION**

Data	Quadrature Coefficients		Composite
Values	$\cos \omega_c t$	$\sin \omega_c t$	Signal
011	0.924	-0.383	$\cos (\omega_c t + \pi/8)$
010	0.383	-0.924	$\cos (\omega_c t + 3\pi/8)$
000	-0.383	-0.924	$\cos (\omega_c t + 5\pi/8)$
001	-0.924	-0.383	$\cos (\omega_c t + 7\pi/8)$
101	-0.924	0.383	$\cos (\omega_c t - 7\pi/8)$
100	-0.383	0.924	$\cos (\omega_c t - 5\pi/8)$
110	0.383	0.924	$\cos (\omega_c t - 3\pi/8)$
111	0.924	0.383	$\cos (\omega_c t - \pi/8)$

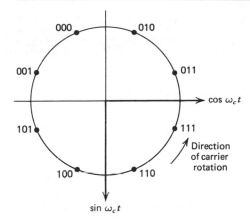

Figure 6.9 Phasor diagram of 8-PSK signal.

for a logic 0 and by not inverting for a logic 1. Some of the basic techniques used for generating multiple phase PSK signals are the following:

1 Synthesis of the desired waveforms using digital signal processing at suitably low carrier frequencies (as in voiceband modems).

2 Generating multiple phases of a single carrier and selecting between the phases depending on the data values.

3 Using controlled delays selected through a switching arrangement to provide the desired phase shifts. Delays are often used to generate the separate signals in method 2.

4 Generating the PSK signals as a linear combination of quadrature signals.

None of the foregoing techniques represents a direct implementation of multiphase PSK modulation as defined in Equation 6.8. In order to implement Equation 6.8 directly, a device is needed that produces carrier phase shifts in direct proportion to the levels in the baseband signal $m_n(t)$. Except for the special case of 2-PSK where a multiplier can be used to produce $\pm 180°$ phase shifts, such devices do not exist.

Direct modulation by baseband signals can produce PSK signals if a quadrature signal implementation is used as indicated in method 4. Actually, two multilevel baseband signals need to be established: one for the inphase (I) signal and one for the out-of-phase (Q) signal. These baseband signals are referred to as $m_I(t)$ and $m_Q(t)$ for the I and Q signals, respectively. The levels chosen for the two baseband signals correspond to the coefficients needed to represent a PSK signal as a linear combination of the I and Q signals. For example, Figure 6.10 shows how an 8-PSK signal, defined in Table 6.3, can be generated by adding two amplitude modulated quadrature signals. A block diagram of the corresponding implementation is provided in Figure 6.11. This particu-

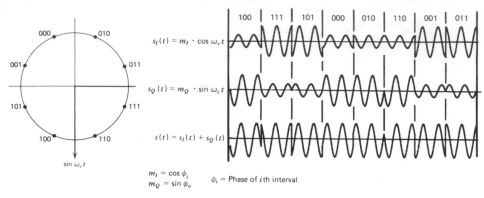

Figure 6.10 Generation of 8-PSK signals by superposition of quadrature amplitude modulated signals.

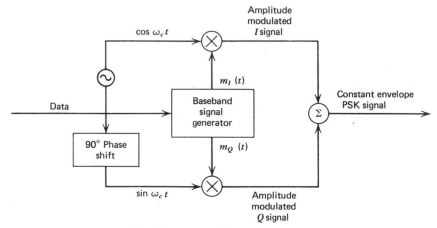

Figure 6.11 Generalized PSK modulator structure.

lar form of modulator is chosen, not so much as a recommendation for actual implementation, but because it demonstrates important modulation concepts and is useful in analyzing the spectrum requirements of PSK signaling.

Demodulator Implementation

Owing to the constant envelope*, all PSK systems must be detected with the aid of a local reference. For 2-PSK systems, the ideal reference is coherent with one of the two possible phases. When this reference is multiplied (mixed) with an in-phase signal, a maximum positive output is obtained. When multiplied by the opposite phase, a maximum nega-

---

*Discussions to this point have not considered the effects of filtering on constant envelope signals. A heavily filtered PSK signal does not have a constant envelope. However, as long as filtering occurs after channel nonlinearities (e.g. power amplifiers), the most harmful effect of a nonlinearity, spectrum spreading, is avoided.

tive output is obtained. In this manner, 2-PSK provides antipodal performance when a local coherent reference is established. The demodulation process for 2-PSK is presented in Figure 6.4 and Equation 6.4, where the mixing and filtering process effectively recovers the baseband signal $m_2(t)$.

The combination of a mixer and a low-pass filter shown in Figure 6.4 is generally referred to as phase detector. The phase detection property is represented mathematically as:

$$y_I(t) = \text{low pass } \{\cos(\omega_c t + \phi) \cdot 2 \cdot \cos \omega_c t\}$$

$$= \text{low pass } \{\cos \phi + \cos \phi \cdot \cos 2\omega_c t - \sin \phi \cdot \sin 2\omega_c t\}$$

$$= \cos \phi \qquad\qquad\qquad\qquad (6.10)$$

where low pass $\{ \cdot \}$ is a low-pass filter function designed to remove twice carrier terms.

When detecting 2-PSK modulation, a single phase detector indicates whether the received phase is closer to $0°$ or to $180°$. The desired information is directly available as the polarity of the phase detector output $\cos \phi$. In multiphase systems, however, the information provided by a single phase detector is inadequate for two reasons. First, a measure of $\cos \phi$ provides no information as to whether ϕ is positive or negative. Second, the output of the phase detector is proportional to the signal amplitude as well as to $\cos \phi$. Hence the magnitude of the phase detector output is meaningless unless referenced to the signal amplitude.

Both of the aforementioned problems are overcome if a second mixer and filter measures the phase with respect to a different reference. As expected, the best performance is obtained when the second reference is orthogonal to the first. If $y_I(t)$ is the output of the first phase detector defined in Equation 6.10, then the second phase detector output $y_Q(t)$ becomes:

$$y_Q(t) = \text{low pass } \{\cos(\omega_c t + \phi) \cdot 2 \cdot \sin \omega_c t\}$$

$$= -\sin \phi \qquad\qquad\qquad\qquad (6.11)$$

The second phase detector not only resolves the positive/negative phase ambiguity, but it also eliminates the need to establish an amplitude reference. All decisions can be based on the polarity of a phase detector output and not on the magnitude. As a first example, consider detection of the 4-PSK signals defined in Table 6-2. Notice that the first data bit in a pair is 0 when the phase angle is positive ($\pi/4$ or $3\pi/4$), and a 1 otherwise. Hence the first data bit is completely specified by the polarity of $\sin \phi$: the output of the second phase detector $y_Q(t)$. Similarly, the second data bit is a 1 when the phase is $\pm\pi/4$, indicating that the polarity of $y_I(t)$ provides all information necessary to detect the second bit. The basic implementation of a 4-PSK (QPSK) demodulator/detector is shown in Figure 6.12. Also shown is a 4-PSK modulator to emphasize the relationships between the modulator and demodulator. A 4-PSK system is presented specifically because it is a popular system, and because it is a useful foundation for describing other modulation techniques.

An important concept to notice in the 4-PSK system shown in Figure 6.12 is that there are, in essence, two separate data streams. The modulator divides the incoming bit stream so that bits are sent alternately to the in-phase modulator I and the out-of-phase modulator Q. These same bits appear at the output of the respective phase detectors in the demodulator where they are interleaved back into a serial bit stream. In this manner two essentially independent binary PSK channels are established using the orthogonal carriers $\cos \omega_c t$ and $\sin \omega_c t$. The channels are usually referred to as the I and Q channels, respectively. This technique of establishing two channels within an existing bandwidth is sometimes referred to as "quadrature multiplexing."

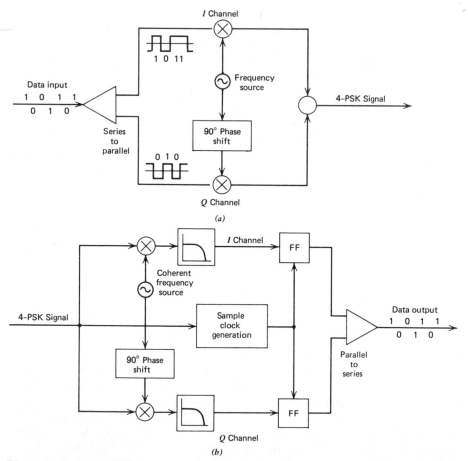

Figure 6.12 Four-PSK modulator/demodulator structure. (*a*) Modulator. (*b*) Demodulator/detector.

As long as the carriers in the modulator and the references in the demodulator are truly orthogonal (coherence is maintained for both channels in the receiver), the I and Q channels do not interfere with each other. Any amount of misalignment in these relationships causes crosstalk between the two quadrature channels. Crosstalk also arises if there is unequalized phase distortion in the transmission channel.

At first thought it might seem that quadrature multiplexing increases the capacity of a given bandwidth by a factor of 2. It must be remembered, however, that the binary PSK signal on each quadrature channel is a double-sideband signal. Hence the bandwidth of the channel, *without quadrature multiplexing*, is only 50% utilized in comparison to a single-sideband system. When quadrature channels are used, a single-sideband operation is no longer possible since the sideband separation process destroys the orthogonality of the two signals. In essence, quadrature multiplexing only recovers the loss of capacity incurred by the double-sideband spectrum. In fact, some single-sideband modulators [5] possess a remarkable resemblance to the QPSK modulator shown in Figure 6.12.

Demodulation and detection of higher-level PSK systems is complicated by the fact that the use of only two references does not provide a simple means of detecting all data bits. There are two basic ways in which simple decisions (positive versus negative) can be established to detect all data. One method is to establish more references in the receiver and measure the phase of the received signal with respect to the additional references. The second method is to use only two references and related phase detectors and generate all additional measurements as linear combinations of the first two.

As an example of the first method, consider the 8-PSK system defined in Table 6.3 and shown in Figure 6.9. Optimum detection for this system can be obtained if two additional phase detector outputs are provided with respect to references at $+\pi/4$ and $-\pi/4$. The two new references are designated A and B as shown in Figure 6.13.

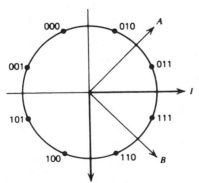

Figure 6.13 Receiver references for 8-PSK detection.

The outputs of four phase detectors corresponding to the four references are determined as follows:

$$y_Q = \text{low pass } \{\cos(\omega_c t + \phi) \cdot 2 \cdot \sin \omega_c t\} = -\sin \phi$$

$$y_B = \text{low pass } \left\{\cos(\omega_c t + \phi) \cdot 2 \cdot \sin\left(\omega_c t + \frac{\pi}{4}\right)\right\}$$

$$= 0.707 \cos \phi - 0.707 \sin \phi$$

$$y_I = \text{low pass } \{\cos(\omega_c t + \phi) \cdot 2 \cdot \cos \omega_c t\} = \cos \phi \qquad (6.12)$$

$$y_A = \text{low pass } \left\{\cos(\omega_c t + \phi) \cdot 2 \cdot \cos\left(\omega_c t + \frac{\pi}{4}\right)\right\}$$

$$= 0.707 \cos \phi + 0.707 \sin \phi$$

As an aid in determining the appropriate decision logic, these equations have been evaluated for each of the eight possible signal phases and listed in Table 6.4. Examination of Table 6.4 indicates that the first data bit is 1 whenever y_Q is positive. Similarly, the second data bit is 1 whenever y_I is positive. The third data bit is a 1 whenever y_A, y_I, and y_B are all positive, or when they are all negative. Hence the third data bit is determined as a logical combination of phase detector outputs. In summary,

$$D_1 = Q$$

$$D_2 = I \qquad (6.13)$$

$$D_3 = AIB + \overline{AIB}$$

where D_i is the ith data bit and Q, I, A, B are logic variables representing positive outputs from y_Q, y_I, y_A, and y_B respectively.

Another method of detecting 8-PSK signals, which avoids two extra references and phase detectors, is revealed in the phase detector Equa-

TABLE 6.4 EIGHT-PSK PHASE DETECTOR OUTPUTS

Data	Phase	y_Q	y_B	y_I	y_A
011	$\pi/8$	-0.383	0.383	0.924	0.924
010	$3\pi/8$	-0.924	-0.383	0.383	0.924
000	$5\pi/8$	-0.924	-0.924	-0.383	0.383
001	$7\pi/8$	-0.383	-0.924	-0.924	-0.383
101	$-7\pi/8$	0.383	-0.383	-0.924	-0.924
100	$-5\pi/8$	0.924	0.383	-0.383	-0.924
110	$-3\pi/8$	0.924	0.924	0.383	-0.383
111	$-\pi/8$	0.383	0.924	0.924	0.383

tion 6.12. The extra measurements y_A and y_Q can be determined as

$$y_A = 0.707\, y_I - 0.707\, y_Q$$
$$y_B = 0.707\, y_I + 0.707\, y_Q$$

(6.14)

Hence the y_A and y_B measurements can be obtained as linear combinations of the quadrature channel phase measurements y_I and y_Q, and no additional phase detectors are required. The resulting 8-PSK demodulator/detector is shown in Figure 6.14. Notice that implementation of the linear combinations in Equation 6.14 can ignore the magnitude of the 0.707 multipliers since only the sign of the result is needed.

The linear combinations in Equation 6.14 essentially represent a rotation of the quadrature channel basis vectors through an angle of $\pi/4$ radians. By changing the angle of rotation, the linear combinations needed for other phase measurements are easily determined. Hence all phase detector based demodulators can be implemented with two phase detectors and as many linear combinations as needed. The general equation for the linear combinations produced by a rotation of α radians is

$$y_A = y_Q \cdot \cos\alpha - y_I \cdot \sin\alpha$$
$$y_B = y_Q \cdot \sin\alpha + y_I \cdot \cos\alpha$$

(6.15)

Reference and Clock Recovery

The demodulators discussed in the preceding paragraphs all require a local, phase coherent, carrier reference for their operation. Furthermore, multiple phase systems require at least one more reference in quadrature with the first one. Recovery of any reference is complicated by the fact that phase shift keying is a double-sideband *suppressed* carrier modulation system. In other words, there is no discrete spectral line

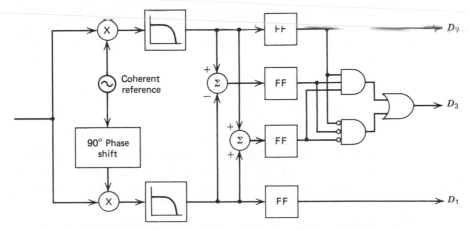

Figure 6.14 Eight-PSK demodulator/detector. Using only two references.

at the carrier frequency as there is with some other types of modulation. In fact, the FCC in the United States has ruled specifically that no spectral lines are allowed in the transmitted signal. The absence of a spectral line at the carrier frequency is overcome by using one of several nonlinear processing techniques [6]. After one coherent reference is established, a quadrature reference is obtained by delaying or differentiating the first.

Clock recovery is obtained in several ways. Once carrier recovery is accomplished and demodulation occurs, the clock can be obtained by locking onto transitions in the baseband signal. Baseband clock recovery occurs as though the signal has never been modulated and demodulated. In contrast, the clock can sometimes be recovered directly from the modulated signal. If a PSK signal is heavily filtered, the envelope is not constant and, in fact, contains amplitude modulation at the signaling frequency. Hence the clock can be recovered, independently of signal demodulation, by envelope detecting a heavily filtered PSK signal [7]. With either method a certain minimum density of clock transitions is required to maintain clock synchronization.

Differential Detection

As an alternative to recovering a coherent reference, some systems merely compare the phase in the present interval to the phase in the previous interval. The signal received in the previous interval is delayed for one signal interval and is used as a reference to demodulate the signal in the next interval. Assuming that the data has been encoded in terms of phase shift, instead of absolute phase positions, we can decode the data properly. Hence this technique, referred to as "differential detection," inherently requires differential encoding.

In general, PSK systems require differential encoding since the receivers have no means of determining whether a recovered reference is a sine reference or a cosine reference. Furthermore, the polarity of the recovered reference is ambiguous. Thus error probabilities for PSK systems are doubled automatically because of the differential encoding process. Differential detection, on the other hand, implies an even greater loss of performance since a noisy reference is used in the demodulation process. Typically, differential detection imposes a penalty of 1 to 2 dB in signal-to-noise ratio [5].

PSK Spectra

By far the easiest way to determine the spectrum of a PSK signal is to analyze the baseband waveforms applied to the quadrature channels. Owing to the orthogonality of the two channels, the signals are uncorrelated, and the composite spectrum is merely the sum of the individual (identical) spectra.

In either 2-PSK or 4-PSK systems the baseband signal is a symmetric

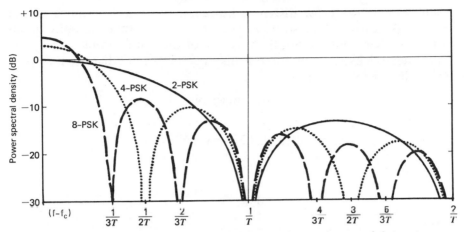

Figure 6.15 Spectrums of unfiltered PSK signals carrying equal data rates.

two-level NRZ waveform. The corresponding spectrum is the common $\sin(x)/x$ spectrum shown in Figure 4.2. High-level systems (8-PSK or greater) use symmetric multilevel NRZ baseband signals similar to that shown in Figure 4.16. The multiphase PSK baseband signal is somewhat different since unevenly spaced levels as defined in Table 6.3 are used.

As mentioned in Chapter 4, a multilevel NRZ signal has the same spectrum as a two-level signal. Hence all conventional PSK systems produce a spectrum that follows the $\sin(x)/x$ response defined in Equation 4.1, but translated to the carrier frequency.* Figure 6.15 shows the PSK spectrum for two-, four-, and eight-phase systems designed to provide the same data rate. Hence the higher-level systems signal at lower rates and have proportionately narrower spectra.

PSK Error Performance

The error performance of any digital modulation system is fundamentally related to the distance between points in a signal space diagram. For example, a 2-PSK system, as represented in the phase diagram of Figure 6.8, is capable of optimum error performance since the two signal points have maximum separation for a given power level (radius of the circle). In other words, one 2-PSK signal is the exact negative of the other. Hence 2-PSK modulation provides antipodal error performance as defined in Chapter 4.

The error performance of a multiphase PSK system is easily compared to a 2-PSK system by determining the relative decrease in the error distance (voltage output of a properly referenced phase detector).

*The baseband levels must be uncorrelated to each other to produce a $\sin(x)/x$ spectrum. If phase transitions from one interval to the next are restricted in some manner, the baseband levels are correlated, and a different spectrum results.

In addition to the error distance, however, the relative values of the noise bandwidths must also be considered. (Recall that the noise bandwidth effectively determines the variance of the noise samples.)

The general expression for the distance between adjacent points in a multiphase PSK system is

$$d = 2 \sin (\pi/N) \tag{6.16}$$

where N is the number of phases.

A general expression of the bit error probability (or bit error rate) of an N-phase PSK system is derived in Appendix C as

$$P_E = \frac{1}{\log_2 N} \, erf(z) \tag{6.17}$$

where

$$z = \sin (\pi/N) \, (\log_2 N)^{1/2} \left(\frac{E_b}{N_0}\right)^{1/2}$$

Equation 6.17 reveals that, with respect to E_b/N_0, 4-PSK provides the same error performance as does 2-PSK. Thus, as mentioned earlier, both systems provide optimum performance, but 4-PSK utilizes half as much bandwidth. The 4-PSK system has an error distance that is 3 dB smaller than the error distance of 2-PSK. However, the shorter error distance is offset by a 3 dB decrease in the noise bandwidth (indicating a 3 dB reduction in noise power at the detector). In order for the 4-PSK system to have the same noise power at the detector it would have to experience a 3 dB greater noise spectral density. Hence conventional signal-power to noise-power ratios (SNRs) can be misleading parameters for comparing digital modulation systems. However, as mentioned in Chapter 4, error rate performances in terms of SNRs are desired when determining the effects of interference or when specifying error rates with respect to measurable quantities. Figure 6.16 displays the error performance of 2-, 4-, 8-, 16-, and 32-PSK as a function of E_b/N_0. Appendix C provides the relationships needed for error rates in terms of SNR.

6.1.4 Quadrature Amplitude Modulation

As described previously, a convenient means of representing phase shift keying with four or more phases involves the use of quadrature signals. In the case of 4-PSK, the quadrature signals represent two separate channels by virtue of the independence of the baseband signals for each quadrature channel. In higher-level PSK systems, the level of a baseband signal for the I channel is not independent of the baseband level for the Q channel (see Table 6.3 or Figure 6.10). After the baseband signals

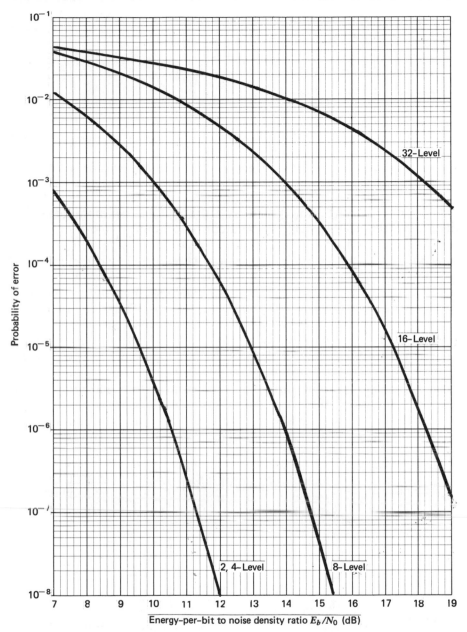

Figure 6.16 Error rates of PSK modulation systems.

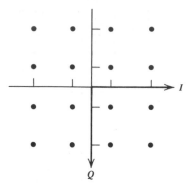

Figure 6.17 Signal constellation of 16-QAM modulation.

have been established, however, the quadrature channel modulation and demodulation processes are independent for all PSK systems.

Quadrature amplitude modulation (QAM) can be viewed as an extension of multiphase PSK modulation wherein the two baseband signals are generated independently of each other. Thus two completely independent (quadrature) channels are established including the baseband coding and detection processes. In the special case of two levels (± 1) on each channel, the system is identical to 4-PSK and is usually referred to as such. Higher-level QAM systems, however, are distinctly different from the higher-level PSK systems. Figure 6.17 shows a signal constellation of a 16-QAM system obtained from four levels on each quadrature channel. The dots represent composite signal points while the hatch marks on the axes represent amplitude levels in each quadrature channel. Figure 6.18 shows a basic QAM modulator and demodulator structure along with a representative waveform for 16-QAM.

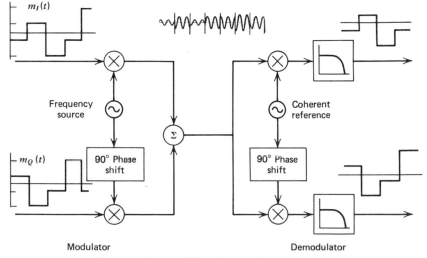

Figure 6.18   QAM modulator/demodulator.

Notice that, in contrast to PSK signals, the QAM signal shown in Figure 6.19 does not have a constant envelope. A constant envelope is maintained with PSK modulation by restricting the combination of levels on the quadrature channels. A QAM system does not restrict the combinations since the levels on each channel are selected independently. Thus QAM systems can not generally be used with saturating devices.

The spectrum of a QAM system is determined by the spectrum of the baseband signals applied to the quadrature channels. Since these signals have the same basic structure as the baseband PSK signals, QAM spectrum shapes are identical to PSK spectrum shapes with equal numbers of signal points. Specifically, 16-QAM has a spectrum shape that is identical to 16-PSK, and 64-QAM has a spectrum shape identically to 64-PSK.

Even though the spectrum shapes are identical, the error performances of the two systems are quite different. With large numbers of signal points, QAM systems always outperform PSK systems. The basic reason is that the distance between signal points in a PSK system is smaller than the distance between points in a comparable QAM system. Figure 6.19 compares the signal points of a 16-QAM system with the signal set of a 16-PSK system using the same peak power.

The general expression for the distance between adjacent signal points in a unit peak amplitude QAM system with L levels on each axis is

$$d = \frac{\sqrt{2}}{(L-1)} \qquad (6.18)$$

Equations 6.16 and 6.18 reveal that an n-ary QAM system has an advantage over an n-ary PSK system with the same peak power level. In terms of average power levels, the QAM system has an even greater advantage. The following equation, derived in Appendix C, provides a

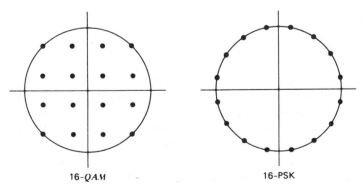

16-QAM   16-PSK

Figure 6.19 Comparison of 16-QAM and 16-PSK signal sets.

general expression for the peak-to-average ratio of a QAM system:

$$\frac{\text{peak power}}{\text{average power}} = \frac{L(L-1)^2}{2 \sum_{i=1}^{L/2} (2i-1)^2} \tag{6.19}$$

where the QAM system has L^2 total points.

EXAMPLE 6.1

Determine the error performance of a 16-PSK system relative to a 16-QAM system with the same peak power level. Also determine the relative performance with respect to identical average powers.

Solution. Since the two systems provide identical numbers of signal points, they signal at the same rate and require the same bandwidth for a given data rate. Thus the relative error performance is completely determined by the relative distances between signal points. (When different rates are used, the effect of different noise bandwidths in the receivers must also be considered.) Evaluating Equations 6.16 and 6.18 indicates that 16-QAM has a 1.64 dB advantage over 16-PSK for a given peak power. Equation 6.19 indicates that 16-QAM has a peak-to-average power ratio of 2.55 dB. Since PSK systems have unity peak-to-average ratios, the advantage of a 16-QAM system over a 16-PSK is 4.19 dB for equal average powers.

The theoretical error rate equation for QAM modulation, derived in Appendix C, is identical to the multilevel baseband error rate equation with the same number of levels. This equation is used to produce error rate curves for 4-, 16-, and 64-QAM modulation in Figure 6.20. Table 6.5 compares the most common forms of digital modulation used in point-to-point microwave radio systems.

Offset Keying

Owing to the interdependence of the baseband signals in high-level PSK modulation, the signal transitions on the quadrature channels necessarily coincide. However, since QAM systems independently modulate the quadrature channels, they are not constrained to align the signal intervals on the two channels. When the signal intervals overlap each other by 50% (Figure 6.21), the mode of operation is referred to as offset keying. Offset keying is commonly used on 4-QAM systems (more commonly referred to as 4-PSK or QPSK systems). The main advantage of offset keying lies in the ability of the reference recovery circuitry to become synchronized to the incoming carrier at lower signal-to-noise ratios than conventional (aligned) QAM and QPSK systems [8, 9].

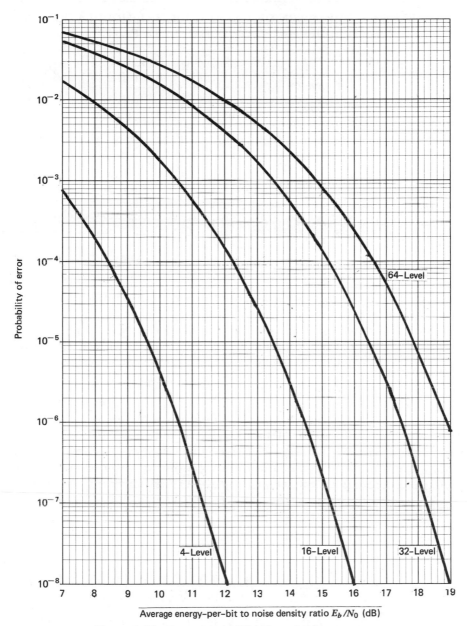

Figure 6.20 Error rates of QAM modulation systems.

TABLE 6.5   COMPARISON OF VARIOUS DIGITAL MODULATION TECHNIQUES BASED ON EQUAL DATA RATES

System Designation	Information Density (bps/Hz)	Signal-to-Noise Ratios for BER = 10^{-6} (dB)		Peak to Average Ratio (dB)[a]
		E_b/N_0 On the Chan	SNR At Decision CKT	
2-PSK	1	10.6	13.6	0.0
4-PSK, 4-QAM	2	10.6	13.6	0.0
QPR	2^b	12.6	17.6	2.0
8-PSK	3	14.0	18.8	0.0
16-QAM	4	14.5	20.5	2.55
16-QPR	4^b	16.5	24.5	4.55
16-PSK	4	18.3	24.3	0.0
32-QAM	5	17.4	24.4	2.3
64-QAM	6	18.8	26.6	3.68

[a] Ratio of maximum steady state signal power to average signal power with random data. Measured on a channel with square root filter partitioning.

[b] In a strict sense, the signal bandwidth of a partial response system is no narrower than the theoretical (Nyquist) bandwidth of a corresponding full response system. As a practical matter, however, the partial response systems require about 17% less bandwidth [2].

QAM Representation of Minimum Shift Keying

The preceding discussions of QAM systems have assumed the use of baseband signals with NRZ level encoding. A more general view of quadrature amplitude modulation allows arbitrary pulse shapes to be used in generating the baseband signals. One particularly interesting

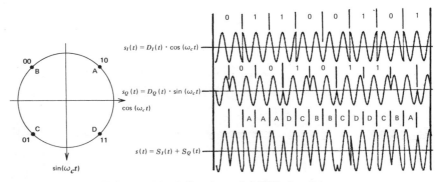

Figure 6.21   Offset keyed 4-PSK signaling.

pulse shape is a half-sinusoid pulse:

$$A(t) = \cos \frac{\pi t}{T} \quad \left(-\frac{T}{2} \leqslant t \leqslant \frac{T}{2} \right) \tag{6.20}$$

where T is the duration of a signal interval.

If half-sinusoidal pulse shaping of duration T is used on both channels of an offset keyed QAM system, the quadrature channel signals $p(t)$ and $q(t)$ can be expressed as follows:

$$p(t) = a_i \cos \left(\frac{\pi t}{2T} \right) \cos \omega_c t \quad \left(-\frac{T}{2} \leqslant t \leqslant \frac{T}{2} \right)$$

$$\tag{6.21}$$

$$q(t) = a_q \sin \left(\frac{\pi t}{2T} \right) \sin \omega_c t \quad (0 \leqslant t \leqslant T)$$

where a_i and a_q are data values (± 1) for the I and Q channels, respectively, $\cos(\pi t/2T)$ is shaping on the I channel, and $\sin(\pi t/2T)$ is shaping on the Q channel. (The sine function is used to account for offset keying.) Adding the two quadrature signals together and applying some fundamental trigonometric identities produces the result:

$$x(t) = p(t) + q(t)$$

$$= a_i \cos \left(\omega_c t + \frac{\pi t}{2T} \right) \quad (a_i = a_q) \tag{6.22}$$

$$= a_i \cos \left(\omega_c t - \frac{\pi t}{2T} \right) \quad (a_i \neq a_q)$$

Equation 6.22 is essentially identical to Equation 6.6 defining minimum shift keying. Hence, except for a logic level transformation of data values, offset QAM with sinusoidal pulse shaping is identical to MSK [4, 10]. The relationship is further demonstrated in Figure 6.22, which shows how an MSK signal is generated by offset keyed quadrature channel modulation with sinusoidal pulse shaping.

These foregoing results show that MSK is very closely related to offset keyed 4-PSK. The only difference is in the use of half-sinusoidal baseband pulse shapes for MSK, and square pulse shapes for 4-PSK. Because of this close relationship it is interesting to compare the amplitude spectra of the two systems in Figure 6.23. As indicated, the MSK spectrum has its first spectral null at a 50% higher frequency than 4-PSK's first null. Other than this, the MSK spectrum is more compact than the 4-PSK spectrum and is significantly lower in amplitude at frequencies outside the main lobe of the spectrum. For this reason MSK is an attractive modulation technique where constant envelopes and little or no filtering are desired. One such application is on digital satellite links with frequency division subchannels [11]. In applications requir-

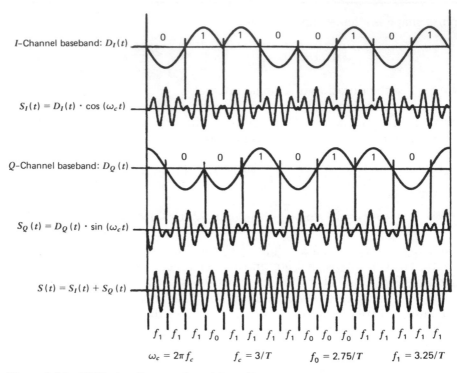

Figure 6.22 MSK signaling produced by offset keyed QAM with sinusoidal pulse shapes.

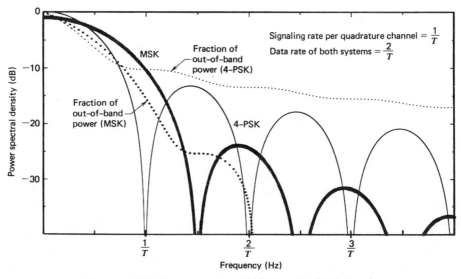

Figure 6.23 Power spectra of MSK and 4-PSK signals.

ing transmitter filters, MSK has no particular advantage over 4-PSK and usually requires a more complicated modulator.

6.1.5 Partial Response QAM

A modulation technique becoming more and more popular is quadrature partial response signaling (QPRS). As commonly implemented [12], a QPRS modulator is nothing more than a QAM modulator followed by a narrow bandpass filter that "overfilters" the quadrature signals and produces controlled intersymbol interference in each channel. The most common application of QPRS involves two levels on each channel before filtering and three levels afterward (see Chapter 4).

This system is essentially a 4-PSK system with partial response filtering to increase the information density. As shown in Figure 6.24, the effect of partial response filtering is to produce nine signal points from the original four. In a similar manner, a 16-QAM partial response system, with four levels on each channel before filtering, has seven levels afterward, and 49 signal points in all.

Although the modulators of a QPRS system can be conventional QAM modulators, the demodulator/detectors must be modified to account for the extra levels in the waveform. After the signal is demodulated, the detection processes for each channel are independent and identical to the baseband PRS detection procedures described in Chapter 4.

Figure 6.25 compares a QPRS system to a QPSK (4-PSK) system in terms of equal peak power out of the modulator. The average transmit powers are different, however, because the partial response system "overfilters" the signal to reduce the transmitted bandwidth. In practice the information density is increased by about 17% [2].

As discussed in Chapter 4, partial response filtering cuts the distance between signal points in half, indicating a 6 dB reduction in error per-

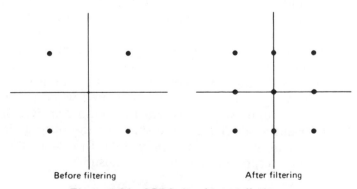

Before filtering   After filtering

Figure 6.24   QPRS signal constellations.

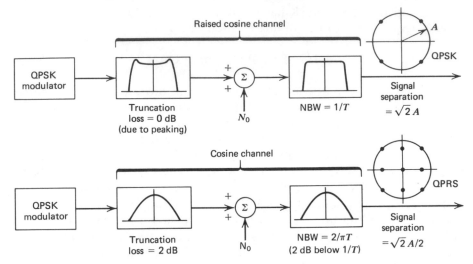

Figure 6.25 Comparison of QPSK and QPRS systems with equal data rates.

formance. However, the noise bandwidth of a PRS receive filter is lower than the noise bandwidth of the corresponding full response system so that some of the error distance degradation is recovered. For the filter systems shown in Figure 6.25, the net performance loss of QPRS system is 4 dB. Notice however, that with respect to power on the channel, the performance loss is only 2 dB.

6.2 FILTER PARTITIONING

The transmitting and receiving equipment of a digital radio system typically contains several filters that limit the signal spectrum to some degree or another. Since the end-to-end frequency response of the channel must conform to certain baseband pulse response objectives, the desired composite filter function must be partitioned among the individual filters. Normally, a number of the filters can be designed to provide their respective functions without significantly impacting the channel pulse response. For example, a mixing process produces a sum and a difference of the input frequencies. Only the sum is wanted when mixing upward, or only the difference is wanted when mixing downward. Usually, the undesired terms can be eliminated by a filter that does not significantly affect the pulse shape of the output signal. The following discussion assumes that only two filters significantly influence the signal response: one in the transmitter and one in the receiver.

6.2.1 Adjacent Channel Interference

One basic purpose of the radio channel receive filter is to minimize the amount of noise present at the detector. A second purpose of this filter is to reject energy in adjacent radio channels. Energy from an adjacent channel that does not get rejected is referred to as "adjacent channel interference." In this discussion, we assume that the spectrum shapes in the adjacent channels are identical to the desired spectra. This situation is shown in Figure 6.26 which depicts a number of frequency division multiplexed digital channels.

As indicated in Figure 6.26, adjacent channel interference occurs as a result of two phenomena. First, the receive filter passes unwanted power P_1 because the adjacent signal is not completely truncated to prevent overlap into the desired channel. The second source of interference P_2 occurs because the receive filter does not provide infinite attenuation of power properly belonging in the adjacent channel. P_1 is minimized by narrowing the transmit filter, while P_2 is minimized by narrowing the receive filter. Since channel pulse response considerations constrain the composite filter function to some minimum width, one component of the interference cannot be reduced without increasing the other. Hence the total filter function must be partitioned in some manner to minimize the sum of P_1 and P_2.

6.2.2 Optimum Partitioning

The optimum filter design for any particular application may depend on a number of factors including: legislated emission specifications,

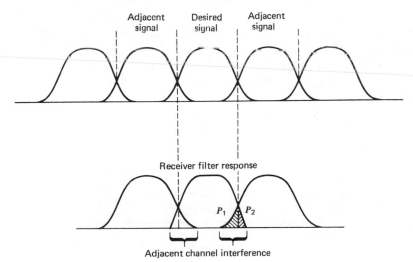

Figure 6.26 Adjacent channel interference.

available technology for power amplifiers, the availability of cross-polarization for adjacent channel isolation, and the relative effects of noise versus adjacent channel interference. In the absence of external constraints, a classical result attributed to Sunde [13] and also presented in References 14 and 15 determines optimum partitioning as one matching the output of the transmit filter to the square root of the desired channel response. Hence if the desired output spectrum $Y(\omega)$ is obtained as follows:

$$Y(\omega) = H_{Rx}(\omega) \cdot H_{Tx}(\omega) \cdot X(\omega) \tag{6.23}$$

where $X(\omega)$ = the channel input spectrum
$H_{Tx}(\omega)$ = the transmit filter response
$H_{Rx}(\omega)$ = the receive filter response

then, the optimum partitioning is obtained as:

$$|H_{Rx}(\omega)| = |Y(\omega)|^{1/2}$$

$$|H_{Tx}(\omega)| = \frac{Y(\omega)^{1/2}}{X(\omega)} \tag{6.24}$$

Equation 6.24 defines the filter amplitude responses to minimize the adjacent channel interference under the condition that the adjacent channels contain identical signals with identical power levels. In addition, if the transmitted spectrum $H_{Tx}(\omega) \cdot X(\omega)$ is the complex conjugate of the receive filter response $H_{Rx}(\omega)$, the receive filter is matched to the channel spectrum, and the best possible error performance is obtained with respect to signal power on the channel.

Partitioning as defined by Equation 6.24 is shown in Figure 6.27 for a pulse input and a raised cosine output (see Appendix C). Even though only baseband signal spectra are shown, the concept is extended easily to passband filter functions.

Although Equation 6.24 provides a sound theoretical basis for determining the optimum filter partitioning, practical considerations of a particular application may require deviation from the optimum. One problem that arises in digital microwave radios is related to the peak at the band edges of the transmit filter in Figure 6.27. With a passive filter

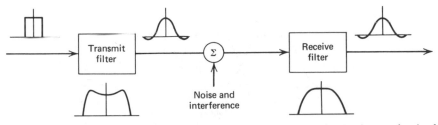

Figure 6.27 Theoretical optimum filtering for square pulse excitation and raised-cosine outputs.

this peak can be obtained only by inserting loss into the midband response. If the transmit power can be increased as compensation, no ill effects result. However, the output level of most digital microwave radios is limited by the technology of microwave-frequency power amplifiers (a few watts typically). Since these radios are device-power limited, midband insertion loss can not be overcome and therefore substracts directly from the received signal level.

In device-power limited applications the optimum transmit filter is one having a flat response in the passband. In order to achieve the desired channel response at the detector, the receive filter must then be peaked at the band edges. Channel insertion losses at this point do not degrade performance since both the signal and the noise is attenuated equally. However, the peaks have the undesirable effect of increasing the receiver noise bandwidth and the P_1 component of adjacent channel interference. Although these increases represent a poorer performance than that obtained by theoretically optimum partitioning, the degradation is not as great as the insertion loss needed to achieve the "theoretical optimum." As one further note, it should be mentioned that not all theoretically optimum transmit filters have a peak at the band edges. In particular, partial response systems do not require band edge peaking. (See Appendix C.)

Another aspect of the optimum design to be kept in mind is that optimum error performance is achieved with respect to signal power on the channel, and not with respect to unfiltered transmitter power. It is this feature that allows any amount of attenuation to be inserted into the channel at the transmitter and not degrade the optimum performance. If performance is measured with respect to unfiltered transmit power, the best transmit filter may be different from that defined in Equation 6.24. Not only would less midband attenuation be desirable, as discussed, but it might also be desirable to widen the transmit filter response. This decreases the truncation loss in the transmit filter so the receive filter can be narrowed to decrease its noise bandwidth.

As an extreme example of how widening the transmit filter can improve error rate performance, consider removing the transmit filter and incorporating it into the receiver. For a given output at the power amplifier, the signal at the detector is unchanged. The receiver noise bandwidth, however, is reduced because the composite filter is narrower than the original receive filter. Hence a higher signal-to-noise ratio is present at the detector.

The penalty incurred for removing the transmit filter, of course, is a greatly increased P_1 component of adjacent channel interference. If adjacent channels do not exist or if they are adequately isolated by cross-polarization (i.e., if the system is noise limited), the performance can be improved by moving some of the transmitter truncation losses to the receiver until adjacent channel interference matches the noise. Keep in mind, however, that if the system is adjacent channel interference limited, the optimum partitioning is indeed defined by Equation 6.24.

6.3 EMISSION SPECIFICATIONS

One unavoidable consideration when determining the filter functions of a transmitter and receiver is the out-of-band emission specifications established by the FCC. In many cases the emission specifications dictate a narrower transmit filter than the theoretical optimum. Thus the partitioning aspect of the filter designs may be predetermined.

It is somewhat ironic that the FCC emission specifications were intended to control adjacent channel interference but, in some cases, actually cause the interference to increase by forcing the use of a wider than optimum receive filter. The FCC specifications, however, were intended to protect adjacent analog radio channels from out-of-band digital emissions. They were not selected with adjacent digital channels in mind.

In January of 1975, when the FCC established the out-of-band emission specifications, two separate specifications were established: for radios operating below 15 GHz and for radios operating above 15 GHz. The emission limitations for operation below 15 GHz are more restrictive than those for operation above 15 GHz because the lower frequencies were heavily used already for analog FDM-FM radios, which are more sensitive to interference. The higher frequencies have not been used extensively because of vulnerability to rain attenuation. These frequencies are generally used for short-distance, special purpose applications and are not as congested as the lower frequencies. The most popular bands for digital microwave have been 11 and 6 GHz.

The emission limitations for operation below 15 GHz are

$$A = 35 + 10 \log_{10} B + 0.8(P - 50) \quad (P > 50) \quad (6.25)$$

where A = the power in a 4 kHz band relative to the mean output power (in decibels)

B = authorized bandwidth in MHz

P = percent removed from carrier frequency ($P = 50$ is the band edge)

In addition, the attenuation A must be at least 50 dB everywhere outside the band but does not need to exceed 80 dB at any point. Notice that the emission limitations are not specified in terms of absolute power levels, but only relative to the transmitted power. Thus these emission limitations do not constrain the output power level of the radio. At the present time microwave output powers are constrained by microwave amplifier technology. (A limit of 10 W of output power does exist when technology can reach it.)

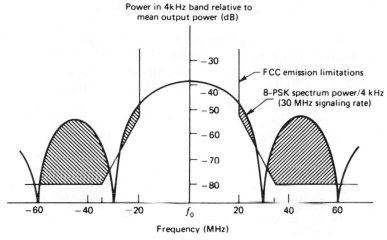

Power in 4kHz band relative to
mean output power (dB)

Figure 6.28   FCC emission mask at 11 GHz and a 90 Mbps 8-PSK spectrum.

Figure 6.28 displays the emission mask for 11 GHz radios using a 40 MHz bandwidth. Also shown is the power spectrum of a 90 Mbps 8-PSK radio using a signaling rate of 30 MHz. To be compatible with the emission specification, the 8-PSK power spectrum is shown in terms of power per 4 kHz band. The difference between the signal spectrum and the FCC mask represents the minimum attenuation required of the transmit filter. The filtered 8-PSK signal thereby provides an information density of 90/40 = 2.25 bps/Hz. This is the basic modulation and filtering format used by Collins/Rockwell in their 90 Mbps digital radio, the MDR-11 [16].

The emission limitations for microwave bands above 15 GHz is defined as

$$A = 11 + 10 \log_{10} B + 0.4 \, (P - 50) \qquad (P > 50) \qquad (6.26)$$

where A = attenuation in a 1 MHz band below mean output power (in decibels)

B = authorized bandwidth in MHz

P = percent removed from the carrier frequency

The attenuation A must be at least 11 dB, but does not have to exceed 56 dB.

In conjunction with the emission limitations, the FCC has stipulated that no discrete lines exist in the transmitted spectrum [17]. Thus no carrier components can exist and no repetitive data patterns are allowed to occur. The repetitive data patterns are effectively eliminated by using a scrambler in the transmitter and a descrambler in the receiver. As mentioned already, PSK and QAM are forms of double-sideband *suppressed* carrier modulation so that the carrier terms are eliminated automatically as long as modulation is continuous.

6.4 RADIO SYSTEM DESIGN

The foremost design requirement of a point-to-point radio system for public telephony is operational dependability, usually referred to as availability. Availability is expressed as the percentage of time that a system provides a specified minimum quality of service. With analog systems, minimum performance is determined by the noise power in the received signal. The performance of a digital system is determined by the bit error rate. Typical objectives for bit error rates range from 10^{-3} for voice traffic to 10^{-6} or 10^{-7} for data traffic. Recall that a bit error rate of 10^{-4} corresponds to the threshold of perceptibility for errors in a PCM voice signal. Typical design objectives for microwave radio systems specify availabilities on the order of 99.98% [18]. Hence the maximum acceptable accumulation of outage, due to all causes, is on the order of 2 hours per year.

Radio system availability is dependent on equipment reliability and path propagation characteristics. The high-availability objectives of a typical radio system mandate redundant equipment and often require redundant paths. The need for redundant paths is determined by the frequency of atmospheric induced outages: (rain attenuation or multipath interference). Rain is a dominant consideration at higher carrier frequencies (above 11 GHz), and multipath must be considered at all frequencies. Multipath fading is dependent on prevailing climate and terrain.

Redundant radio equipment typically operates in either a backup or a hot-standby mode with automatic protection switching. The transmission path is backed up with spare channels (frequency diversity) or spare paths (space diversity receivers). In extreme cases a backup route may even be utilized.

6.4.1 Fade Margins

The main technique used to circumvent atmospheric induced outages is to provide an extra strong signal at the receiver during normal path conditions. The difference between the normal received power and the power required for minimum acceptable performance is referred to as the fade margin. Greater fade margins imply less frequent occurrences of minimum performance levels. Radios operating in higher frequency bands generally require greater fade margins because they are more susceptible to rain attenuation. A 50 dB fade margin is typical for a digital radio at 11 GHz, while a 40 dB fade margin is typical for lower microwave frequencies.

The amount of fade margin actually required for a particular route depends on the probability of multipath induced fades and heavy rainfall occurrences. Thus drier climates permit lower fade margins—thereby allowing greater repeater spacings. In some mountain-based microwave links in the western United States, microwave hops can be 100 miles long. By comparison, the average hop in other parts of the country is less than 30 miles long.

When large fade margins are provided, the received signal power during unfaded conditions is so strong that bit errors are virtually non-existent. Nevertheless, problems with extra strong signals do exist. Namely, automatic gain control in a receiver must operate over a wide dynamic range. If the maximum signal level into the demodulation and detection circuitry is not controlled, saturation is likely to degrade performance—especially in high-density modulation formats such as 16-QAM, where information is encoded into the signal amplitudes.

6.4.2 System Gain

One of the most important parameters used to characterize digital microwave system performance is the system gain A_s. System gain is defined to be the difference, in decibels, of the transmitter output power and the minimum receive power for the specified error rate:

$$A_s = 10 \log_{10} \left(\frac{P_T}{P_R} \right) \qquad (6.27)$$

where P_T = transmitter output power
P_R = receive power for specified error rate

The minimum acceptable receive power is sometimes referred to as the threshold power and is primarily dependent on the receiver noise level; the signal-to-noise ratio required by the modulation format; and various system degradations such as excess noise bandwidth, signal distortions, intersymbol interference, carrier recovery offsets, timing jitter, and coupling and filter losses that either attenuate the signal or increase the noise level.

Noise power in a receiver is usually dominated by thermal noise generated in the front-end receiver amplifiers. In this case, the noise power can be determined as follows:

$$P_N = (F)(N_0)(B)$$
$$= F(kT_0)B \qquad (6.28)$$

where F = the receiver noise figure
N_0 = the power spectral density of the noise
B = the receiver bandwidth*
$k = 1.38(10)^{-23}$ is Boltzmann's constant
T_0 = the effective receiver temperature in degrees Kelvin

---

*Typically, B is assumed to be the minimum theoretical bandwidth for the particular modulation format in use. Excess bandwidth required by practical implementations is then incorporated into a system degradation factor.

Equation 6.28 essentially states that the receiver noise power is determined by the spectral noise density of the receiver input resistance and the additional noise introduced by the amplification process (noise figure F). Normally, a reference temperature of 290°K is assumed so that the thermal noise density (kT_0) is $4(10)^{-21}$ W/Hz.

The noise figure of any device is defined as the ratio of the input signal-to-noise ratio to the output signal-to-noise ratio:

$$F = \frac{(S/N) \text{ in}}{(S/N) \text{ out}} \qquad (6.29)$$

In effect, the noise figure specifies the increase in noise power relative to the increase in signal power. Since all physical devices introduce noise, the noise figure of any system is always greater than one, and is usually expressed in decibels. If a system has no gain or attenuation, the noise figure is exactly equal to the ratio of output noise to input noise. Noise figures of low-noise microwave amplifiers typically range from 2 to 5 (3 to 7 dB). Radio receiver noise figures are typically 6 to 10 without a low-noise amplifier.

Combining Equations 6.27 and 6.28 and incorporating a term D for the degradation from ideal performance produces the following general expression for system gain:

$$A_s = 10 \log_{10} \left(\frac{P_T}{(SNR)\,(F)\,(kT_0)\,(B)} \right) - D \qquad (6.30)$$

where SNR is the theoretical signal-power to noise-power ratio required for the maximum acceptable error rate, and D includes all degradations from ideal performance.

Notice that the SNR term in Equation 6.30 refers to signal-power to noise-power ratios and not E_b/N_0. The relationship between SNR and E_b/N_0 is provided in Appendix C.

The system gain, in conjunction with antenna gains and path losses, determines the fade margin:

$$\text{fade margin} = A_s + G_T + G_R + 20 \log_{10} \lambda - A_f - A_0 \qquad (6.31)$$

where G_T = the transmitter antenna gain (dB),
   G_R = the receive antenna gain (dB),
   λ = the transmitted wavelength,
   A_f = the antenna feeder and branching loss (dB),
   $A_0 = 20 \log_{10} (4\pi d)$ is the free space attenuation. (The distance d must be in the same units as λ.)

The directivity, and consequently the gain, of an antenna is directly proportional to the size of its aperture and inversely proportional to the square of the transmitted wavelength. In determining the receive power, however, it is actually only the area of the antenna that is important and not the directivity or gain. Thus radio system designers conveniently consider transmit and receive antenna gains as contributing to signal power, but include a wavelength normalization ($20 \log_{10} \lambda$) to relate the gain of the receiving antenna back to the size of its aperture.

In addition to providing increased antenna gain, greater directivities also reduce multipath problems. The longer secondary paths, which arise from greater emanation angles, have lower power levels when the directivity is increased. Unfortunately, antenna gains are limited by practical considerations in several regards: economically sized towers can only support limited sized antennas, mechanical alignment is difficult, and directional stability of both the antenna and the path is limited.

The feeder and branching losses A_F included in Equation 6.31 arise because single antenna systems typically carry several channels for separate radios. Furthermore, reliability considerations usually dictate that spare transmitters and receivers be available for protection switching. The process of combining signals for transmission or distributing them after reception inherently introduces various amounts of attenuation or splitting of signal power.

EXAMPLE 6.2

Determine the system gain of a 10 Mbps, 2 GHz digital microwave repeater using 4-PSK modulation and an output power of 2.5 W. Assume the excess bandwidth of the receiver is 30% and that other departures from ideal performance amount to 3 dB degradation. Assume a noise figure of 7 dB for the receiver, and the desired maximum error rate is 10^{-6}. Also determine the fade margin assuming antenna gains of 30 dB each and a path length of 50 kilometers. The branching and coupling losses are 5 dB.

Solution. From Figure 6.16, the required value of E_b/N_0 for 4-PSK modulation can be determined as 10.7 dB. Using Equation C.42 (in Appendix C), it can be determined that the signal-power to noise-power ratio at the detector is 3 dB higher than E_b/N_0. Thus, the required SNR is 13.7 dB.

Since 4-PSK modulation provides 2 bits per baud, the signaling rate is 5 MHZ which is the theoretical minimum (Nyquist) bandwidth. Equation 6.30 can now be used to determine the system gain:

$$A_s = 10 \log_{10} \frac{2.5}{4(10)^{-21} (5)10^6} - 13.7 - 7 - 3 - 10 \log 1.3$$

$$= 116 \text{ dB}$$

At a carrier frequency of 2 GHz, the wavelength is $3(10)^8/2(10)^9 = 0.15$ meters. Thus the fade margin can be determined from Equation 6.31:

fade margin $= 116 + 60 + 20 \log_{10} (0.15) - 5 - 20 \log_{10} [4\pi 5(10)^4]$

$= 38.5$ dB

6.4.3 Frequency Diversity

As mentioned previously, neither the transmitting and receiving equipment nor the path is normally reliable enough to provide an acceptable level of system availability. Frequency diversity is one means of providing backup facilities to overcome both types of outages. A deep multipath induced fade occurs when a signal from a secondary path arrives out of phase with respect to the primary signal. Since the phase shift produced by a path is proportional to frequency, when one carrier fades it is unlikely that another carrier fades simultaneously. Frequency diversity involves the use of a spare transmitter and receiver operating in a normally unused channel. Since separate hardware is used, frequency diversity also provides protection against hardware failures.

The simplest means of implementing frequency diversity is to use one-for-one $(1:1)$ protection switching as indicated in Figure 6.29. One-for-one protection switching implies that one spare channel is provided for each assigned message channel. When high-spectrum efficiency is required, it is generally necessary to have only one spare for a group of N channels ($1:N$ protection switching). In fact, the FCC in the United States has stipulated that, in some frequency bands, a system must be implementable in one-for-N configurations. The main impact

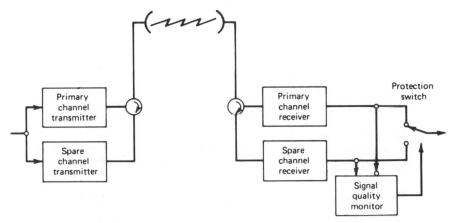

Figure 6.29 Protection switching with one-for-one frequency diversity.

of a 1:N protection system is the complexity of the switching unit and the need to switch back to the assigned channel in the first available hop so that a single spare channel can be reused repeatedly on a long route.

Frequency diversity normally does nothing to alleviate rain outages since all channels in a particular frequency band are simultaneously affected. When rain is a particular problem, it can only be overcome by using: higher transmit powers, shorter hops, or backup channels in lower-frequency bands.

6.4.4 Space Diversity

Space diversity is implemented, as shown in Figure 6.30, by vertically separating two receive antennas on a single tower. The resulting difference in the two paths is normally sufficient to provide independent fading at the two antennas. Space diversity is the most expensive means of improving the availability, particularly if separate receivers for multiple channels are used for each antenna. The cost can be minimized, however, by combining the two received signals in a phase coherent manner for input to a common receiver [18]. This technique provides less hardware backup than when completely separate receivers are used for each antenna.

6.4.5 Adaptive Equalization

Since multipath effects are frequency dependent, not all frequencies in a particular channel simultaneously experience the same amount of fading. Thus multipath fading can not only produce a general attenuation of the received signal, but it can also produce the equivalent of in-band amplitude (and phase) distortion. Distortion of the spectrum amplitude produces a general degradation in the error performance over

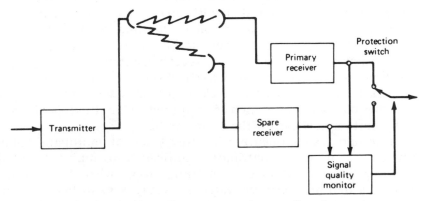

Figure 6.30 Protection switching with space diversity receivers.

and above the penalty incurred by the general attenuation of the signal. In wideband digital radios, frequency selective fading (as opposed to flat fading) has proved to be the dominant source of multipath outages. Fortunately, better performance can be achieved if the spectrum amplitude is equalized (adjusted to a uniform level).

Since atmospheric induced multipath varies with time, removal of multipath induced amplitude distortion requires adaptive equalization. The most common approach to adaptive amplitude equalization merely samples the energy at selected frequencies in the received signal spectrum. A compensating filter shape is then inserted into the signal path to adjust all energy samples to a common level. This basic technique is used in digital radios manufactured by Bell Northern [18] and Rockwell International [19]. Reference 19 reports a 7 dB improvement in effective fade margin can be achieved with an adaptive equalizer.

6.4.6 Route Design

The layout of a point-to-point microwave relay system involves numerous considerations of the local terrain, prevailing atmospheric conditions, radio frequency interference with other systems, and interference from one hop to another within a single system.

Path Length

The foremost consideration in setting up a single hop of a microwave system is that line-of-sight transmission is required. Using antenna heights of about 60 m, the curvature of the earth limits the transmission distance to approximately 50 km. Longer line-of-sight distances are possible if taller antenna towers are used. However, practical considerations of mechanical stability limit the height of a tower, particularly since longer distances imply larger and more directive antennas.

When antennas are rigidly mounted on the top of a building or the side of a mountain, mechanical stability may not be a problem but the stability of the path itself may become a limiting factor. Under some atmospheric conditions, radio waves can be refracted to the point that a narrow beam completely misses the receiving antenna. Bending of the propagation path can also cause a fixed obstacle to intermittently obstruct transmission, implying that clearance between the normal path and nearby physical obstructions is required. Thus there is always a practical limit on how narrow the beam can be.

Local terrain also influences the prevalence of multipath fading. Nearby bodies of water contribute significantly to multipath conditions during late evening or early morning hours when there is little wind. Direct transmission over water is usually very difficult because of reflections off the water surface. Transmission over water often requires higher antennas and some means of blocking direct reflections.

Cochannel Interference

Since there are a limited number of channels available for point-to-point microwave systems, the same channels must be used over and over again. Reuse of microwave frequency bands is enhanced by the directivity of the antennas and the general need for line-of-sight reception. In many metropolitan areas, however, there is so much traffic converging into one particular area it is impossible to completely isolate two systems using the same channel. This type of interference is referred to as cochannel interference.

Cochannel interference results from converging routes or from overreach of one hop into another hop of the same system reusing a channel. Sometimes reflections or atmospheric refractions can contribute to overreach, even when direct line of sight is not present. Overreach is usually controlled by zigzagging the hops so that a beam from one transmitter misses all subsequent receive antennas in a route.

6.4.7 Limitations of Technology

The choice of a modulation system for a digital microwave radio is not only dependent on the desired information density, but also on the available technology. As already mentioned, power amplifiers at microwave frequencies have limited output power and typically display significant saturation when operating near their maximum levels. For this reason constant amplitude modulation formats such as phase shift keying are very desirable.

When a modulation format with multiple amplitudes such as 16-QAM is selected, it is usually necessary to back off the power amplifier to obtain a more linear characteristic. Thus, multiple amplitude systems typically trade raw output power for more efficient utilization of average and peak powers with multiple amplitude signals.

Another important aspect of microwave power amplifiers that must be considered is the effect of varying signal amplitudes on the amplifier delay. Microwave amplifiers typically exhibit lower delay times when amplifying higher amplitude signals. From a modulation point of view, the variable delay times represent amplitude dependent phase shifts. Hence this phenomenon is referred to as amplitude modulation to phase modulation (AM to PM) conversion. Correspondingly, saturation effects are sometimes referred to as AM to AM distortion.

In addition to idiosyncrasies of microwave amplifiers, other factors that strongly influence the choice of a modulation format are linearity and speed of the modulators, filter and equalization requirements, average power and peak power, carrier and clock recovery, and susceptibility to all forms of noise and interference.

REFERENCES

1 Federal Communications Commission, *Report and Order*, Docket No. 19311, FCC 74-985, released September 27, 1974.

2 J. W. Bayless, R. D. Pedersen, and J. C. Bellamy, "High Density Digital Data Transmission," *National Telecommunications Conference Record*, 1976, pp 51.3-1 to 51.3-6.

3 Y. Yoshida, Y. Kitahara, and S. Yokoyama, "6G-90Mbps Digital Radio System with 16-QAM Modulation," *IEEE International Conference on Communications*, 1980, pp 52.4.1–52.4.5.

4 S. Pasupathy, "Minimum Shift Keying: A Spectrally Efficient Modulation," *IEEE Communications Magazine*, July 1979, pp 14–22.

5 R. E. Ziemer and W. H. Tranter, *Principles of Communications*, Houghton Mifflin Company, Boston, 1976.

6 F. M. Gardner, "Rapid Synchronization: Carrier and Clock Recovery for High Speed Digital Communication," *Microwave Systems News*, Vol. 6, No. 1, 1976, p 57.

7 C. R. Hogge, Jr. "Carrier and Clock Recovery for 8-PSK Synchronous Demodulation," *IEEE Transactions on Communications*, May 1978, pp 528–533.

8 S. A. Rhodes, "Performance of Offset-QPSK Communications With Partially Coherent Detection," *National Telecommunications Conference Record*, November 1973, pp 32A-1 to 32A-6.

9 M. K. Simon and J. G. Smith, "Offset Quadrature Communications with Decision-Feedback Carrier Synchronization," *IEEE Transactions on Communications*, October 1974, pp 1576–1584.

10 S. A. Gronemeyer and A. L. McBride, "MSK and Offset QPSK Modulation," *IEEE Transactions on Communications*, August 1976, pp 809–820.

11 Y. Morihiro, S. Nakajima, and N. Furuya, "A 100 Mbit/s Prototype MSK Modem for Satellite Communications," *IEEE Transactions on Communications*, Oct. 1979, pp 1512–1518.

12 C. W. Anderson and S. G. Barber, "Modulation Considerations for a 91 Mbit/s Digital Radio," *IEEE Transactions on Communications*, May 1978, pp 523–528.

13 E. D. Sunde, "Ideal Binary Pulse Transmission by AM and FM," *Bell System Technical Journal*, November 1959, pp 1357–1426.

14 W. R. Bennet and J. R. Davey, *Data Transmission*, McGraw-Hill, New York, 1965.

15 R. W. Lucky, J. Salz, and E. J. Weldon, Jr., *Principles of Data Communications*, McGraw-Hill, New York, 1968.

16 P. R. Hartmann and J. A. Crossett, "A 90 MBS Digital Transmission System at 11 GHz Using 8-PSK Modulation," *IEEE International Conference on Communications*, 1976, pp 18-8 to 18-13.

17 Federal Communications Commission, Memorandum Opinion and Order modifying FCC *Report and Order* of Docket 19311, released January 29, 1975.

18 G. H. M. de Witte, "DRS-8: System Design of a Long Haul 91 Mb/s Digital Radio," *IEEE National Telecommunications Conference*, 1978, pp 38.1.1–38.1.6.

19 P. R. Hartmann and E. W. Allen, "An Adaptive Equalizer for Correction of Multipath Distortion in a 90 MB/S 8-PSK System," *IEEE International Conference on Communications*, 1979, pp 5.6.1–5.6.4.

PROBLEMS

6.1 In order to prevent the transmission of line spectra, digital radio terminals use data scramblers to randomize the data patterns. Furthermore, differential encoding is normally required for proper data detection. Both functions introduce error multiplication. If the combined effect of these operations causes an average of five decoded errors for every channel error, what is the effective penalty in transmit power for a 4-PSK system at a $BER = 10^{-6}$? At a $BER = 10^{-3}$?

6.2 If a digital radio receiver is experiencing thermal noise induced errors at a rate of 1 per 10^6 bits, what is the new error rate if the path length is decreased from 30 to 25 miles?

6.3 Derive Equation 6.18.

6.4 What is the minimum theoretical bandwidth of an 8-PSK signal carrying 4800 bps?

6.5 A 32-QAM signal set is implemented by eliminating the four corner points of a 36-QAM signal. What is the minimum error distance in terms of the peak signal power? How does this answer compare to the error distance for antipodal signaling? (Express answers in decibels.)

6.6 What is the peak-to-average ratio of a 32-QAM signal (Problem 6.5) assuming transmission of random data?

6.7 How does 32-QAM (Problem 6.5) error performance compare to 32-PSK error performance?
(a) In terms of average signal powers?
(b) In terms of peak signal powers?

6.8 A carrier transmission system using 4-PSK modulation provides an error rate of 10^{-6}. If the modulation is changed to 16-PSK to narrow the transmission bandwidth (data rate unchanged), how much must the transmit power be increased to maintain the same error rate?

6.9 Repeat Problem 6.8 assuming that the bandwidth is unchanged and the data rate is doubled.

6.10 What is the error rate of an ideal 16-QAM signal with a signal to Gaussian noise ratio (SNR) of 18 dB?

6.11 What is the error rate of Problem 6.10 if interference is present at a level that is 21 dB below the received signal? (Assume the effect of the interference is identical to Gaussian noise at the same rms power.)

6.12 How much must the transmit power in Problem 6.11 be increased to offset the effect of the interference? (Assume the interference level is fixed.)

6.13 Repeat Problem 6.12, but assume that the interference increases in proportion to the increase in transmit power. (Signal powers in the adjacent channels are increased along with the desired signal power.)

CHAPTER SEVEN

NETWORK SYNCHRONIZATION, CONTROL, AND MANAGEMENT

In Chapters 4 and 6 some synchronization requirements of transmission systems are discussed. These requirements involve: carrier recovery for coherent detection of modulated signals, clock recovery for sampling incoming data, and framing procedures for identifying individual channels in a TDM signal format. All of these considerations are inherent in digital transmission systems and, for the most part, operate independently of other equipment in a network. An instance of one system's dependency on another was noted for T1 lines. T1 source data must include minimum densities of 1's to maintain timing on the transmission link. In contrast, other line codes are described that maintain clock synchronization independently of the source data.

This chapter discusses more general synchronization considerations for interconnecting various digital transmission and switching equipment. Foremost among these considerations is coordinating the operations of the synchronous transmission and switching equipment. When individual synchronous equipments are interconnected to form a network, certain procedures need to be established that either synchronize the clocks to each other or provide for their interoperability when each subsystem uses independent clocks. Following the discussion of network clock synchronization, the concept of synchronization is extended to higher-level aspects of network control. These considerations include connection control, traffic routing, and flow control.

7.1 TIMING INSTABILITIES

All digital systems inherently require a frequency source or "clock" as a means of timing internal and external operations. Operations timed from a single frequency source do not require particularly stable

321

sources since all commonly clocked elements experience timing variations in common. A different situation occurs when transfers are made from one synchronous equipment to another (as from a transmitter to a receiver). Even if the clock of the receiving terminal is "synchronized" to the transmitting terminal on a long-term or average basis, short-term variations in either clock may jeopardize the integrity of the data transfer. Thus it is generally necessary to use frequency sources (oscillators) in both the transmitter and the receiver that are as stable as is economically feasible.

7.1.1 Timing Jitter

No matter how stable the clocks are at both ends of a digital transmission system, certain amounts of instability inevitably occur in the received signal because of external electrical disturbances and changing physical parameters of the transmission link. The resulting instability in the line clock is referred to as "timing jitter." The following are main sources of timing jitter:

1 Noise and interference disrupting synchronization circuitry in a receiver.
2 Changes in the path length.
3 Changes in the velocity of propagation.
4 Doppler shifts from mobile terminals.
5 Irregular timing information.

Noise and Interference

A common means of synchronizing a receiver clock to a transmitter clock uses a phase-locked-loop as shown in Figure 7.1. A phase detector continuously measures the phase difference between a locally generated clock and one derived from the incoming signal. The output of the phase detector is filtered to eliminate as much receive noise as possible, and then the phase measurement adjusts the frequency of the voltage-controlled-oscillator (VCO) to reduce the phase difference. Some

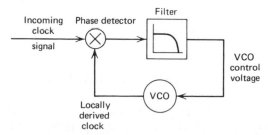

Figure 7.1 Phase-locked-loop clock recovery circuit.

amount of noise or interference inevitably passes through the phase detector and the filter causing erroneous adjustments in the VCO frequency. As time passes, however, frequency offsets produce ever-increasing phase shifts. When the phase difference builds up, it is easier to detect, and the appropriate changes in the VCO occur. Hence the local clock maintains the desired average frequency but inherently produces a certain amount phase jitter as it continuously hunts the underlying clock frequency of the source.

At high signal-to-noise ratios, the phase offsets are small and no ill effects result. As the relative noise level increases, the phase jitter increases and suboptimum sample times result that contribute to the error rate. At very low signal-to-noise ratios, the phase-locked loop may lose synchronism altogether. Even if the recovered clock stays unsynchronized for only a few cycles, the effects can be severe. Not only do bit errors occur, but also there may be a loss of bit count integrity which destroys framing at all levels. When the recovered clock slips a count with respect to the line clock, the transmission link is essentially broken until loss of synchronization is detected, and resynchronization at all levels is reestablished. Normally, a transmission link produces intolerable error rates before clock recovery circuits lose synchronization.

An important consideration in the design of a digital transmission link is the build up of jitter at tandem clock recovery circuits. If a recovered clock is used to time the transmission of outgoing data, as in a regenerative repeater, the incoming jitter is imbedded into the outgoing clock. The clock recovery circuit in the next receiver tracks its incoming clock but introduces even more jitter due to noise and interference on the second section. Thus jitter accumulates at every regenerative repeater using its received line clock as its transmit clock. If there is a large number of regenerative repeaters, the jitter can accumulate to a point where subsequent clock recovery circuits have a difficult time tracking the receive timing, produce sampling errors, and possibly lose lock.

Changes in Path Length

Path length changes occur as a result of thermal expansion or contraction of guided transmission media, or as a result of atmospheric bending of a radio path. While a path is increasing in length, the effective bit rate at the receiver is reduced because more and more bits are being "stored" in the medium. Similarly, as the path shortens, the bit rate at the receiver increases because the number of bits stored in the transmission link is decreasing. After the path has stabilized, the receive signal returns to the nominal data rate. The most significant changes in path length occur with communications satellites. Current geostationary satellites produce path length variations of approximately 200 miles, which cause propagation time variations of approximately 1 ms [1].

EXAMPLE 7.1

Determine the change in path length, the change in the number of bits in the path, and the relative change in received data rate of a 500 mile long T2 transmission link using 22 gauge copper wires. Assume the temperature changes by 20°C over a period of 1 hour.

Solution. The coefficient of thermal expansion for copper and the velocity of propagation of 22 gauge wire can be obtained from Reference 2 as $16.5(10)^{-6}/°C$ and 29,400 miles per second, respectively.* Thus the change in path length is determined as

$$\Delta d = (500)(16.5)(10^{-6})(20) = 0.165 \text{ miles}$$

The change in the number of bits in the path is determined as

$$\Delta B = \frac{(6.312)(10)^6(0.165)}{29,400} = 35.4 \text{ bits}$$

Assuming that the temperature changes at a constant rate, we determine the change in the data rate as

$$\Delta R = \frac{35.4}{3600} = 9.8(10)^{-3} \text{ bps}$$

Hence the relative change in the received data rate (instability) is

$$\frac{\Delta R}{R} = \frac{9.8(10)^{-3}}{6.312(10)^6} = 1.56(10)^{-9}$$

Although the value of received clock instability obtained in Example 7.1 is a very small number, it is comparable to the clock accuracy objective of 1.7 parts in 10^9 for the Bell System digital network [3]. It is important to note, however, that instability due to path length variations is only a transient phenomenon while accuracy objectives represent a maximum offset in steady state clock rates.

Changes in the Velocity of Propagation

Temperature changes not only cause expansion and contraction of wireline transmission media, they can also change those propagation constants of the media that determine the velocity of propagation. The resulting change in received clocked stability, however, is much less than that produced by the change in path length [4].

The propagation velocity of radio waves in the atmosphere also changes with temperature and humidity. Although these velocity

---

*These parameter values are obtained for conventional 22 gauge cable pairs and do not necessarily represent values for low-capacitance cable used in T2 systems.

changes are more significant than those occurring in wire lines, they are still smaller than the path-length-induced variations derived in Example 7.1 [5]. Notice that a change in propagation velocity is effectively equivalent to a change in path length since the number of bits "stored" in the transmission path is changed.

Doppler Shifts

The most significant source of potential timing instability in a received clock occurs as a result of Doppler shifts from airplanes or satellites. For example, a Doppler shift induced by a 350 mph airplane amounts to an equivalent clock instability of $5(10)^{-7}$. Again, Doppler shifts occur, in essence, as a result of path length changes.

Irregular Timing Information

As discussed in Chapter 4, a fundamental requirement of a digital line code is that it provide sufficient timing information to establish and maintain a receiver line clock. If the timing information is data dependent, jitter in the recovered clock increases during periods of relatively low-density timing marks. The magnitude of the jitter is not dependent solely on the density of timing marks, but also on the timing (data) patterns. In an ideal repeater, only the density would matter. In practice, however, various imperfections lead to pattern dependent jitter [6].

As discussed later in this chapter, higher-level digital multiplexers necessarily insert overhead bits into a composite data stream for various purposes. When the higher-rate data stream is demultiplexed, the arrival rate of data within individual channels is irregular. This irregularity produces timing jitter when generating new line clocks for the lower-rate signals. This source of jitter (waiting time jitter) is often the most troublesome and is discussed in more detail later.

7.1.2 Elastic Stores

The timing instabilities described in the preceding paragraphs essentially represent changes in the number of bits stored in a transmission link. In the case of noise and interference induced jitter, the change in "bits stored" occurs because data is sampled a little earlier or a little later than nominal. Since the outgoing data of a regenerative repeater is transmitted according to the recovered clock, a phase offset in the clock means the delay through the repeater is different from when there is no misalignment in timing.

If phase offsets in successive regenerative repeaters coincide, a net change of several bits of storage in a long repeated transmission link occurs. Since these extra bits enter or leave the transmission link over relatively short periods of time, the accumulated jitter may represent a relatively large, but short lived, instability in the receive clock.

Because regenerative repeaters use incoming sample clocks as output clocks, sustained timing differences between inputs and outputs do not exist. The endpoints of a transmission link, however, may interface to a local clock. In this case a difference between a received and a relatively fixed local clock must be reconciled with an elastic store. An elastic store is a data buffer that is written into by one clock and read from by another. If short-term instabilities exist in either clock, the elastic store absorbs the differences in the amount of data transmitted and the amount of data received. An elastic store can compensate only for short-term instabilities that produce a limited difference in the amounts of data transmitted and received. If sustained clock offsets exist, as with highly accurate but unsynchronized clocks, an elastic store will eventually underflow or overflow.

TDM—Switch Interface

A typical need for an elastic store occurs when digital transmission links are interfaced to a digital time division switch. As shown in Figure 7.2, the elastic store is placed between the incoming digital transmission link and the inlet side of the switch. In most instances the digital switch provides timing for all outgoing TDM links so that no timing discrepancies exist between these links and the switch. For the time being, assume that the far end of the digital link derives its clock from the receive signal and uses that clock to time digital transmissions returning to the switch. This is the situation that arises when a remote channel bank is connected to a digital switch through T1 lines and is commonly referred to as channel bank "loop timing." When loop timing is used, the line clock on the incoming link is nominally synchronized to the switch clock. However, for reasons discussed previously, a certain amount of instability in the incoming clock necessarily exists. The elastic store absorbs these instabilities so that purely synchronous data is available for the switch.

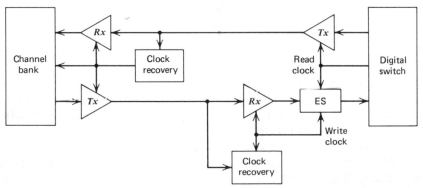

Figure 7.2 Interface between TDM transmission link and a digital switch using an elastic store.

In essence, the loop formed by the transmission links and the elastic store maintains a constant and integral number of clock intervals between the inlet and outlet of the switch. Thus, from a timing point of view, the inlets and outlets operate as though directly connected to each other using a common source of timing.

Removal of Accumulated Jitter

Another application for an elastic store is shown in Figure 7.3 where it is used to remove transmission induced timing jitter in a regenerative repeater. Normally, a regenerative repeater establishes the transmit timing directly from the locally derived sample clock. In Figure 7.3, however, the transmit timing is defined by a separate local clock. The elastic store absorbs the short-term instabilities in the receive clock, but the long-term frequency of the transmit clock is controlled by maintaining a certain "average level of storage" in the elastic store. Thus the transmit clock is synchronized to the line clock on a long-term basis, but not on a short-term basis. If the elastic store is large enough to accomodate all transient variations in the data rate, high-frequency instability of the output clock is independent of the input clock.

All regenerative repeaters, regardless of the mechanism used to recover timing, derive their output clocks by averaging the incoming timing information over a period of time. Tuned circuits average the incoming clock for relatively few signal intervals, phase-locked loops for many intervals. In all cases a certain amount of storage or delay is implied. An elastic store is merely a mechanism to increase the available delay so that output timing adjustments can be made more gradually. As discussed later, elastic stores cannot remove arbitrarily low-frequency jitter, but low-frequency jitter is no problem if the output clock is derived from the input clock.

Elastic Store Implementations

The required size of an elastic store varies from a few bits to several hundred bits for high-speed, long-distance communications links. Figure 7.4 shows one means of implementing a small elastic store utilizing a series-to-parallel converter, a register, and a parallel-to-series con-

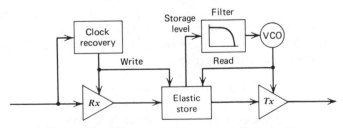

Figure 7.3   Jitter-removing regenerative repeater.

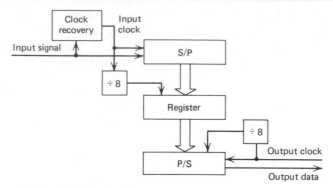

Figure 7.4 Basic implementation of an elastic store.

verter. As indicated, incoming data are transferred into the register as soon as each word is accumulated in the series-to-parallel converter. Some time later data in the register are transferred to the output parallel-to-series converter as a complete word is shifted out. Notice that transfers to the parallel-to-series converter are independent of the incoming clock. As long as output transfers occur between input transfers no data is lost, and short-term jitter is absorbed by varying delays through the elastic store.

Normally, some control circuitry (not shown) is needed to initialize the elastic store so that the first transfer into the register occurs midway between output transfers. This process means some incoming data are initially discarded by the series-to-parallel register until the desired transfer time occurs.

The relative times of the parallel transfers into and out of the holding register provide a direct indication of the relative phase of the input and output clocks. Thus the parallel transfer clocks contain the information needed to generate VCO control voltages if the elastic store is being used to remove accumulated transmission jitter.

The basic structure shown in Figure 7.4 can be extended to implement larger elastic stores as shown in Figure 7.5. The only change involves the substitution of a first-in, first-out (FIFO) buffer for the holding register of Figure 7.4. This data buffer is designed specifically to allow input transfers under the control of one clock while outputs are controlled by a different clock. Normally, the FIFO is initialized by inhibiting output transfers until the buffer is half full. In fact, some commercially available FIFOs have an output signal specifically indicating when the buffer is half full or greater.

7.1.3 Jitter Measurements

A simple circuit for measuring timing jitter is shown in Figure 7.6. As indicated, it is nothing more than a phase-locked-loop (PLL) with the

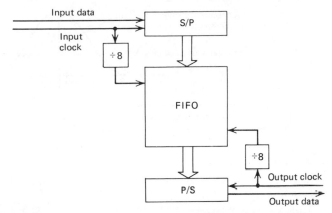

Figure 7.5 FIFO implementation of an elastic store.

output of the phase comparator providing the measurement of the timing jitter. Normally, the bandwidth of the low-pass filter (LPF) is very small so the VCO ignores short-term jitter in the timing signal. If there is no jitter at all, the output of the phase comparator is constant and no signal is passed by the high-pass selection filter (HPF).

Very low-frequency jitter can not be measured by the circuit in Figure 7.6 because the VCO tracks slowly changing phase shifts. Low-frequency jitter may be of no concern, however, because it can be tracked by a PLL. Higher-frequency jitter, on the other hand, is more apt to cause sampling errors or a loss of lock in the clock recovery of a repeater. Thus the spectral content of the jitter as well as its magnitude is of interest.

Phase jitter is commonly specified as a power measurement in units of radians squared or cycles squared. As indicated in Figure 7.7 phase jitter power is a measure of the variance of the number of time slots stored in the transmission link. In this case the jitter is expressed in units of clock cycles squared. In a physical sense jitter "power" has little meaning since it represents timing variations—not power. Some

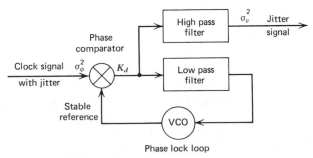

Figure 7.6 Circuit for measuring timing jitter.

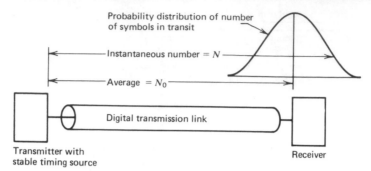

Figure 7.7 Phase jitter modeled as the variance in the number of symbols "stored" in the transmission link.

physical justification for expressing jitter as a power can be obtained by observing that the rms power σ_v^2 of the phase detector output signal is proportional to the rms phase jitter σ_ϕ^2 :

$$\sigma_v^2 = K_d^2 \sigma_\phi^2 \quad \text{(volts}^2\text{)} \tag{7.1}$$

where K_d is the phase-detector gain factor in volts/radian.

EXAMPLE 7.2

Given an rms phase jitter of 10.7 dB relative to one slot-squared, what is the standard deviation of the phase offset?

Solution. The variance of the signal phase is determined as $\sigma_\phi^2 = 10 \exp{(^{10.7}/_{10})} = 11.76$ slots-squared. Hence, the standard deviation is $(11.76)^{1/2} = 3.43$ symbol intervals. Since 68% of a normal probability distribution lies within one standard deviation, the phase of this signal is within ±3.43 symbol intervals for 68% of the time. One percent of the time the signal phase will be outside 2.6 standard deviations or ±8.9 symbol intervals.

If phase jitter arises as a result of additive Gaussian noise on a stable signal, the phase noise can be approximated as

$$\sigma_\phi^2 = \frac{\sigma_n^2}{2 \cdot P_s} \quad \text{(radians}^2\text{)} \tag{7.2}$$

where σ_n^2 is the additive noise power, and P_s is the signal power.

Equation 7.2 is the basic equation of phase jitter produced by additive noise on a continuous sinusoid [7]. When timing is extracted from a data signal, the timing information is usually not continuous. The distinction is not important because jitter produced by additive noise is normally insignificant compared to other sources [8]. For an analysis of phase jitter produced by regenerative repeaters operating on randomly occurring timing transitions see reference [9].

7.1.4 Systematic Jitter

An original analysis of jitter in a chain of digital regenerators was reported by Byrne, Karafin, and Robinson in Reference 6. Figure 7.8 shows the basic model of their analysis. Each of the regenerative repeaters in a T-carrier line extracts timing from the received waveform and passes that timing on to the next regenerator as a transmit clock. Owing to implementation imperfections (primarily intersymbol interference) in the timing recovery circuits (tuned filters of T1 lines), jitter produced by repeaters is dependent on the data patterns. One worst case pattern produces an extreme phase lag. Another pattern produces an extreme phase lead. When the data pattern shifts from one worst case to another, a phase ramp occurs. Because every repeater has the same basic implementation, the jitter produced by individual repeaters tends to be coherent. The systematic nature of this jitter makes it the most significant source of accumulated line clock jitter at the end of a chain of repeaters.

Accumulated line clock jitter is primarily a concern of a timing interface at a higher-level multiplexer or a switch. In both of these cases elastic stores synchronize an incoming data stream to a local clock. A switch interface using channel bank loop timing (Figure 7.2) is particularly concerned with accumulated phase jitter since the local clock is fixed and the elastic store must absorb all jitter. The elastic store of a higher-level multiplexer can be somewhat smaller because the effective output channel rate can be varied using a technique referred to as "pulse stuffing" (described later in this chapter).

At an individual repeater, accumulated jitter is not as important as

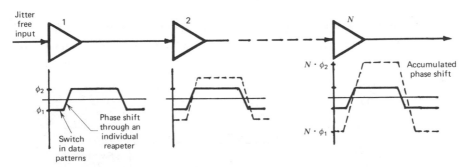

Figure 7.8 Model of systematic jitter in a string of regenerative repeaters; ϕ_2 = phase shift produced by worst case data pattern for phase lead; ϕ_1 = phase shift produced by worst case data for phase lag.

the jitter between the incoming clock and the local sample/transmit clock. This latter jitter can lead to sampling errors and is usually referred to as "misalignment jitter." The portion of accumulated jitter that gets tracked by the local clock recovery circuit is of no significance. Only at a timing interface where the local clock has little or no adjustment capability is accumulated jitter important.

7.2 TIMING INACCURACIES

In the preceding section the nature of certain instabilities or transient variations in timing were discussed. Although these variations represent shifts in the frequency of a line clock, the shifts are only temporary and can be absorbed by elastic stores. In some instances digital communications equipment using autonomous frequency sources must be interconnected. When this happens, the clock rates of the two system are never exactly the same, no matter how much accuracy is designed into the frequency sources. An offset in the two clocks, no matter how small, produces interconnection requirements not fulfillable by elastic stores alone.

In the preceding section, channel bank loop timing was mentioned as an example of how remote terminals are synchronized to a digital switch. When the remote terminal is another digital switch using its own frequency source as a reference, a different situation results. As shown in Figure 7.9, the outgoing clock for each direction of transmission is defined by the local switch clock. Thus the incoming clock at each switch interface not only contains transmission line induced jitter but also a small and unavoidable frequency offset. CCITT recommends that the 2.048 Mbps primary signal have a bit rate tolerance of 50 parts per million [10].

7.2.1 Slips

As indicated in Figure 7.9, the interface of each incoming digital link necessarily contains an elastic store to remove transmission link timing

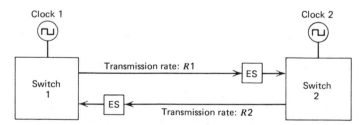

Figure 7.9 Communication between autonomously timed digital switches.

jitter. The elastic store at the first digital switch is *written into* by the recovered line clock, but *read from* at the local rate R_1. If the average rate of the recovered line clock R_2 is different from R_1, the elastic store will eventually underflow or overflow, depending on which rate is larger. When R_2 is greater than R_1, the elastic store at the first digital switch overflows causing a loss of data. If R_2 is less than R_1, the same elastic store underflows, causing extraneous data to be inserted into the bit stream entering the switch. Normally, the extraneous data is a repetition of one or more data bits already transferred into the switch. Disruptions in the data stream caused by underflows or overflows of an elastic store are referred to as "slips."

Uncontrolled slips represent very significant impairments to a digital network. Slips are most detrimental if they cause a loss of frame synchronization. Therefore, slips are allowed to occur only in prescribed manners that do not upset framing. One general approach to controlling the slips is to ensure that they occur only in the form of a repetition or deletion of an entire frame. Thus the time slot counters and framing logic associated with the multiplex group remain synchronized. Controlled slips comprising entire frames can be assured by using elastic stores with at least one frame of storage. As a slip occurs, the storage level in the elastic store is effectively increased or decreased by a full frame. Rather than actually inserting or deleting a frame of information, the desired effect is achieved more easily by indexing address pointers in a random access memory. Such a system is shown in Figure 7.10.

The elastic store in Figure 7.10 operates by sequentially writing input information into memory addresses corresponding to individual TDM channels. Data for individual output channels are obtained by

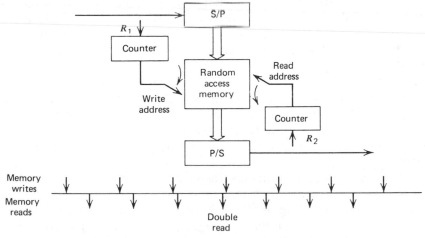

Figure 7.10 Elastic store operation with a one-frame memory.

reading the same addresses in the same sequential manner. Ideally, if there is no offset between the clock rates, the read times of each channel occur midway between write times for corresponding channels. The elastic store then has the capability of absorbing transmission delay variations up to one-half of a frame time.

The timing diagram in Figure 7.10 depicts an exaggerated timing offset in which the switch clock R_2 is greater than the incoming clock R_1. As indicated, the read times catch up gradually with the write times until a "double read" occurs. At that time the information retrieved for each channel is a repetition of the information retrieved for the previous outgoing frame. Although write and read times for only one channel are shown, the corresponding times for all other channels have the same relationship. Thus all channels slip together. Notice that if R_1 is greater than R_2 a slip occurs when a "double write" on all channels causes the information in the previous incoming frame to be lost.

The elastic store operation depicted in Figure 7.10 is very similar to the operation of a time slot interchange memory described in Chapter 6. This relationship is exploitable in a TST switch where the inlet memory can provide both the elastic store function and the time switching function [5]. When the two functions are combined, slips generally occur at different times for different channels. Nevertheless, individual channels maintain proper frame alignment since each channel is transferred through the inlet memory using dedicated memory addresses.

One attractive feature of using the inlet memory as an elastic store is that, when setting up a new connection, an internal switching time slot can be chosen so that the inlet memory read is halfway between inlet memory writes for the particular channel. Thus a slip in that connection will not occur for a long time—probably not until long after the connection is released. (With clock accuracies of only one part in 10^8, the time between slips in any one channel is 3.5 hours.)

One potential problem with the elastic store in Figure 7.10 occurs when write and read times nearly coincide. When both accesses to a single channel occur one after the other, transient timing instabilities can cause the two accesses to cross back and forth with respect to each other. Thus slips caused by double reads may follow slips caused by double writes and vice versa. To remedy this situation, some amount of hysteresis is needed in the counter adjustment process. The hysteresis, in turn, implies that additional storage is needed to defer the occurrence of one type of slip after a slip of the other type has recently occurred.

One means of implementing an elastic store with the desired hysteresis is to use two frames of storage as shown in Figure 7.11. For convenience the elastic store is divided into an A-frame memory and a B-frame memory. The counter logic again accesses the memories in sequential fashion except that frames are written alternately into the A

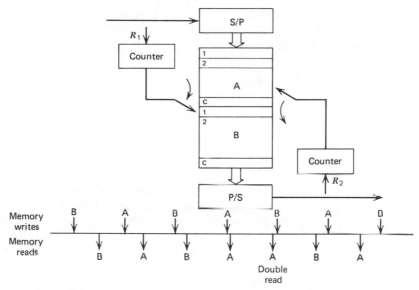

Figure 7.11 Elastic store with a two-frame memory.

and B memories. Under normal operation, the memories are accessed in the same way for output data. When a slip is imminent, however, control logic causes the output channel counter to be reset so that the A memory is read twice in a row. This situation is depicted in the timing diagram of Figure 7.11, which again assumes that R_2 is greater than R_1. The important point to be noticed in the timing diagram is that after the counter adjustment produces a double read of memory A, the write and read times of each individual memory are approximately one frame time apart. Thus another adjustment can be deferred until the write and read accesses again drift one full frame time with respect to each other. The structure and mode of operation shown in Figure 7.11 describe the elastic store used for DS-1 signal interfaces to the Bell System No. 4ESS [11].

Slip Rate Objectives

If the difference between an elastic store's input data rate and its output data rate is ΔR,* the time between slips is

$$\Delta T = \frac{N}{\Delta R} \tag{7.3}$$

*Typically, clock offsets are specified in relative terms (e.g., one part in 10^9). A clock that is accurate to P parts per million (PPM) has a maximum offset of $\Delta R = R \cdot P/10^6$.

where N is the number of bits that get dropped or repeated whenever a slip occurs. Normally, a slip involves a full frame of data, in which case the slip rate is determined as

$$\text{slip rate} = \frac{1}{T_s}$$
$$= \Delta F \qquad\qquad (7.4)$$

where ΔF is the difference in frame rates.

As long as slips are controlled so that they do not disrupt higher-level synchronization processes, their only effect is an infrequent repetition or deletion of the information within affected TDM channels. The audible effect of slips on a digitized voice signal is an occasional "click." Only one slip in 25 produces an audible click in PCM voice [1]. If we assume that a duration of 5 minutes between audible clicks is acceptable, the tolerable slip rate for voice is 300 slips per hour. With an 8 kHz frame rate, 300 slips per hour occur when clocks differ by 1 part in (8000) (3600)/300 = 96,000—an easily achievable accuracy.

Encrypted traffic (voice or data) is more susceptible to slips since the encryption/decryption process usually relies on bit synchronous scramblers and unscramblers. When the bit count is altered by insertion or deletion of bits in a time slot, counters in the source and destination become unsynchronized. At best, the decryption process causes every slip to be audible. At worst, unintelligible speech or data result until the unscrambler is resynchronized.

A more subtle, and troublesome, aspect of slips occurs when a digitized channel carries voiceband data. High-speed data modems for the analog telephone network use phase shift modulation with coherent detection in the receiver. Since these modems are particularly sensitive to phase shifts, they are particularly vulnerable to slips. An 8 bit slip in a digitized modem signal using a carrier of 1800 Hz causes an instantaneous phase shift of 81°. Obviously, a phase shift of this size causes a data error, but more importantly, it upsets the carrier recovery circuitry in the receiver and causes multiple errors. A single slip can upset the operation of some voiceband modems for several seconds [3].

When a digital transmission link is being used to transmit data directly, the effect of a slip may not be any more significant than a single channel error. Most data communications receiving equipment requests a complete retransmission of any block of data not satisfying certain redundancy checks. Thus one error is as bad as many errors or a complete loss of data. The effect of the slip, will be more significant, however, if the communications protocol (e.g. Digital Equipment Corporation's DDCMP) relies on byte count procedures to delimit message blocks. Insertion or deletion of data by the network causes the receive counter to become unsynchronized, and the normal exchange of information is disrupted until the loss of synchronization is recognized.

From the foregoing considerations for data transmission, the slip rate objective for the evolving digital network in North America has been set at one slip in 5 hours for an end-to-end connection [3]. Since slips can occur at multiple points within a network, the objective for slips at individual trunk and switching interfaces is set at one slip every 20 hours.

EXAMPLE 7.3

Determine the relative accuracy requirements of two independent clocks to maintain a mutual slip rate objective of one slip in 20 hours. Assume a frame rate of 8 kHz as in PCM voice signals.

Solution. The slip rate objective implies that the frame rate produced by one clock can be different than the frame rate produced by the other clock by no more than

$$\Delta F = \frac{1}{20(60)(60)} = 1.39(10)^{-5} \text{ slips per second}$$

Since there are 8000 frames per second, the relative accuracy is determined as

$$\frac{1.39(10)^{-5}}{8000} = 1.7(10)^{-9} \text{ slips/frame}$$

Hence the clocks must be accurate to greater than one part in 10^8.

7.2.2 Pulse Stuffing

In the preceding section certain aspects of network synchronization were discussed that implied the need for clock synchronization to prevent a loss of data by way of slips. In this section we discuss a procedure referred to as "pulse stuffing" that avoids both slips and clock synchronization. The term pulse stuffing can be somewhat misleading since it implies that pulses are inserted into the line code to make timing adjustments. Actually, pulse stuffing is a purely logical operation that is independent of the line code or modulation system in use. Pulse stuffing is a term commonly used in North America while the same concept is referred to as "justification" in Europe.

The basic concept of pulse stuffing involves the use of an output channel whose rate is purposely higher than the input rate. Thus the output channel can carry all input data plus some variable number of "null" bits or "stuffed bits." The null bits are not part of the incoming

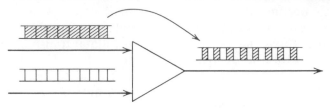

Figure 7.12 Two-channel multiplexer showing equal output data rates for each input.

data. They are inserted, in a prescribed manner, to pad the input data stream to the higher output rate. Naturally, the extraneous null or "stuffed" bits must be identifiable so that "destuffing" can recover the original data stream.

Pulse stuffing is applied most often to higher-level multiplex signals when each of the incoming lower-level signals is unsynchronized to each other. In this situation pulse stuffing is the only possible means of avoiding slips. The fundamental problem is shown in Figure 7.12, which portrays a simple two-channel multiplexer. Notice that the output rates for both channels are necessarily identical, since each channel is assigned an equal number of time slots over a period of time. Hence if the two input channels have slightly different rates, at least one of these rates is different from its respective output channel rate.

When pulse stuffing is used, each subchannel of the multiplexer operates at exactly the same rate. However, the rate of information flow within each subchannel can vary by inserting noninformation-bearing null pulses as needed. Thus the stuffing operations in each channel are necessarily independent. An example of pulse stuffing in a two-channel multiplexer with exaggerated timing differences on the input signals is shown in Figure 7.13. Notice that within each subchannel the number of information bits per frame can vary. Channel 2 contains more "stuffed" bits because its input rate is lower than that of channel 1. A word-interleaved format is shown for conceptual purposes only. Multiplexers working on serial inputs normally use a bit-interleaved format.

Each frame in the multiplexer output signal of Figure 7.13 contains a fixed number of time slots. However, not all time slots contain information. In addition to framing bits (not shown), the frame contains

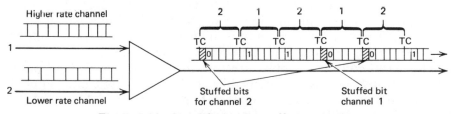

Figure 7.13 Simplified pulse stuffing example.

a stuffing specification bit (C) and a special timing bit (T) for each channel. When a control bit is a 1, the T bit contains information. When a C bit is a 0, the respective T bit is a null (stuffed) bit. (The logical value of a stuffed bit can be either a 0 or a 1 since its only purpose is to make timing adjustments. By using a specified value, however, channel errors or misframe detection is possible when the logical value of a received stuffed bit contradicts the definition.)

An important point to notice about a higher-level multiplexer using pulse stuffing is that the output frame structure is unrelated to the frame structure of the lower-level inputs. As far as the higher-level multiplexer is concerned, each input signal is merely a serial bit stream with no particular structure assumed. Framing bits in the lower-level multiplex signals are transmitted right along with the information bits. After the higher-level signal is demultiplexed (and destuffed), framing of the lower-level signals must be established for further demultiplexing.

Although pulse stuffing can be implemented with a variety of higher-level framing formats, the generally most desirable features of a pulse stuffing format can be identified as follows:

1 The use of fixed length master frames with each channel allowed to stuff or not to stuff a single bit in the master frame.
2 Redundant stuffing specifications.
3 Noninformation bits distributed across a master frame.

Timing offsets are generally quite small, so only small adjustments from a single occasional timing bit are required. Thus a large number of multiplex frames can be combined into a master frame with one specific bit position identified as the T bit. In unstuffed master frames this bit position contains information. In others this bit position is a null bit. Nominally, approximately one-half of the master frames contain the maximum number of information bits N_M, and the other half contain $N_M - 1$ information bits.

The purpose of a pulse stuffing operation is to prevent a loss of data when two interconnected digital transmission links are unsynchronized with respect to each other. If single bit errors can cause a null bit to be interpreted as information (or vice versa), the basic objective is lost. Furthermore, notice that an erroneous interpretation of a stuff code causes lower-level multiplex signals to lose framing. For these reasons, the interpretation of T bits must be encoded redundantly. If we assume that channel errors are random, the probability of misinterpreting a stuff code is

$$P_F = \binom{2n + 1}{n + 1} p^{n+1} (1 - p)^n \tag{7.5}$$

where p is the probability of a channel error, and n is the number of correctable stuff code errors ($2n + 1$ bits in a stuff code).

Noninformation bits should be distributed across a master frame for several reasons. First, by separating these bits as much as possible, errors in redundant stuffing specification bits (C bits) are more likely to be independent. If the specification bits are too close together and burst errors are prevalent, the redundant encoding is of little use. Second, by distributing noninformation bits, the irregularity of information flow is minimized. When higher-level multiplex signals are demultiplexed, a clock for each individual lower-level signal needs to be derived from the irregular information rate in each channel. Generation of a suitably stable clock synchronized to the information rate is simplified if information bursts or gaps are minimized. Furthermore, elastic stores needed to smooth out the information rates are smaller when the tendency toward bursts is minimized.

An example of a higher level multiplexing format is provided in Figure 7.14. This is the format used for 6.312 Mbps DS-2 signals in the North American digital hierarchy. As indicated, a DS-2 signal is derived by bit interleaving four DS-1 signals and adding the appropriate overhead bits.

A DS-2 masterframe is 1176 bits long. Of these there are 1148 information bits (287 per channel), 12 framing bits (M_0, M_1, F_0, F_1), 12 stuffing control bits (C_1, C_2, C_3), and 4 T bits (T_1, T_2, T_3, T_4). Since a T bit can be a null bit or an information bit, each channel can send 287 or 288 bits in a master frame. A T bit is designated as an information bit if all three of the corresponding C bits are 1. The T bit is a null (stuff) bit if all three corresponding C bits are 0. Obviously, this encoding procedure allows for single error correction in the stuffing control bits.

The first level of framing is established by the alternating F_0, F_1,

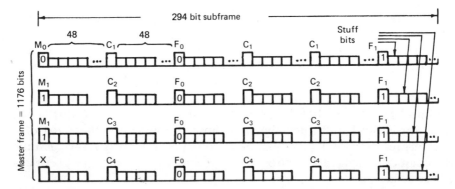

Figure 7.14 Frame format of DS-2 digital signal. Stuffing occurs in channel i when the previous C_i bits = 111. X is an alarm bit that equals 1 for no alarm condition. Framing is established by the $F_0 F_1 F_0 \ldots$ sequence with 146 intervening bits.

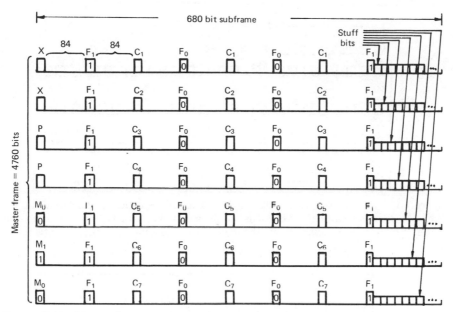

Figure 7.15 Frame format of DS-3 digital signal. P is even parity over all message bits in the previous master frame. Stuffing occurs in channel i when the previous C_i bits = 111. The X (alarm) bits and the P bits must be 11 or 00 so the $M_0 M_1 M_0$ sequence can identify the end of the master frame.

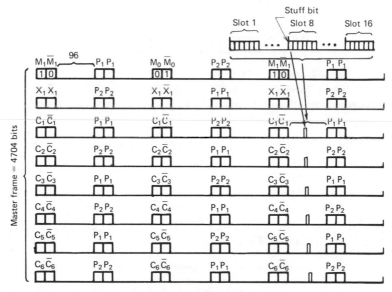

Figure 7.16 Frame format of DS-4 digital signal. $C_i = 111$ implies stuff the eighth message bit position for channel i following the last C_i. P_1 is even parity over the 192 previous odd numbered message bits. P_0 is even parity over the 192 previous even numbered message bits.

341

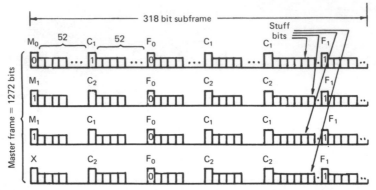

Figure 7.17 Frame format of DS-1C digital signal. Stuffing occurs in channel i when the previous C_i bits = 111. X is an alarm bit that equals 1 for no alarm condition. Framing is established by the $F_0 F_1 F_0$ ... sequence with 158 intervening bits.

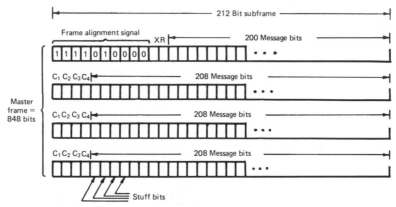

Figure 7.18 Frame format of second level digital signal of CCITT. Stuffing occurs in channel i when the previous C_i bits are 111.

F_0 ... pattern. Notice that exactly 146 bits separate the F_0 and F_1 bits. Another level of framing for identifying the C and T bits is established by the M_0 and M_1 bits. A fourth M bit (M_x) is not used for framing and therefore can be used as an alarm service digit. Similar frame structures exist for other higher level digital signals. Figures 7.15 through 7.18 show the structures for DS-3, DS-4, DS-1C, and the second level digital signal of CCITT, respectively.

EXAMPLE 7.4

Determine the minimum and maximum input channel rates accommodated by an M_{12} multiplexer. Also determine the rate of DS – 1 misframes caused by an erroneous interpretation of a stuffed bit. Assume the bit error rate is 10^{-6}.

Solution. The maximum information rate per channel is determined as

$$\frac{6.312(288)}{1176} = 1.5458 \text{ Mbps}$$

The minimum information rate per channel is determined as

$$\frac{6.312(287)}{1176} = 1.5404 \text{ Mbps}$$

Since there are three possible combinations of two errors in the C bits, the probability of misinterpreting a T bit is

$$3(10^{-6})^2 = 3 \cdot 10^{-12}$$

The duration of each master frame is $1176/6.312 = 186$ μs. Thus the rate of misframes per DS-1 signal is

$$\frac{(3) \, 10^{-12}}{(186) \, (10)^{-6}} = 0.016(10)^{-6} \text{ misframes per second}$$

which is equivalent to one misframe every two years.

Example 7.4 demonstrates that the tolerance of a 1.544 MHz DS-1 clock is -3.572 to $+1.796$ kHz. Thus the relative accuracy between the DS-1 and DS-2 clocks must be $1.796/1544$ or only 1 part in 860. This relatively large timing tolerance is greater than what is normally required for reasonable clock and line instabilities. The timing adjustment capabilities and unsymmetric tolerance range were chosen out of a desire to: (1) minimize DS-2 reframe times; (2) provide a line clock that is a multiple of 8 kHz; and (3) minimize waiting time jitter [12].

Elastic Store Size Requirements

A functional model of an M_{12} multiplexer is shown in Figure 7.19. Associated with each lower level (DS-1) input is an elastic store to hold incoming data until it is tranferred to the higher level (DS-2) output. The elastic stores serve two purposes: to remove the arrival jitter of the incoming data and hold data for the proper time slots. In addition to generating framing, the control logic of the multiplexer monitors the storage level (which serves as a phase comparator) of each elastic store and initiates a stuffing operation whenever the elastic store is less than half full. Conversely, no stuffing occurs when the elastic store is more than half full.

Since stuffing can occur at only certain times and only at a certain maximum rate, the elastic store must be at least as large as the peak jitter (peak phase offset) of the incoming signal. As discussed at the

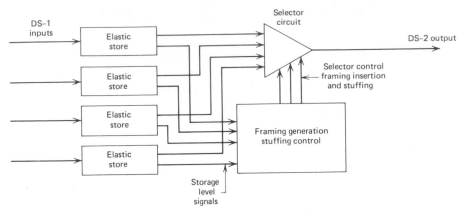

Figure 7.19 Functional diagram of a M12 multiplexer.

beginning of this chapter, jitter accumulates along the entire length of a repeatered transmission link. Thus longer links require larger elastic stores if slips are to be prevented.

As an example of the relationship between line length (number of repeaters) and number of bits of elastic storage needed by an M_{12} multiplexer refer to Figure 7.20. This figure was submitted by AT&T to the CCITT special study group on jitter [13]. The analysis is an extension of the systematic jitter analysis of Reference 6.

The abscissa of Figure 7.20 is the maximum phase slope produced by a clock recovery circuit in the presence of a worst case shift in data patterns (worst case systematic jitter). Since phase slope is nothing more than frequency offset, the required elastic store size can be determined as the maximum phase slope times its maximum duration. Since the total phase slope is proportional to the number of repeaters, Figure 7.20 displays the maximum number of repeaters per storage cell in an elastic store versus the jitter slope of an individual repeater when making a worst case timing transition.

As an example, current T_1 repeaters produce a worst case slope of 2.4 kHz. M_{12} multiplexers allocate 5 bits of storage to input phase jitter (3 more bits are included for implementation ease and waiting time jitter). From Figure 7.20 it can be seen that the ratio of N_{max} to A is 56, which implies that $N_{max} = (56)(5) = 280$ repeaters.

7.2.3 Waiting Time Jitter

When demultiplexing a TDM data stream, it is necessary to generate a clock for each derived subchannel. Because the subchannels are transferred (or transmitted) as synchronous data, the derived clock must be continuous. Derivation of synchronous subchannel clocks is complicated by the insertion into TDM data streams of overhead bits that

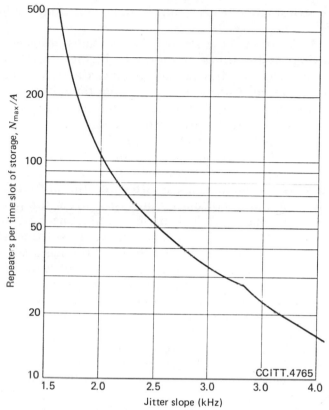

Figure 7.20 Maximum number of regenerative repeaters as a function of elastic store size and jitter slope. A = Number of cells in elastic store.

create gaps in the bit arrival times. Irregularity in the data arrival rate caused by these gaps is referred to as "waiting time jitter."

Most of the overhead bits (e.g., framing bits, parity bits, stuffing specification C bits) occur on a regular and predictable basis. The waiting time jitter caused by these gaps can be eliminated easily with an elastic store and a output clock derived from the incoming line clock. For example, a single PCM channel clock at 64 kbps can be derived from a 1.544 Mbps T_1 line clock by multiplying by 8 and dividing by 193.

In contrast, waiting time jitter produced by pulse stuffing is more difficult to deal with. The difficulty arises because waiting times produced by stuffed pulses are irregular and unpredictable. In general, no rational relationship exists between one autonomous clock and another, so that stuffing occurrences arise in relation to a nonrepetitive decimal expansion (much like leap years, which do not occur every 100 or 1000 years). For this reason the subchannel output clocks derived from a pulse stuffed TDM line must be derived independently and only from the average arrival rate of each channel's data—not from the higher-level TDM rate!

Output clocks from pulse stuffed data streams are generated using jitter removing elastic stores as shown in Figure 7.3. If large elastic stores and very slowly adjusted output clocks are used, most of the jitter can be removed. Unfortunately, waiting time jitter has frequency components down to zero frequency so the jitter can never be eliminated entirely. However, the jitter can be confined to as low a band of frequencies as desired by using a large enough elastic store.

Waiting time jitter is basically a function of how often pulses are stuffed, but it is also dependent on the ratio of actual stuffs to stuffing opportunities. If the input clock is jitter-free the output jitter peaks when one-half of the opportunities are used. From the point of view of maximum tolerance for clock offsets, a stuffing ratio of one-half is ideal. To reduce the waiting time jitter, however, stuffing ratios of approximately one-third are often used. For a thorough analysis of waiting time jitter see Reference 14. As an example of waiting time jitter dependence on stuffing (justification) ratios, see Figure 7.21 obtained from Reference 13. The abscissa of Figure 7.21 represents the ratio of stuffs to opportunities while the ordinate is jitter power produced by a single pulse stuffing process. The jitter power is expressed in decibels relative to one time slot squared (0 dB implies an rms jitter value of 1 bit-time squared, i.e., 2π radians squared). Curve A shows the output jitter produced when the input jitter is -20 dB (0.1 slot rms). Curve B shows the output jitter when the input jitter is -12 dB (0.25 slot rms). Figure 7.21 shows jitter produced by a single multiplexer. From measured data [13], a good order-of-magnitude estimate of the waiting time jitter accumulated by N_m tandem M_{12} stuffing/destuffing oper-

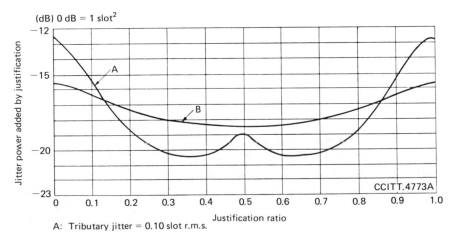

A: Tributary jitter = 0.10 slot r.m.s.
B: Tributary jitter = 0.25 slot r.m.s.

Figure 7.21 Waiting time jitter dependence on justification ratio.

ations is

$$\sigma_w^2 = \frac{N_m}{100} \text{ slots}^2 \tag{7.6}$$

7.3 NETWORK SYNCHRONIZATION

As discussed in the preceding section, whenever a digital transmission link is connected to a digital switch it is desirable to synchronize the two systems by having the transmission link obtain its timing from the switch. An obvious exception to this mode of operation occurs when a digital transmission link is connected to a digital switch on both ends. Generally, a transmission link in an all-digital network derives its timing from just one of the switches to which it is connected. If the other switch is not synchronized to the first in some manner, an unsynchronized interface necessarily results. This section is concerned with network synchronization as a whole, not simply the synchronization of a single interface. Basically, network synchronization involves synchronizing the switches of the network. The transmission links can then be synchronized automatically by deriving timing directly from a switching node. An exception occurs on higher-level multiplexers that at present use unsynchronized higher-rate clocks with pulse stuffing to overcome clock inaccuracies and fluctuations. The use of unsynchronized clocks for higher-level TDM links does not preclude network synchronization of lower-level (e.g., DS-1) signals.

There are two basic reasons for paying assiduous attention to the timing requirements of a digital network. First, the network must prevent uncontrolled slips which could produce misframes, inadvertent disconnects, and crossconnects. It is generally very difficult, or very expensive, to prevent slips altogether. Thus a second aspect of a network timing plan requires establishing a maximum rate of controlled slips as part of the end-to-end circuit quality objectives. As mentioned previously, the slip rate objective of the North American network was established primarily in consideration of data and encrypted traffic requirements.

There are six basic approaches used, or considered for use, in synchronizing a digital network:

1 Plesiochronous.
2 Networkwide pulse stuffing.
3 Mutual synchronization.
4 Network master.
5 Master-slave clocking.
6 Packetization.

7.3.1 Plesiochronous

A plesiochronous network does not synchronize the network, but merely uses highly accurate clocks at all switching nodes so the slip rate between the nodes is acceptably low. This mode of operation is probably the simplest to implement since it avoids distributing timing throughout the network. A plesiochronous network, however, implies that the smaller switching nodes carry the cost burden of highly accurate and redundant timing sources. For this reason, Bell System engineers decided not to use independent clocks for network timing in the United States.

Plesiochronous timing is planned, however, for international digital network interconnections. In recommendation G-811 [15], the CCITT has established the stability objectives for clocks of all international gateway digital switches. The stability objective of one part in 10^{11} implies that slips between international gateway switches will occur at a rate of one per 70 days. (This assumes one clock is positive one part in 10^{11} and another clock is negative one part in 10^{11}, and there are 8000 frames per second.)

7.3.2 Networkwide Pulse Stuffing

If all internal links and switches of a network were designed to run at nominal rates slightly higher than the nominal rates of the voice digitization processes, all voice signals could propagate through the network without slips by stuffing the information rate up to the local channel rate. None of the clocks would have to be synchronized to each other, and relatively coarse clock accuracies could be tolerated. At every interface between systems running under different clocks, however, the individual channels would have to be unstuffed from the incoming rate and stuffed up to the local or outgoing rate. In essence, the TDM links of the network would provide TDM channels through which user data flows at lower and variable rates—the differences being absorbed by internal pulse stuffing.

In contrast to pulse stuffing operations of higher-level multiplexers where all channels in a lower-level digital signal are stuffed as a group, switching operations imply that each channel must be stuffed independently. The need for separate pulse stuffing operations is illustrated in Figure 7.22, which depicts two voice signals being switched into a common TDM outlet link. Obviously, the bit rate R_3 of both output channels is identical. If the two channels originate in portions of the network running under different clocks R_1 and R_2, the pulse stuffing adjustments must be made separately for each channel.

The need to independently stuff each channel at every switching interface precludes pulse stuffing as an economically feasible solution

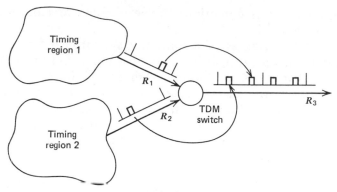

Figure 7.22 Switching two channels with different rates onto a common TDM output.

to networkwide timing problems. Pulse stuffing is being used at the present time for higher-level multiplexers primarily because other synchronization procedures have not yet been established. As the digital network evolves and the ability to synchronize the clocks of all lower-level digital signals arises, pulse stuffing may become unnecessary. One instance in which pulse stuffing might always be desirable is over digital satellite links that are timed by free running clocks in the satellite [16].

7.3.3 Mutual Synchronization

The two preceding sections discuss modes of operation for the network that do not involve synchronization of individual clocks. This section, and the next two, describe network timing plans that synchronize each individual clock to a common frequency. The first method, mutual synchronization, establishes a common network clock frequency by having all nodes in the network exchange frequency references as shown in Figure 7.23. Each node averages the incoming references and uses this for its local and transmitted clock. After an initialization period, the network clock normally converges to a single stable fre-

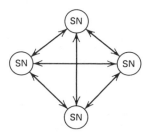

Figure 7.23 Mutual synchronization. SN = switching node.

quency. Under certain conditions, however, the averaging process can become unstable [17].

The main attractiveness of a mutually synchronized network is its ability to remain operational in spite of a clock failure in any node. The main disadvantages are the uncertainties of the exact average frequency and unknown transient behavior. For these reasons, mutual synchronization is not being considered for the North American telephone network. In Great Britain, however, a hierarchical timing structure is being developed that utilizes mutual synchronization within some of the levels [18].

7.3.4   Network Master

Another method of synchronizing the network is shown in Figure 7.24. With this method a single master clock is transmitted to all nodes enabling them to lock onto a common frequency. As indicated, all network nodes are directly connected to the network master implying the need for a separate transmission network dedicated to the distribution of the reference. Reliability considerations also imply that alternate paths be provided to each node. Because of cost considerations for the separate network, a network master with direct transmission to each node is undesirable.

7.3.5   Master-Slave Synchronization

The main drawback to network master synchronization as described in the preceding section is its need for separate and reliable transmission facilities to every node. Figure 7.25 shows a network configuration that disseminates a master reference by way of the message links themselves. A network reference frequency is transmitted to a few selected higher-level switching nodes. After these nodes synchronize their clocks to the reference, and remove transmission link induced timing jitter, the reference is passed on to lower-level switches by way of existing digital links. The next lower-level switches, in turn, synchronize to an incoming link from the higher level and pass timing on

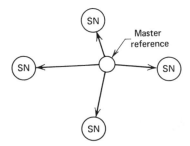

Figure 7.24 Network master synchronization. SN = switching node.

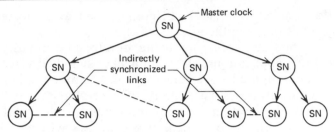

Figure 7.25   Master-slave synchronization.

to another level of switches by way of their outgoing digital links. The process of passing the reference downward from one level to the next is referred to as "master-slave synchronization."

Since all switching nodes in the network are synchronized either directly or indirectly to the same reference, they all run at the same nominal clock rate. Thus slips should not occur normally. However, owing to the different paths through which timing is disseminated, short-term frequency differences can occur between some nodes. If these nodes are synchronized indirectly, as shown in Figure 7.25, infrequent slips might occur. Furthermore, reliability considerations imply that backup clocks must be provided in all switches should the clock distribution system fail. When this happens, slips become more likely, but only after relatively stable backup clocks have had enough time to drift from the common reference frequency.

The Bell System and the United States Independent Telephone Association (USITA) have selected master-slave synchronization for the evolving switched digital network in the United States [19]. The reference frequency is located in Hillsboro, Missouri from which selected No. 4ESS switching centers receive their timing by way of dedicated transmission facilities. Synchronization of all other switches occurs by way of existing digital transmission links.

7.3.6 Packetization

The synchronization discussions of the five preceding sections have assumed implicitly that a synchronous, circuit-switched network was being considered, since prevailing digital voice networks operate in that manner. For completeness, however, another form of network must be mentioned—a packet-switched network.

As discussed in Chapter 8, packet-switched networks break up measages into identifiable blocks (packets) of data. In between the blocks, the transmission links carry either idle codes or control messages. If all messages (control and data) are separated by a nominal interval of idle transmission, elastic stores can be reset in preparation for the next block. As long as each block is limited in length, the elastic stores can absorb clock differences and avoid losses of data. (In essence, slips occur in the idle codes only.)

Packet-switched networks are being developed primarily for data communications applications, although packet-switched voice networks have also been proposed. (See references in Chapter 8.) This mode of operation demonstrates that purely packet-switched (or message-switched) networks have an inherent tolerance for clock offsets. Each node in the network can be timed autonomously with relatively inaccurate clocks. Notice, however, that transmission link efficiency is degraded because of greater overhead between blocks. Furthermore, eventual generation of a decoder sample clock for voice reconstruction is complicated by the bursty arrivals (waiting time jitter).

7.4 NETWORK CONTROL

In essence, the synchronization procedures described in the preceding section represent methods for controlling the timing between transmission and switching systems. In this section we discuss synchronization in a more general sense. Instead of just time or frequency control, the synchronization concept is extended to higher-level functions of connection control and network control as a whole. Fundamental to the control concept is the interaction between two processes (e.g. the exchange of information from one switching machine to another to set up or monitor a connection).

A particularly useful means of defining the interaction of two processes is a state transition diagram. The main purpose of the state transition diagram is to abstract the operational states of a process and indicate what events cause transitions from one state to another. When these events are messages (signaling tones) from another process, the state transition diagram effectively defines how two communicating processes interact.

7.4.1 Hierarchical Communication Process

As an example of a more generalized concept of synchronization, Figure 7.26 has been included to demonstrate three distinct levels of control for a conventional telephone connection using digital transmission and multiplexing. The lowest-level process shown in Fig. 7.26a depicts nothing more than the clock synchronization process required to transmit and receive digital information. There are only two states to the process in both the transmitter and the receiver. The purpose of clock synchronization is to cause transitions between the two states in the receiver to coincide with transitions in the clocking process of the transmitter. To accomplish this a certain amount of transmission capacity is required in the form of line code transitions.

Figure 7.26b depicts a higher-level synchronization process involving the framing of a time division multiplexer. Both processes represent a modulo-C counter where C is the number of channels in the TDM frame. The two processes are synchronized (framed) by utilizing some of the transmission capacity to send framing patterns. Once the receiver acquires framing, the counter in the receiver counts in synchronism with the counter in the transmitter so that individual TDM channels are properly identified.

Figure 7.26c provides state transition diagrams of a somewhat more complicated, but easily understood process. The figure depicts the connection control of a conventional telephone call. The state transition

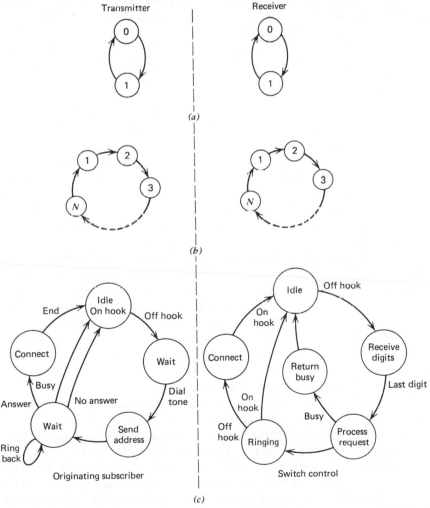

Figure 7.26 State transition diagram of synchronization processes. (*a*) Timing process. (*b*) Framing process. (*c*) Telephone connection process.

diagram of the first process represents a subscriber placing a call (going off-hook). The second state transition diagram represents the sequence of states the control element in the local switch goes through to set up the connection.

As indicated, the process begins by the originating subscriber going off-hook and waiting for the dial tone. When the switch recognizes the off-hook signal (current flow in the line), it connects the subscriber line to a digit receiver which returns dial tone. The subscriber then dials the address of the desired telephone and enters another wait state. Upon receiving the last digit of the address, the switch control processes the request. If a local connection is desired, a busy tone or a ringback tone is returned to the originating subscriber. A busy tone prompts the sub-scriber to hang up (go on-hook) while a ringback signal causes the sub-scriber to stay in the wait state until the called party answers, or until the caller "times out" and hangs up.

When the called party answers, both processes enter the connected state and communication between the end users begins. The end users then get involved in yet another level of "synchronization." Voice tele-phone users begin by exchanging greetings and identities to establish a "connection" between their thought processes to communicate on a mutually understood subject. The message exchange process also requires synchronization so that only one person talks at a time. Hence various forms of control signals are needed to "turn the line around." Although being somewhat subtle in nature, these control signals repre-sent transmission overhead in the same sense as control signals within the network. A talker may indicate his end-of-transmission by asking a question, by inflections in the voice, by the message itself, or more commonly by a pause.

Data communications equipment goes through the same basic pro-cedures in order to establish connections and exchange information. In this case, the procedures are defined more formally and, conse-quently, are more restrictive. The formal rules of communication between data communications equipment are usually refered to as a "protocol." Data communications protocols typically include a def-inition of certain control codes, code interpretations, message framing, turn-around procedures for half-duplex lines, error control, message sequencing, fault control, and recovery.

Automated fault control and recovery procedures for communica-tions networks can become quite involved and difficult to implement reliably. When individual voice circuits malfunction (e.g. become noisy or disconnected), the recovery procedures are left to the users. They merely redial the connection and take up where they left off. However, large trunk groups or switching systems must be designed for higher levels of dependability and maintainability. The dependability criterion ensures that failures or malfunctions rarely occur, or that they are circumvented automatically by protection switching. High levels of maintainability ensure that failures are repaired quickly when they occur. Within switching systems, most of the instructions and memory words of the processor are typically dedicated to hardware and software performance monitoring, recovery procedures, and maintenance diagnostics.

7.5 NETWORK MANAGEMENT

In addition to controlling individual connections and equipment, a communications network must also manage its facilities on more macroscopic levels. The basic goal of network management is to maintain efficient operations during equipment failures and traffic overloads. The main considerations are the following:

1 Routing control.
2 Flow control.

7.5.1 Routing Control

Routing control refers to procedures that determine which paths in a network are assigned to particular connections. If possible, connections should use the most direct routes at the lowest levels of the network. The direct routes are obviously desirable because they use fewer network facilities and generally provide better transmission quality. However, economic considerations often limit the capacities of the direct routes so that alternate routes are needed to maintain suitably low-blocking probabilities between one end office and another.

If a trunk group between two switching offices contains enough circuits to provide an acceptably low-blocking probability, a significant number of the circuits in the group are idle during average traffic loads. A more economical design allocates a limited number of heavily utilized trunks in the direct route and provides alternate routes for overflow (alternately routed) traffic. In this manner the users are able to share larger portions of the network. Chapter 9 presents basic examples of how a network can be engineered to minimize the transmission facilities while providing a given grade of service (blocking probability).

7.5.2 Flow Control

In the preceding section, alternate routing is discussed as one aspect of managing traffic in a communications network. Routing algorithms are concerned only with the utilization of paths or directions of travel within a network. Another requirement of network management is to control the amount of traffic in a network. Managing the rate at which traffic enters a network is referred to as flow control. A network without effective flow control procedures becomes very inefficient or ceases to function entirely when presented with excessively heavy traffic conditions.

The generalized performance of a large, uncontrolled network is shown in Figure 7.27 as a function of the offered traffic. As indicated, when light traffic conditions exist, the network carries all traffic requests presented. As the load increases, however, some of the offered traffic is rejected because no appropriate circuits are available for particular connections, that is, blocking exists. As the input load increases even further, a network with no flow control eventually begins to carry less traffic than it does when lighter loads are presented. If the offered load increases even more, the network may even cease to carry any traffic at all.

The reason that the volume of carried traffic decreases when the offered traffic exceeds some critical value is that partially completed requests tie up network resources while trying to acquire other resources tied up by other partially completed requests. Thus a form of deadlock occurs. As a specific example, consider the public telephone network and the occurrence of extremely heavy traffic between two distant cities. Since each request requires an end-to-end connection, a sequence of intermediate transmission and switching facilities is needed to set up each circuit. Under normal operations both sources of traffic are serviced by successively seizing trunks to intermediate switching nodes until the destination is reached. If heavy traffic exists, requests emanating from both sides of the network encounter congestion somewhere near the middle. At that time, the partially completed connections are tying up facilities needed by other requests. In the limit, when

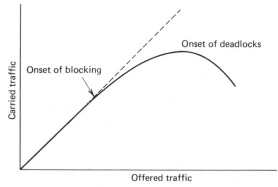

Figure 7.27 Traffic carried versus traffic offered for a network with no flow control.

extremely heavy traffic exists, all network resources are held by partially completed requests and no complete connections can be established!

Another example of the need for flow control, is automobile traffic at a metropolitan intersection. Have you ever encountered an intersection in which your lane of traffic was blocked by cross traffic backed up into the intersection? In essence, the driver blocking your direction of travel seized a common resource (the middle of the intersection) without being able to obtain the next resource required (the other side of the intersection). Bumper-to-bumper traffic in one direction can significantly degrade the throughput in other directions. With heavy traffic in all directions, total throughput can grind to a halt until the congestion is relieved from the periphery inward.

The fundamental principle demonstrated by these examples is that efficient use of the common resources of a heavily loaded network requires some form of flow control. In the automobile example, smooth operation of an intersection depends on each individual driver looking ahead and not entering an intersection unless he can get all the way across. A telecommunications network must use the same basic principle (it is hoped with more discipline). The control elements at the periphery of the network must be aware of the internal status of the network and control the flow of traffic from its sources.

More than one level of flow control is sometimes implemented within a network. In a data communications link, some form of flow control is usually required to keep a source terminal from overloading the terminal at the other end of the link. The receiving terminal uses a reverse channel (sometimes with a lower bit rate) to inform the source when to cease or begin transmissions. This level of flow control in a circuit-switched network involves the terminals themselves and is of no concern to the network since the traffic flows within an established connection. Of more concern to a circuit-switched network is how to control the flow of connection requests into the interior of the network. In setting up a long-distance connection the first few circuits required should not be seized unless there is a reasonable chance that all of the circuits necessary to complete the connection can be obtained. Partially completed circuits only degrade the network capacity by increasing congestion without satisfying a service request. The following paragraphs describe some of the basic mechanisms used in the present telephone network to control the flow of connection requests.

Dial Tone Delay

A natural consequence of a busy network at the local level is that common equipment such as digit receivers are much in demand. The effect of the heavy use is to produce queues for service and concomitant delays in returned dial tones. Although dial tone delays are not intended to do so, they provide an effective means of slowing the flow of requests into the interior of the network. Since flow control for local connections is normally unnecessary, dial tone delay is undesirable as an intentional means of delaying service requests at the periphery of the network. A preferred means of flow control is first to receive the digits of a service request, and then to reject the request, if necessary, by returning a reorder tone (busy tone) or recorded message to the originating subscriber. The subscriber can effectively place himself in a queue for service, if he continually repeats the request. This approach has the one drawback that subscribers with serviceable requests experience increased dial tone and processing delays because of the load produced by unserviceable retries.

Trunk Directionalization

The operation of trunk circuits can be classified according to two different ways of controlling seizures for particular calls. Two-way trunks can be seized at either end. One-way trunks, on the other hand, can be seized only at one end. (Notice that this has nothing to do with the direction of message transfer on established connections, which is always in both directions.) When one-way trunking is used, the trunk group is usually partitioned into one group that can be seized at one end and one group that can be seized at the other end. Two-way trunk groups are obviously more flexible in servicing fluctuating traffic patterns, but they are more difficult to control since the possibility of simultaneous seizures (called glare) at both ends must be resolved.

A useful feature to incorporate into two-way trunks is the ability to directionalize them by marking them busy at one end and effectively creating one-way trunks. With this mechanism, a distant, overloaded switching node can be relieved of additional incoming traffic while providing it sole access to the trunk group for outgoing traffic. Thus the overloaded node relieves its congestion while inhibiting new arrivals.

When the network as a whole experiences heavy traffic loads, trunk directionalization can be used to reduce the flow of connect requests into the interior of the network while establishing one-way trunks from the interior to the periphery. Thus connect requests that manage to get to the interior have a much better chance of obtaining all facilities required for the complete connection.

Cancellation of Alternate Routing

Alternate routing of traffic is primarily useful for accommodating localized overloads by transferring traffic to underutilized routes. During networkwide overloads, however, alternate routing is undesirable for two reasons. First, alternate routes imply that a greater number of transmission and switching facilities are needed for a connection. If these same facilities could be assigned to two or more direct connections, the total number of links per call could be reduced and the network could carry more traffic.

Second, the probability that an alternately routed call can acquire all the necessary resources is relatively low. Trying to set up a connection with a large number of facilities is undesirable if the probability of getting all of the facilities is low (particularly so, if some facilities are fruitlessly tied up while less demanding requests are pending).

Code Blocking
Code blocking refers to artificially blocking calls intended for specific destination codes. If the calls are blocked at originating end offices before they acquire internal network facilities, the destinations are relieved of incoming traffic without tying up facilities that may be needed for outgoing requests from the specified areas.

This method of flow control is particularly useful in times of natural disasters, which typically stimulate large numbers of calls both into and out of the area of the disaster. In these events a network control center can initiate code blocking for all, or a large percentage, of the calls into the area. The principle of giving preference to the outgoing calls serves two purposes. First, no network facilities are seized unless there is a reasonable chance of obtaining all facilities necessary. It is the trunks into or out of the disaster area that are the focal point of network congestion. Once one of these trunks is seized the rest of the connection can probably be established. Second, code blocking is useful because outgoing calls are probably more important than incoming calls.

Centralized Connection Control
All of the flow control procedures described previously are designed to eliminate seizures of common resources if the desired connection has a low completion probability. Because of the distributed nature of network control implied by these operations, these control procedures are necessarily probabilistic. In order to maintain a certain amount of network efficiency, the network is purposely operated at less than maximum capacity.

A more desirable mode of operation, from a throughput point of view, is to allocate network facilities from a single centralized control node. Since this central node has access to the status of all network resources, no facilities are assigned to a particular request unless all facilities needed for the complete connection are available. In this manner network transmission links are assigned in an analogous fashion as internal links of common control switches.

Complete centralized control of a network as large as the public telephone network is obviously infeasible from the point of view of maintaining status of all local links and telephones, and from the point of view of survivability of the network when the control node fails. However, certain aspects of centralized control are being implemented in North America and around the world with common channel interoffice signaling (CCIS). For example, INWATS call requests can be routed to a central node that determines if the destination is busy or not. If the destination is busy, the originating end office is instructed to return the busy tone without any of the internal transmission links ever being seized [20]. This mode of operation is particularly useful for 800 numbers (INWATS) that occasionally experience very heavy traffic flow because of national television announcements.

Without CCIS, the normal mode of operation is to return a busy tone all the way through the network from the place at which the busy circuit or subscriber is located. Thus the path through the network is tied up during the time the busy tone is being returned. CCIS allows the originating office to return the busy tone so internal network facilities can be released and reassigned immediately upon detecting the busy condition.

REFERENCES

1   M. Decina and U. deJulio, "International Activities on Network Synchronization for Digital Communication," *IEEE International Communications Conference*, 1979.

2   *Reference Data for Radio for Radio Engineers*, Howard Sams Inc., Indianapolis, Indiana, 1972.

3   J. E. Abate, L. H. Brandenburg, J. C. Lawson, and W. L. Ross, "The Switched Digital Network Plan," *Bell System Technical Journal*, September 1977, pp. 1297–1320.

4   J. R. Pierce, "Synchronizing Digital Networks," *Bell System Technical Journal*, March 1969, pp. 615–636.

5   A. A. Collins and R. D. Pedersen, *Telecommunications, A Time for Innovation*, Merle Collins Foundation, Dallas, Texas 1973.

6   C. J. Byrne, B. J. Karafin, and D. B. Robinson, "Systematic Jitter in a Chain of Digital Repeaters," *Bell System Technical Journal*, November 1963, pp 2679–2714.

7   F. M. Gardner, *Phaselock Techniques*, 2nd ed., John Wiley & Sons, New York, 1979.

8   E. D. Sunde, "Self-Timing Regenerative Repeaters," *Bell System Technical Journal*, July 1957, pp 891–938.

9   D. L. Duttweiler, "The Jitter Performance of Phase-Locked Loops Extracting Timing from Baseband Data Waveforms," *Bell System Technical Journal*, January 1976, pp 37–58.

10   "Characteristics of Primary PCM Multiplex Equipment Operating at 2048 kbit/s," CCITT Recommendation G.732, *Orange Book* Vol. 3-2, 1976, p 425.

11 J. F. Boyle, J. R. Colton, C. L. Dammann, B. J. Karafin, and H. Mann, "No. 4ESS: Transmission/Switching Interfaces and Toll Terminal Equipment," *Bell System Technical Journal*, September 1977, pp 1057–1097.

12 Members of Technical Staff, Bell Telephone Laboratories, *Transmission Systems for Communications*, Bell Telephone Laboratories, 1971.

13 "Impact of Jitter on the Second Order Digital Multiplex at 6312 kbit/s," AT&T Submittal to CCITT study group on jitter, *Green Book*, Vol. 3, pp 861–869.

14 D. L. Duttweiler, "Waiting Time Jitter," *Bell System Technical Journal*, January 1972, pp 165–207.

15 "Plesiochronous Operation of International Digital Links," CCITT Recommendation G. 811, *Orange Book*, Geneva, Switzerland 1976.

16 T. Muratani and H. Saitoh, "Synchronization in TDMA Satellite Communications," *IEEE International Conference on Communications*, 1979, pp 11.4.1–11.4.6.

17 J. P. Moreland, "Performance of a System of Mutually Synchronized Clocks," *Bell System Technical Journal*, September 1971, pp 2449–2464.

18 P. A. Mitchell and R. A. Boutler "Synchronization of the Digital Network in the United Kingdom," *IEEE International Conference on Communications*, 1979, pp. 11.2.1–11.2.4.

19 C. A. Cooper, "Synchronization for Telecommunications in a Switched Digital Network," *IEEE Transactions on Communications*, July 1979, pp 1028–1033.

20 George Ebner and Lawrence Tomko, "CCIS: A Signaling System for the Stored Program Controlled Network," *Bell Laboratories Record*, February 1979, pp 53–59.

PROBLEMS

7.1 Determine the size of an elastic store needed to accommodate a velocity shift of ± 1000 km/hr that lasts for 10 seconds if the data rate is 10 Mbps. (The speed of light is $3(10)^8$ m/sec.)

7.2 How many bits are needed in an elastic store designed to interface the primary digital signal of CCITT to a digital switch? What is the maximum slip rate if the line clock and switch clock differ by +50 parts per million to −50 parts per million (the maximum recommended offsets)?

7.3 What is the maximum phase offset (in signal intervals) produced by a jitter power of +10 dB relative to 1 radian squared?

7.4 Determine the rate at which DS-1 signals in a DS-2 multiplex lose framing because stuff codes of a DS-2 signal are incorrectly interpreted. Assume the channel BER is 10^{-3}. Assume the BER is 10^{-6}.

7.5 What would the rate of incorrect DS-2 stuff code occurrences be if a 5 bit stuff code were used instead of the 3 bit stuff code? Assume the channel BER is 10^{-3}.

7.6 A digital transmission link is to be used to transmit blocks (packets) of data without slips. If the transmission link is autonomously timed with respect to the receiving terminal, what is the maximum allowable block length if the clocks vary by ±50 parts per million each and an elastic store of 16 bits exists in the receiving terminal? Assume the elastic store is initialized to half full between each block.

7.7 Assume that worst case systematic jitter from a single repeater produces a phase slope of 300 radians per second for 1 ms. What is the peak-to-peak jitter in decibels relative to a signaling interval at the end of a line with 200 such repeaters?

7.8 Repeat Example 7.4 for an M_{23} multiplexer.

7.9 A jitter power of 100 decibels relative to one radians-squared is observed at the receiving end of a digital microwave link. What is the probability that the phase offset will exceed ±3 symbol intervals?

CHAPTER EIGHT

DIGITAL NETWORKS

Although this book emphasizes the use of modern digital electronics as a means of supplanting traditional analog implementations of voice communications networks, it should be remembered that communications by digital techniques is actually an ancient practice. In early times communication over distances greater than the range of the human voice was provided by sight or sound of discrete signals (e.g., heliographs, smoke signals, flags, drum beats, bugles, etc.). Furthermore, the first practical communication system to use electricity, the telegraph, is inherently digital.* As telegraphy evolved from the original manually based systems to fully automated systems, they developed into what are commonly referred to as message switching networks. Modern message switching is discussed in the first section of this chapter.

More recently, digital communications networks have evolved to satisfy data communications needs. Because of its widespread availability it is only natural that the public telephone network is being used to provide most data transmission services. The availability overshadows numerous technical shortcomings of a network designed primarily for voice communications services. The main deficiencies of the telephone network for data transmission are

1   Need for signal transducers (modems).
2   Low-data rates.
3   High-error rates.
4   Inefficient circuit utilizations.

As data communications requirements have increased, so has the justification for more cost-effective data communications networks. One solution to reducing data transmission costs is to improve circuit utilizations through the use of packet switching networks. The technology of packet switching was pioneered by the Advanced Research Projects Agency of the U.S. government. This agency developed a network re-

---

*It is also interesting that early attempts to transmit voice over an electrical circuit used vibrating relays as transducers onto and from a telegraph circuit. Hence these attempts were quasi-digital in nature.

ferred to as the ARPANET [1]. ARPANET has served as a test bed for evaluating numerous packet switching concepts and is the most-cited example of packet switching in the United States. One of the greatest contributions of ARPANET is the wealth of publications describing its operation. Particularly illuminating are descriptions of some of the problems that arose during its development [2, 3].

In addition to ARPANET, which is used only by government, educational, and industrial research institutions, a number of public packet switching networks have been developed in the United States and throughout the world. Two notable examples are DATAPAC [4] in Canada and TRANSPAC [5] in France. In the United States, the two most prominent public networks are TELENET [6] and TYMENET [7]. Because these public networks use leased lines from common carriers and add switching facilities before selling their services, these networks are sometimes referred to as value added networks.

Packet switching networks are an outgrowth of the need to utilize the capacity of a data communications circuit more efficiently. Circuit inefficiencies arise because the statistics of most data communications traffic vary significantly from voice traffic statistics. A subscriber voice terminal is typically involved in one or two calls per busiest hour of the day. The average telephone call lasts for 3 to 4 minutes. Thus, a typical voice terminal is in use only for 10% of the busiest hour of the day. In contrast, data communications terminals generally produce more frequent and shorter messages. In common applications a data terminal uses a voice-grade circuit for extended periods of time, even though actual use of the transmission capacity is very low. Data communications usage patterns represent abnormal loads to the telephone network whether infrequent and long holding times are used or whether frequent, short holding times occur: Short calls overload the call processing elements of a switch, and a few calls with extended holding times overload interoffice trunks and paths through switching matrices. Because of increasing amounts of "abnormal" traffic, and because of computational capabilities of stored program control switches, telephone operating companies are beginning to employ usage sensitive accounting for local calls in lieu of traditional flat rate service.

From a customer's point of view, the most inefficient aspect of data communications occurs on long-distance connections (or measured local service) that incur charges based on total holding times and not on actual utilization of channel capacity. In a typical interactive computer session, the transmission capacity is less than 1% utilized.

A second approach to improving data communications involves developing separate networks specifically designed for digital transmission (no analog circuits with modems). The first major enterprise of this type in the United States was a nationwide digital microwave network developed by Digital Transmission Corporation (DATRAN). Unfortunately for data communications users, DATRAN ran into financial difficulty and declared bankruptcy in 1977.

Following the DATRAN service introduction, digital transmission services became available from AT&T by way of their Dataphone Digital Service (DDS) [9]. More recently, Satellite Business Systems (SBS) was formed by IBM, Aetna Life Insurance Company, and Comsat General to implement satellite based digital transmission. A similar system referred to as XTEN was announced by Xerox [10] but subsequently withdrawn.

Except for the SBS network, which includes voice services, the aforementioned networks are digital by the nature of their application: either messages or data. Since voice networks can be implemented quite effectively with digital techniques, it is natural to consider designing digital networks that service both types of traffic. A later section of this chapter discusses basic considerations to be made when designing a single digital network to provide *both* voice and data services. The last section of this chapter discusses future prospects for all-digital telephone networks in the United States.

8.1 MESSAGE SWITCHING

As one telegraph system after another was installed in the countries around the world, nationwide communications networks evolved. A message could be sent from one point to another even if the two points were not serviced by a common telegraph line. In this case, telegraph operators at intermediate points would receive a message on one line and retransmit it on another. When a telegraph office had several lines emanating from it, the process of transferring a message from one line to another was, in essence, a switching function.

The process of relaying, or switching, a message from one telegraph line to another became semiautomated when teletypes with paper tape punches and readers were developed. An incoming message could be punched automatically onto a paper tape by one teletype and subsequently read by another teletype for transmission on the appropriate outgoing line. The process of transferring a message from one line to another in this manner led to these systems being referred to as torntape message switches. Some torn-tape message switches still exist today.

One of the world's largest message switches was completely automated in 1963 when Collins Radio Company of Cedar Rapids, Iowa installed a computer based message switch for the airline companies of North America. This system, and its more recent successors, eliminate paper tape transfers by storing incoming messages directly into a computer memory (disk file) and forwarding them automatically to the appropriate output line when available. Hence this mode of operation is often referred to as "store-and-forward" message switching.

Included with each message is a header containing an address and possibly routing information so the message processor at each node can determine which output line to switch the message to. As indicated in Figure 8.1, the processor in each node maintains message queues for each outgoing link. These queues are normally serviced on the first-come first-served basis. However, priority information can sometimes be included in each header to establish different classes, or grades of service—thereby allowing time-critical messages to be placed at the head of a queue.

A message-switching network is fundamentally different from a circuit-switching network in that the source and destination do not interact in real time. In fact, most message-switching networks can deliver a message on a delayed basis if a destination node is busy or otherwise unable to accept traffic. There is no need to determine the status of the destination node before sending a message, as there is in circuit switching.

Message-switching networks are also fundamentally different from circuit-switching networks in their response to traffic overloads. A circuit-switching network blocks or rejects excess traffic, while a message-switching network normally accepts all traffic but provides longer delivery times as a result of greater average queue lengths.

Another important distinction of a message-switching network is that the transmission links are never idle while traffic is waiting to use them. In a circuit-switching network, a circuit may be assigned to a

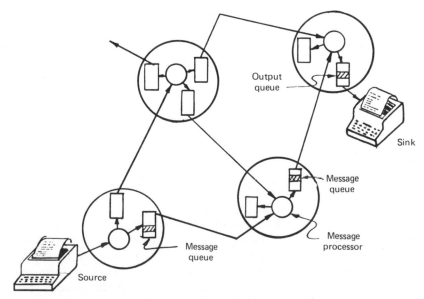

Figure 8.1　Message switching network.

particular connection but not actually carrying traffic. Thus some of the transmission capacity may be idle while some users are denied service. In contrast, utilization of the transmission links of a message-switching network is directly related to the actual flow of information. Arbitrarily high-utilization efficiencies are possible if increased store-and-forward queueing delays are acceptable. Chapter 9 provides basic results of queueing theory that relate utilization efficiency to queueing delay.

8.2 PACKET SWITCHING

The circuit-switched telephone network is ill-suited to interactive data traffic because it is fundamentally designed for less frequent service requests with comparatively long holding times (3 to 4 minutes on average). Neither the control elements in the switches nor the capacity of the signaling channels are capable of accommodating frequent requests for very short messages. The result is that connection setup time (up to 15 seconds for long distances but being reduced by CCIS) is many times greater than the holding time of a data message. Obviously, more efficient utilization of the network requires greater control channel bandwidth and increased call processing capacities in the switches. Beyond this, however, interactive data traffic with low-activity factors requires a network operation that is fundamentally different from a conventional circuit-switched network. The most appropriate mode of operation for traffic that comes in bursts is more closely related to a message-switched network than to a circuit-switched network.

Figure 8.2 depicts both the conceptual structure and the conceptual operation of a packet-switched network. A single message at the source

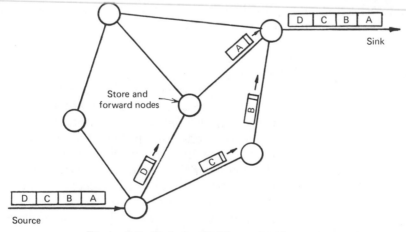

Figure 8.2 Packet switching network.

is broken up into "packets" for transmission through the network. Included in each packet is a header containing address and other control information. Each packet is relayed through the network in a store-and-forward fashion similar to a message switching network. At the destination node, the packets are reassembled into the original contiguous message and delivered to the sink.

The main feature of a packet switching operation is the manner in which the transmission links are shared on an as-needed basis. Each packet is transmitted as soon as the appropriate link is available, but no transmission facilities are held by a source when it has nothing to send. In this manner, a large number of relatively inactive sources can share the transmission links. In essence, link utilization is improved at the expense of storage and control complexity in the nodes. A circuit-switched network has control overhead associated with connection setup, but very little control thereafter. In contrast, packet-switching nodes must process the header information in each packet as it arrives. Thus a long message in a packet-switched network requires more control overhead than if it were serviced in circuit-switched network. Considering the declining cost of digital memory and processing, the increased control complexity becomes less and less significant as time goes on.

As the traffic load in a packet-switched network increases, the average transmission delay increases correspondingly. In contrast, a circuit-switched network either grants service, or rejects it. There is no graceful degradation in service. Conversely, when only a few circuits are in use in a circuit-switched network, much network transmission capacity is idle. When there is a light load on a packet-switched network, the active users benefit by shorter than usual delay times. Hence from a user's grade-of-service point of view, the two network types are fundamentally different.

Using automatic repeat request (ARQ) error control, packet-switching networks normally provide essentially error free transmission for each node-to-node transfer. This process requires the receiving nodes to monitor redundant check bits appended to each packet to determine if the packet was received correctly. When errors are detected, a retransmission is requested. Hence transmitting nodes must hold all transmitted packets in memory until a positive response (ACK) is returned by the receiving terminal. Furthermore, an entire packet must be received and checked for errors before forwarding it to another node.

As commonly implemented, the nodes of a packet switched network are interconnected by leased analog lines from the common carriers. Voice-grade lines are usually conditioned to provide the maximum data rates possible using voiceband modems (9600 bps typically). Wideband channels providing 56 kbps are also commonly used.

Customers access packet networks by way of local leased lines or dial-up connections. Dial-up connections are used by infrequent users,

while leased lines are preferred by heavy users to achieve constant availability, higher data rates, and often lower error rates. Most computers
are normally connected by way of leased lines, whereas terminals are
more apt to use dial-up connections.

If flat rate accounting is in effect for local dial-up access, packet
switching customers have long-distance data communications capabilities at standard monthly rates from the local operating company. A
packet switching company typically charges only for the amount of
data actually transmitted. Thus customers realize considerable savings
if they transmit and receive data using low-activity factors.

Despite the similarity to a message-switching operation, a packet-
switching network is different in two important respects:

1 The store-and-forward delay through a packet-switched network
   is relatively short. Thus interactive communications can occur in
   much the same manner as if a dedicated, end-to-end circuit is
   established.

2 A packet-switched network does not provide storage of messages
   —except in an incidental manner while relaying packets from
   one node to another. The network is designed to provide switched
   communication between two nodes, both of which are actively
   involved in the communications process. A basic packet-switching
   network does not store a message for later delivery to an inactive
   or busy terminal.

One reason for breaking messages into packets is to allow transmission of the first segment of a long message while other segments are in
transit. If the entire message had to be received at each node before
forwarding it to the next node, the delays through the nodes might be
too large. Another reason for breaking the messages into packets arises
from operational simplifications derived from storing, processing, and
transmitting fixed-length blocks of data. In addition, if long messages
are transmitted intact, short messages experience excessive delays when
queued behind long messages. Packetization allows short messages to
get through a transmission link without waiting behind long messages.
This same principle occurs in multiprogrammed computers, which use
time slicing to allow short jobs the opportunity of finishing before
previously started long jobs. One more motivation for packetization is
that when a transmission block is too long, it is unlikely that the entire
message will be received correctly. Packetization provides a means
of retransmitting only those portions of a message that need to be
retransmitted.

8.2.1 Packet Formats

The format of a packet in a packet-switching network can vary significantly from one network to another. Some formats include numerous

fields for control information while other systems rely more heavily on special control packets to transmit control information. Generally speaking, the control information associated with a particular message or link is included in the header of a message packet. Less frequent, networkwide control information is communicated through special control packets. The structure of a packet format sometimes depends on the type of computer used to process the packets and the available communications protocols.

As indicated in Figure 8.3, a packet contains three major fields: the header, the message, and the redundancy check bits. Some packets may not contain a message field if they are being used strictly for control purposes. Although a variety of techniques for generating redundancy checks are possible, the most popular technique uses cyclic redundancy codes (CRCs). Basically, a CRC is nothing more than a set of parity bits that cover overlapping fields of message bits. The fields overlap in such a way that small numbers of errors are always detected and large numbers of errors are detected with a probability of 1 in 2^M where M is the number of bits in the check code [11]. Standard CRCs can now be generated and checked using cyclic shift registers implemented in a single intergrated circuit.

A header typically contains numerous subfields in addition to the necessary address field. Additional fields sometimes included in a header are the following:

1 An operation code to designate whether the packet is a message (text) packet or a control packet. In a sense this field is a part of the destination address, with the address specifying the control element of a switching node.

2 A source address for recovery purposes or identification of packets at a destination node that is capable of simultaneously accepting more than one message.

3 Sequence number to reassemble messages at the destination node, detect faults, and facilitate recovery procedures.

4 Length code to indicate the length of a packet when less than a standard size packet is transmitted. Some protocols insert special

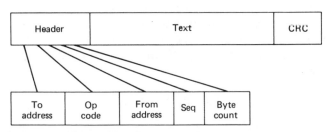

Figure 8.3 Typical packet format.

delimiters (flags) at the end of a packet and therefore do not use a length count.

8.2.2 Statistical Multiplexing

The digital time division multiplexing techniques described in the previous chapters provide multiple channels by periodically assigning a time slot to each channel. The time slots are assigned whether or not the respective sources have anything to send. Channel assignments may be altered on a connection-by-connection basis, but for the duration of a "call" a particular time slot is dedicated to a respective connection. Because of the continuous manner in which time slots occur for each channel, this form of multiplexing is sometimes referred to as synchronous time division multiplexing (STDM). In this section we describe another form of multiplexing, variously referred to as asynchronous time division multiplexing (ATDM) or statistical time division multiplexing (STATDM or STAT-MUX). This form of multiplexing is mentioned here because of its close relationship to packet-switching techniques.

Asynchronous time division multiplexers operate with framing formats that are basically identical to STDM framing formats. The major difference is that an ATDM system periodically redefines the length of its frames to change the number of time slots and, hence, the number of channels. Whereas an STDM system permanently assigns a time slot to each of its sources, an ATDM system assigns a time slot only when a source becomes active. A time slot is eliminated (the frame shortened) when the respective source becomes inactive. Notice that the terms "synchronous" and "asynchronous" in this context have nothing to do with the mode of operation of the transmission link. The transmission links of either an STDM or an ATDM system normally operate with synchronous transmission and detection but can conceivably operate with asynchronous (start/stop) transmission.

ATDM or STAT-MUX systems are typically used to provide line sharing for a multiple number of interactive terminals communicating with a host computer. When only a few sources (terminals) are active, each source has a relatively high data rate channel available to it. As the number of active sources increases, the frame length increases so individual channel rates decrease. Some systems limit the number of active sources to ensure certain minimum data rates.

The purpose and performance of STAT-MUX systems is very similar to the purpose and performance of a packet-switching link. The main difference is that a packet-switching link transmits larger blocks of data with a header included in each block. Each time slot of a STAT-MUX system is shorter and contains only source data. Figure 8.4 contrasts the basic operation of message switching, packet switching, and statis-

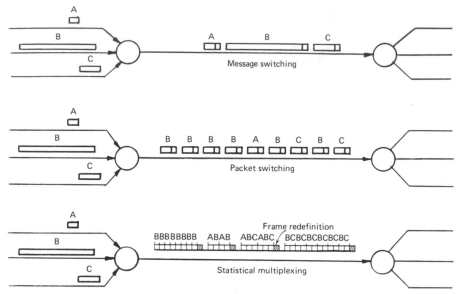

Figure 8.4 Comparison of message switching, packet switching, and statistical multiplexing.

tical time division multiplexing. The message switch transmits each message in its entirety in a first-come first-served manner. Packet switching breaks messages up to allow interleaving of packets from other sources. Thus short messages never get queued behind long messages such as file transfers. The STAT-MUX system breaks the messages up into even finer blocks (words) of data and adds periodic frame definition messages so that receiving terminals can properly identify the individual time slots and switch the incoming data accordingly.

As indicated in Figure 8.4, a packet-switching operation becomes very similar to a STAT-MUX operation if the size of the packets is small. In fact, commercially available statistical time division multiplexers can be used to build up a packet-switching network. In addition to the multiplexers, however, a packet-switching node needs sufficient storage to accumulate an entire packet, check it for errors, process it, and retransmit it. Strictly speaking, a statistical multiplexer does not accumulate an entire message or packet. It switches the incoming data as they are received. Error control functions (redundancy checking and requests for retransmission) are implemented between endpoints of a "connection" instead of between nodes. In summary, a basic ATDM or STAT-MUX system is strictly a multiplexer/demultiplexer. A packet-switching node provides multiplexinglike functions, but also provides message-level and network-level control functions as well.

8.2.3   Routing Control

Much discussion and experimentation has been undertaken regarding various procedures for routing packets from sources to sinks through a packet-switching network [12]. All routing techniques allow for a certain amount of adaption or alternate routing to circumvent line or node failures. The various techniques differ, however, in how fast they respond to failures and whether or not they circumvent network congestion as well as equipment failures. The basic techniques are the following:

1 Dynamic routing.
2 Virtual circuit routing.
3 Fixed path routing.

Each of these techniques can be implemented in a variety of ways and can assume some of the characteristics of the other routing control procedures.

Dynamic Routing

Dynamic routing is implemented on a distributed basis with network nodes examining the destination address of each received packet to determine the appropriate outgoing link. The outgoing link is selected by processing locally stored information to determine which path provides minimum delay to the destination. The routing criteria are routinely updated to include both the operational status (health) and the size of the queues in the neighboring nodes. The routing decisions are adjusted rapidly enough that individual packets of a single message may follow *different* paths through the network. Dynamic routing, with its ability to respond quickly to changes in network topology or traffic conditions, is one of the original features espoused for packet switching. In fact, dynamic routing is sometimes considered to be inherent in the definition of a packet-switching network.

In spite of the obvious attraction of being able to adjust to rapid fluctuations in traffic patterns, dynamic routing has a number of significant drawbacks. One implication of allowing successive packets in a message to follow different routes through the network is that packets may arrive at a destination out of sequence. Although sequence numbers are used to rearrange the packets properly, the reassembly process is complicated—particularly since the destination does not know if a missing packet is merely delayed, or lost entirely.

Another drawback to dynamic routing is the possibility of oscillation occurring in the routing decisions. If the bandwidth assigned to updating the routing control algorithms is too small, a lightly loaded node will attract more traffic than it can handle before neighboring nodes are informed of the change in the traffic flow. In this instance, a packet might even wind up at a node from which it has previously been sent.

Purely distributed control, in general, and dynamic routing, in particular, also presents difficulties with respect to flow control in a packet network. As mentioned in Chapter 7, alternate routing in the switched telephone network is suspended when the network as a whole becomes overly congested (alternate routes require more resources). Obviously, the same principle applies to a packet-switching network. Flow control in packet networks is discussed in later sections.

Dynamic routing is most appropriate for small networks or in a military environment where survivability of the network in the presence of multiple-node failures is a requirement. A military network is typically more concerned with reliable and timely completion of a few important messages than with achieving the highest possible throughput from a given amount of resources. Dynamic routing has been used successfully in the ARPANET [13], but not without associated problems [2, 3].

Virtual Circuits

A virtual circuit network embodies some of the basic aspects of both circuit switching and packet switching. The transmission capacity is dynamically assigned on an "as needed" basis, but all packets of a multipacket message follow the *same* route through the network. Before interactive communication begins, a route is established through the network, and all participating nodes are informed of the "connection" and how to route the individual packets that follow. From then on, all packets flowing between the two end points follow the same route through the network. In essence, a virtual circuit is a logical concept involving addresses and pointers in the nodes of the network, but no dedicated transmission facilities. At the end of a connection (or "session" in data communications terminology), a virtual circuit is released by a "disconnect" message propagating through the network.

Separate "connections" or sessions involving the same two end points do not necessarily use identical paths through the network. Each virtual circuit is established during the call establishment phase depending on the traffic patterns at the time. Thus a virtual circuit network can respond to network failures or changing traffic patterns, but in a longer time frame than a dynamically routed network. When virtual circuits are changed from one "connection" to the next, the mode of operation is sometimes referred to as a "switched virtual circuit network" by direct analogy to conventional circuit switching.

Virtual circuits can be established using either distributed or centralized control. When distributed control is used, the call establishment message propagates through the network with each node making a local decision as to which outgoing link should be selected. In this manner, distributed path selection is basically identical to the routing process used in the public telephone network.

As discussed in Chapter 7 concerning conventional circuit switched networks, centralized call establishment has the basic advantage of being able to set up circuits with a networkwide view of network status and traffic patterns. The TYMNET network of the United States [14] and the DATAPAC network of Canada [15] are examples of virtual circuit, packet-switched networks with centralized control.

Since the call establishment phase of a virtual circuit represents excessive overhead for single-packet messages, a virtual circuit mode of operation is obviously most useful when a network services a relatively large number of multiple-packet messages or sessions. Thus a dual mode network is suggested requiring virtual circuits for longer messages, and immediate transmission with dynamic routing for single-packet messages. In this instance, the single-packet messages are usually referred to as "datagrams."

One of the main advantages of a virtual circuit operation is its ability to provide more orderly control of packet delivery. If a node in a virtual circuit never forwards a packet pertaining to a particular "connection" until the previous packet has been acknowledged, packets can not arrive out of order. A second advantage of a virtual circuit is the reduced addressing requirements of individual packets. Once a virtual circuit has been established, complete destination addresses are no longer needed in the packets. In place of destination addresses, virtual circuit identifiers can be used that are local to each link. In essence, virtual circuit identifiers are pointers to memory addresses in the call processors of the packet-switching nodes. The designated memory addresses contain the pertinent information concerning the actual source, destination, and routing of the packets. Another important feature of a virtual circuit operation is its inherent ability to aid in flow control, as discussed in the next section.

The main disadvantage of a virtual circuit operation is the possibility of greater transmission delays. When a path for a virtual circuit is established, it is chosen to minimize the delay through the network under the traffic patterns at that time. If the traffic patterns change, packets pertaining to a particular virtual circuit may experience long queueing delays on some links while alternate links are more lightly loaded. Yum and Schwartz [16] report that analyses of routing techniques for several small network configurations indicate a packet delay improvement of 10 to 27% is possible when adaptive routing is used in lieu of a fixed routing rule.

When first considered, call establishment might seem to be a disadvantage of a virtual circuit network. Actually, however, flow control considerations require some type of query/response message to determine the status of the destination nodes before a source is allowed to begin sending a packet stream. Thus the control overhead and the delay associated with setting up a "connection" is usually a fundamental requirement, even in a dynamically routed network.

Fixed Path Routing

Fixed path routing embodies the same basic concepts of a virtual circuit network except successive "connections" or sessions between any two end points always use the same path. In essence a fixed path network is one that assigns permanent virtual circuits to each pair of end points.

One attractive feature of a fixed path network is the absence of the call establishment phase of a virtual circuit network. However, unless the necessary resources are permanently allocated, a "clear-to-transmit" message is needed before sending packets. Inactive virtual circuits do not tie up network resources in the same manner as conventional circuits, but each virtual circuit does require transmission capacity and store-and-forward buffers in a probabilistic sense. If minimum grades of service (delay times) are to be guaranteed, a network must limit the number of virtual circuits existing at any particular time. Hence clear-to-transmit signals are needed before a fixed path circuit becomes active. Of course, a network can provide two modes of operation: permanently active (hot line) virtual circuits and virtual circuits activated and deactivated as needed.

An obvious disadvantage of a purely fixed path network, as described so far, is its vulnerability to node or link failures. To circumvent this problem, a network control center can assign semipermanent paths through the network that are changed only when necessary for failure survivability or maintenance. Some message-switching networks are implemented with fixed path routing in the strict sense. This mode of operation is more appropriate to message-switching networks since message delivery is less time critical and can be deferred while repairs are undertaken for inoperative equipment.

8.2.4 Flow Control

As discussed in Chapter 7 concerning conventional circuit-switched networks, routing and flow control are two closely related operational requirements of any communications network. The same basic principle for controlling the flow in circuit-switched networks also applies to packet-switched networks. Namely, whenever the network is experiencing excessive stress, either from a loss of capacity due to failures or from an abnormally heavy demand for services, new service requests must be blocked at the periphery of the network before they tie up common resources and compound the congestion. In a packet-switched network, the common resources are store-and-forward buffers and transmission links.

Flow control in a packet network is primarily concerned with buffer management. For example, if all store-and-forward buffers in adjacent nodes become filled with packets destined to each other, the nodes are unable to receive additional packets, and a deadlock exists. Recall that a packet-switching node does not release a buffer as soon as it transmits a packet. The buffer is released when an acknowledgment is returned from the receiving node. If the receiving node has no available buffers, it cannot accept a new packet and therefore cannot acknowledge it.

Flow control requirements imply that all interface nodes in a packet-switched network are aware of overload conditions and refuse new requests for service until the congestion is relieved. A particularly attractive feature of a virtual circuit network is that the call establishment phase provides an automatic means of determining whether or not a particular request should be serviced. If no path through the network can be established because the interior nodes are too congested, the request is rejected. On the other hand, if a virual circuit is established, there is reasonable expectation that the entire request will be serviced in a timely manner.

Unless network nodes are very conservative in accepting requests for new virtual circuits, the ability to set up a circuit does not guarantee the avoidance of excessive congestion or deadlocks. A node accepts a new virtual circuit based on an expected load and its capacity to service that load. If the traffic volume and patterns happen to exceed the expected load, excessive delay or congestion is possible.

Virtual circuits are an effective means of controlling the flow of multiple-packet messages, but they represent too much overhead for single-packet or datagram flow control. If a virtual circuit network must support a significant number of single-packet messages, it can allow immediate transmission of these messages and forego the call establishment phase. In terms of transmission overhead, a datagram is not much greater than a call-establishment packet. Thus, from this point of view, the packet might as well be sent immediately and be considered its own circuit-setup message. In terms of store-and-forward buffers, however, a single-packet message is much different from a call-establishment packet. A call-establishment packet requires a certain amount of storage and processing by the call processor of each network node it reaches, but it does not compete for store-and-forward buffers as does a message packet. If necessary, a call-establishment message can be ignored by an overloaded node, and no ill effects result. The originating node merely times-out waiting for the network response and reissues the request. The time out should be long enough that the network has had a chance to relax.

In contrast to call-establishment packets, if a message packet is ignored by an overloaded node, congestion migrates to the node that last transmitted the packet, since this node is holding a copy of the rejected packet in a buffer. The buffer can not be released until an acknowledgment is received. Hence datagrams cannot be allowed to enter the network unless a reasonable chance for complete passage exists. When the network is heavily congested, attempts to set up virtual circuits might also be suspended.

A conventional circuit-switched network is unconcerned with flow control between the end users of a connection since, once the circuit is established, the activity or inactivity of the users has no effect on other connections or on the network as a whole. End users necessarily administer flow control between themselves so the source does not overrun the receive buffers of a sink. These procedures concern only the end users.

In contrast, the very nature of a packet-switching network implies direct involvement with user activity. If one user pumps excessive traffic into a network, other users experience degraded performance. Hence interfaces to a packet-switching network necessarily include flow control for respective sources. Source flow control establishes a maximum data rate for a network. If a sink accepts data at a lower rate for a sustained period of time, this fact must be communicated to the network node serving the source in order to slow the source down.

Store-and-forward buffers of a packet-switching network are needed for communications purposes and are not used as a storage medium for messages. If a packet-switching network also provides message-switching services, message storage functions should be implemented separately from the communications buffers. Then, the packet-switching network is used to transfer data to the message storage facility.

Another implication of maximizing store-and-forward buffer utilization is the need to wait until a sink is ready to accept data before a source begins sending. If a sink is not ready, packets get stuck in store-and-forward buffers at the far end of a network and cause congestion. Thus some form of request-to-transmit/clear-to-transmit sequence is needed before message transmission begins. This requirement is independent of the routing algorithms employed in the network. Hence in actual practice the setting up of a virtual circuit does not represent a time penalty.

Single-packet messages can be an exception. If they are transmitted without a clear-to-transmit signal, they may be discarded at the destination node if the sink is inactive or has no receive buffers available. The destination node then needs to return a rejection message to the source indicating the status of the sink. In this manner, the message itself is a request-to-transmit signal. The ARPANET uses a flow control strategy wherein single-packet messages serve as their own requests for buffer storage, but multiple-packet messages require preallocated buffers in groups of eight at the destination [2, 13].

Flow control in TYMENET is implemented in a different manner because of its exclusive use of virtual circuits for all messages.* Before any node in the network can send a packet to a neighboring node, it must receive a clear-to-transmit signal from the neighboring node for the particular virtual circuit. The clear-to-transmit signal is an indication that a specified number of store-and-forward buffers are being held in reserve for that particular virtual circuit. After a node sends the specified number of packets, it sends no more until another clear-to-transmit signal is received (indicating the previous packets have been forwarded or more buffers have been allocated to the virtual circuit). By using the same node-to-node method of flow control at the interface between the network and the users, a very solid end-to-end flow control strategy is established.

The TYMENET flow control strategy is somewhat conservative in that it may allocate store-and-forward buffers to one virtual circuit while another virtual circuit has more use for them (possibly causing a decrease in line utilization). This conservatism, however, provides a number of useful features:

1   Networkwide flow control is established automatically by the flow control within each virtual circuit.

2   Under light traffic conditions, packet flow within each virtual circuit adjusts automatically to the maximum rate that the source and sink can support. If only a few virtual circuits exist, a relatively large number of buffers can be assigned to each circuit— allowing the return of more frequent clear-to-transmit signals.

3   If a sink stops accepting packets for some reason, this fact propagates back to the source by way of cessation of clear-to-transmit signals. Thus the source stops sending when all allocated buffers are full.

4   As long as a node never overcommits its buffers, store-and-forward lockups can not occur. If several sinks stop accepting data, the store-and-forward buffers assigned to the particular virtual circuits get filled and become unavailable. Other virtual circuits, however, can maintain transmission through their own assigned buffers.

5   The mechanism is fail-safe in the sense that positive indications (clear-to-transmit signals) are needed before packets are for-

*TYMENET is oriented almost entirely to interactive terminal users that access remote computers somewhere on the network. A virtual circuit is established for each session. Thus single-packet messages occur only within the confines of an established virtual circuit.

warded to neighboring nodes. When network links are overloaded, flow control bandwidth requirements are minimal. If flow control signals stop altogether, packet transmission stops.

Both the routing and flow control strategies of TYMNET are radically different from those of the ARPANET. For this reason TYMENET is sometimes referred to as a virtual circuit network as opposed to a packet-switching network. With this point of view, "packet switching" implies something about internal network operations other than just packetization of the messages.

8.2.5 Ancillary Benefits

The main motivation for the development of packet-switching networks was the desire to improve the utilization of the transmission links for traffic with high peak-to-average ratios of data rates. A peak-to-average ratio is a quantitative way of expressing the "bursty" characteristic of most computer-based communications. Besides accomodating bursty traffic, packet-switching networks can provide the following additional features:

1 Resource sharing.
2 Load sharing.
3 Interface standardization and translation.
4 Speed conversion.
5 Code conversion.
6 Error control.
7 Alternate routing.
8 Virtual subnetworks.

Resource Sharing

The computational requirements of an individual or group of individuals generally vary from simple and routine tasks to rather sophisticated and involved computations. To support this range of applications, a comensurate range of hardware and software resources is required. If a computer system is sized to support the most demanding tasks, but is used primarily for the simpler tasks, significant cost penalties result. The inefficiencies are compounded because different individuals and institutions have similar needs but end up duplicating the same, infrequently used, resources. A general solution to these inefficiencies is a communications network allowing sharing of a common set of resources. A packet-switching network is an effective means of providing such a network.

Any individual with access to the network can use the processing facilities of any computer (commonly called a host) that is also connected to the network. Thus a user can access one computer for one type of job or a different computer for a different type of job. In this manner special processors, programs, or data bases can be shared by a large number of users.

Load Sharing

The resource sharing aspects of a computer network imply that a specific task is best suited to a specific processor. For many tasks any one of several machines might be appropriate. Load sharing is the process of directing general tasks to any one of several processors, depending on which one is most available. Ideally, load sharing should be implemented as a network operating system that schedules tasks according to needs and availability. When a user "logs on" and indicates what his processing requirements are, the network operating system directs his tasks to the appropriate facility, unbeknownst to the user. If a task requires a special facility, it is sent accordingly. If it is a general task, it is sent to the least-loaded general purpose processor.

A load sharing computer network also has important benefits when expanding the capabilities of a network to meet increased loads. When the capabilities of a single centralized computer are exceeded, upgrading the system can be a traumatic experience—particularly if a new system requires modification of existing software or operating procedures. In a load sharing system, however, the processing capabilities can be expanded by merely adding another identical computer to the network. The users are oblivious to the additional facilities.

The current trend in the data processing industry is to move toward load-sharing networks under the name of distributed data processing. In essence, distributed data processing embodies the same basic concepts of network organization that have allowed the telephone network to grow in a graceful manner. From a communications viewpoint, the computing facilities connected to a distributed computer network represent processing channels, and the network operating system is a call processor routing the tasks through the appropriate channels until a job is completed.

Interface Standardization and Translation

To say that the evolution of data communications technology has been an enigma to the users is an understatement. Basically, chaos in data communications evolved because each communications application was developed in a more or less ad hoc manner (possibly so that mainframe computer manufacturers could lock customers into their peripherals). The result is a myriad of communications products, both hardware and software, from various manufacturers performing similar functions but incompatible with each other. Even the communications product line of single manufacturers contains many incompatible equipments. For example, a plotter may have communications support in one processor or operating system, but not in another.

The obvious solution to the problem is a set of interface and communications standards around which all data communications hardware and software is designed. The basic electrical and mechanical aspects of an interface are well standardized by the Electronic Industries Association (EIA) RS-232-C standard in the United States and CCITT's comparable standard V.24.

Communications procedures have not been nearly as well standardized. Some de facto standards have arisen by emulating common peripheral equipment such as IBM's 2780 and 3780 remote job entry terminals. More recently, packet-switching companies have developed interface standards for their networks that encompass both the electrical/mechanical aspects of an interface (RS-232-C) and the higher-level operations of line control, code interpretation, message formatting, error control, flow control, and so on.* Most notable of these standards is CCITT's recommendation X.25 link access protocol [17]. Although this standard is specifically intended for accessing a communications network, its importance transcends its application to packet-switching networks. If all computer and terminal manufacturers provide support for this interface (or any other single interface), the means for mixing equipment from the various manufacturers becomes available for any application.

Recommendation X.25 actually contains specifications for several types of functional interfaces. One of these is referred to as a virtual circuit interface and is designed to simulate a direct connection between two communicating devices by way of a full time circuit. This implies that messages crossing the interface can be of arbitrary length. Hence the nature of the network (packet-switched or circuit-switched) is transparent to the communication devices. Notice that a virtual circuit in this context refers to the nature of the interface and does not imply anything about the internal operations of the communications network.

A packet-switching network does not have to provide a single interface to help alleviate data communications incompatibilities. In fact, present networks necessarily provide a variety of interfaces so the most common computers and terminals can use the network. In essence, the interfaces emulate the communications aspects of commonly used computers and terminals. Figure 8.5 depicts a packet-switching network supporting a variety of interfaces for various computer types. Notice

*Computer manufacturers are also undergoing a transition from ad hoc communications products to communications equipments that function within network architectures. Basically, a network architecture is a set of communications standards (software modules) that all products comply with so that arbitrary network configurations are possible.

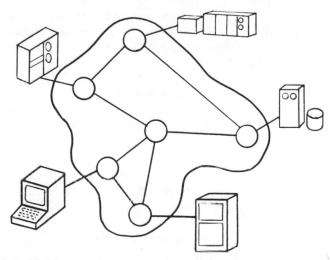

Figure 8.5 Multiple machine access through a packet-switching network.

that a single terminal with a single network interface can communicate with any one of the host computers. Furthermore, the computers can send messages to each other.* In contrast, if dial-up lines were used to access the computers, each terminal would have to support the communications protocol of each computer. In summary, a packet-switching network is capable of providing interface translation services as well as communications functions.

Speed Conversion

Since two devices communicating with each other by way of a packet-switched network are not communicating directly with each other, there can be numerous differences in their respective interfaces. One difference can be the speed at which the interfaces exchange data. For example, a computer interface may be designed to operate at 9600 bps, while teletype terminals may operate at only 110 bps. The packet-switching network can accumulate messages at 110 bps from the teletype, transfer them through the network, and then deliver them to the computer in bursts at 9600 bps.

Code Conversion

An additional aspect of interface translation is the ability of a packet-switching network to provide code conversion. Since there are numer-

---

*The packet-switching interface is a message level interface allowing a message in a buffer of one machine to be transferred to a buffer in another machine. Proper interpretation of this message is another matter. Hence to complete the communications hierarchy, compatibility between higher-level communications processes and operating system commands is needed.

ous codes in use by various data communications equipments (e.g. Baudot, ASCII, EBCDIC, Transcode, etc.) code incompatibility is another data communications idiosyncrasy that can be alleviated by a packet-switching network.

Error Control

As mentioned already, node-to-node transfers in a packet-switching network generally use automatic repeat request (ARQ) error control. If an equipment is directly connected to the network such that the interface does not require error control, the data terminal equipment is relieved of having to provide error control functions itself. However, if the equipment accesses the network by way of noisy lines, the interface requires error control anyway.

Alternative Routing

Depending on the nature of the application, the availability (reliability) of a communications facility can be all important. The cost of providing backup communications for leased lines is quite significant, particularly if the backup facilities are selected automatically. Packet-switching networks provide alternate routing capabilities with the routing cost being shared by all users. If access lines to the network are reliable (e.g. dial-up lines with alternate routing between switching office provided by the telephone company), users do not need to provide backup facilities.

Virtual Subnetworks

Many corporations have a large variety of data communications needs: remote data processing, order entry, inventory control, accounting, design documentation, and various forms of inquiry/response applications (e.g. credit card or check verification). To meet these needs large corporations have often developed numerous independent networks. Smaller corporations have either relied on special service networks of commercial concerns or done without the desired communications altogether.

To add to the diversity of data communications applications, significant new requirements are evolving for electronic funds transfer, high-rate facsimile, distributed data processing, communicating word processors, teleconferencing, and electronic mail. These applications can be supported by a communications network with both low- and high-speed interfaces. Significant economies of scale result if a single network is designed to accomodate all applications. Figure 8.6 depicts a single packet-switching network with a number of virtual subnetworks to service specific applications. Each subnetwork is a distinct logical entity. The major network keeps track of which users are "connected" to which subnetworks and controls the accesses accordingly. Furthermore, net-

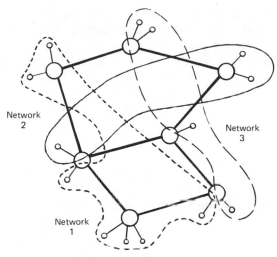

Figure 8.6 Virtual packet-switched networks.

work call processors translate all logical addresses of particular subnetworks into physical network addresses. Even though the subnetworks are functionally independent, they share the costs of transmission links, switching nodes, maintenance, and backup facilities.

A particularly exciting prospect for small and medium users of data communications services in the United States is the proposed Advanced Communications Service (ACS) planned by AT&T [18]. ACS will be implemented as a packet-switching network designed specifically to provide virtual subnetworks. Thus small data communications users, who can not afford to develop and maintain their own networks, will be able to establish virtual networks within the confines of ACS. Furthermore, ACS will provide the means for establishing interorganizational networks as well.

A full scale, nationwide ACS system represents a mammoth undertaking that few corporations, other than AT&T, could support. It would have a significant impact on both the data communications and the data processing industry. Thus significant regulatory issues must be resolved before ACS can be realized.

8.3 DIGITAL DATA SYSTEM

In December of 1974 the Bell System began offering a new transmission service specifically designed for data communications users. In contrast to previous Bell System data services such as Dataphone® service, using analog facilities for transmission, the new service utilizes digital transmission links almost exclusively. Hence the new service is referred to as Dataphone® Digital Service (DDS). The digital transmission facilities used to provide the service are referred to as the Digital Data System.

DDS relies primarily on T1 lines for local area digital transmission links. Except for special channel banks, DDS lines are implemented with the same equipment used for digital voice links. In some cases, DDS channels even share a TDM link with digitized voice signals. Functionally, however, the voice network and DDS are completely independent. In no instance are the digital channels sometimes used for the voice network and sometimes used for DDS.

Dataphone® Digital Service is primarily intended for point-to-point full-duplex transmission at data rates of 2.4, 4.8, 9.6, 56, 1344, and 1544 kbps [19]. However, multipoint service (a single channel with more than two stations connected to it) is also available [20]. All transmission is synchronous in nature with the customer equipment deriving its transmit rate from the received data clock. Access to DDS is normally provided by four-wire leased lines that have been specially conditioned to carry the digital signals. Sometimes, however, analog transmission with voiceband data sets (modems) is necessary to access the nearest DDS serving office. Figure 8.7 depicts the basic implementation of the DDS network.

Although T1 lines are readily available for interoffice digital transmission within metropolitan areas, there are presently no digital transmission links interconnecting greatly separated metropolitan areas. To meet this requirement, the 1A-Radio Digital System (data under voice) was developed so a single DS-1 signal could be added to existing TD and TH analog radio channels. (A fully developed analog route has 18 radio

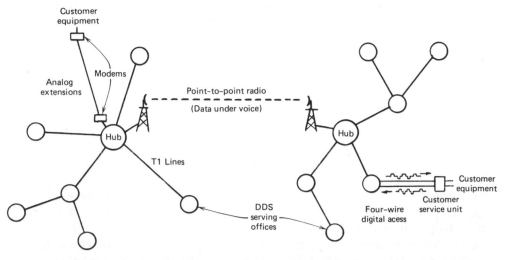

Figure 8.7 Digital data service network.

channels, thereby allowing a maximum of 18 DS-1 signals per route.) The Bell System does not use digital radios for long-haul voice traffic, but digital radios with one or two DS-3 signals are being used more and more for short-haul applications. These facilities, as well as other transmission systems in the digital hierarchy, are being used for DDS distribution.

Cost and Performance

At the present time rates for Dataphone® Digital Service have been set by the FCC to be comparable to the cost of similar services obtainable by data transmission over analog channels. Thus there is no particular cost advantage for using DDS services. This is somewhat paradoxical since as many as twenty 2.4 kbps channels are multiplexed into the equivalent of a single 64 kbps digital voice channel. Despite the lack of a cost advantage, DDS channels are much in demand by data communications users. The reason is that the digital channels have a much lower-error rate than a typical voiceband data circuit. Hence the throughput of a DDS channel is higher because block retransmissions are rare. The design objective for DDS channels is to provide at least 99.5% error-free seconds at 56 kbps, and even better quality at the lower-data rates [19].

Network Access

As indicated in Figure 8.7, a local DDS serving area is organized in a treelike hierarchy with a hub office at the center of the serving area. Every DDS circuit must enter a hub office where it is a) crossconnected to another incoming circuit for local point-to-point service, or b) cross-connected to an interhub circuit for long-distance point-to-point service.

When customer equipment is close enough, it is connected directly to a hub office by way of a four-wire loop designed to carry a bipolar coded digital signal at the selected rate. The four-wire loops necessarily have no loading coils, no build-out capacitors, and only limited-length bridged taps. If the service requirements are large enough, DDS circuits can be extended to satellite offices by way of T1 lines. The satellite offices multiplex the signals from individual customer loops into a DS-1 format for transmission to the hub office.

The maximum acceptable attenuation for a DDS digital access signal is 31 dB. As indicated in Figure 4.11, a bipolar spectrum has its peak amplitude at a frequency equal to one-half the data rate. The maximum length of a customer loop is therefore determined by the attenuation of available cable pairs at one-half the desired data rate. Table 8.1 shows the maximum loop length of various cable pairs as a function of data rate.

A major drawback of Dataphone® Digital Service is its limited availability. The service is provided only in major metropolitan areas where T1 lines are used extensively. Even in a metropolitan area, a customer

TABLE 8.1   CABLE LOSSES AND CORRESPONDING MAXIMUM
LOOP LENGTHS FOR DDS ACCESS CIRCUITS[a]

Wire Gauge	DDS Bit Rate (kbps)							
	2.4		4.8		9.6		56	
	Loss[b] (dB/kft)	Length (kft)	Loss (dB/kft)	Length (kft)	Loss (dB/kft)	Length (kft)	Loss (dB/kft)	Length (kft)
19	0.27	114.8	0.36	86.1	0.46	67.4	0.76	40.8
22	0.42	73.8	0.55	56.4	0.72	43.1	1.28	24.2
24	0.55	56.4	0.73	42.5	0.96	32.3	1.80	17.2
26	0.74	41.9	0.96	32.3	1.25	24.8	2.40	12.9

[a]Information in Table 8.1 was obtained from Reference [21], p 169.
[b]Loss figures are for frequencies equal to one half the data rate.

must be within the distances indicated in Table 8.1 to get direct digital access to a designated DDS office. In lieu of digital access, analog extension of any length can be used as indicated in Figure 8.7. These extensions use standard voice-grade lines with modems at both ends and therefore are limited to 9.6 kbps.* In addition to cost and data rate penalties imposed by analog extensions, end-to-end quality may suffer appreciably since the over-all error rate of a high-rate channel is most likely determined by the analog extension facility and not the DDS channels.

Another drawback of DDS service is the absence of voice communication as an alternative (unless a user supplies his own codecs). Analog data circuits can always be used as voice circuits by bypassing the modulation and demodulation circuitry of the data sets. The voice mode is particularly useful when manually setting up a connection to business machines or for coordinating the work of maintenance personnel.

8.4   DIGITAL SATELLITE SYSTEMS

The first communications satellites were used to provide communications to remote areas of the world where no feasible alternatives existed. The technology quickly developed, however, so satellite systems began competing with terrestial-based communications facilities—first for overseas traffic, and then for domestic communications.

Development of satellite systems in the United States, has not progressed as rapidly as technology and economics would allow. In the interest of encouraging competition, the FCC expressly forbade AT&T

*The higher rate, for example, 56 kbps, can be obtained on voiceband analog channels only with multiple circuits.

from offering domestic satellite services. Instead, other companies were given an invitation to develop systems free from initial competition with the Bell System. The first company to respond to what has been referred to as the "Open Skies" policy of the FCC was Western Union with their Westar satellites in 1974. Various other companies have since developed satellite systems—primarily for television program distribution and data transmission.

The first generation of satellite systems used analog transmission with frequency division multiplexing for channelization. Systems currently under development, however, are using digital transmission with time division channelization. One such system has been developed by Satellite Business Systems (SBS), a company founded as a consortium of International Business Machines, Aetna Life Insurance Company, and Comsat General. Another satellite-based system was proposed by Xerox and is referred to as Xerox Telecommunication Network (XTEN). Even though the XTEN project was canceled in 1981 it is worth describing because it introduced a new concept for local digital transmission service (DTS).

8.4.1 Satellite Business Systems

The SBS system is designed to provide large, geographically dispersed corporations or government organizations with wideband digital channels for applications such as:

1 Conventional voice/data communications.
2 Teleconferencing.
3 High-rate facsimile.
4 Processor to processor communications.

SBS transmits voice using 32 kbps delta modulation. Since it is assumed that most data transmission will use the digital channels directly, the voice coders have not been designed to support high-data rates in a voiceband signal. If a modulated data signal is passed through the network in a voice channel, the delta modulator effectively restricts the transmission rate to 1800 bps [22]. Higher-rate voiceband data must be demodulated and applied to a digital circuit for SBS transmission.

The last three applications mentioned are possible only because high data rate channels are available. Teleconferencing is provided with a range of data rates. At the low end, SBS supports fixed-frame video for charts and high-resolution images. Full motion video is possible with higher data rates. The facsimile service provides image transmission and reproduction at rates approaching one page per second. (As opposed to 6 minutes per page using voiceband facsimile.) At the present time

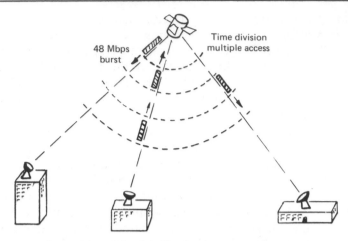

Figure 8.8 Satellite business system.

(1981) interprocessor communications address an undefined market since economical high-rate data circuits to support such applications are only now becoming available. (The first SBS satellite was launched November 15, 1980.)

The SBS system concept is illustrated in Figure 8.8. As indicated, each user location has its own earth station to access the satellite. The earth stations are owned and maintained by SBS. Transmission to and from the satellite uses a time division multiple access (TDMA) technique that requires each ground station to transmit bursts of data on a common carrier frequency in prescribed time slots so that bursts from separate ground stations do not overlap each other. Functionally, the satellite acts as a transponder—merely transferring the incoming bursts of data to a different down-link carrier frequency. The earth stations monitor the down-link data stream and extract whatever data is intended for their respective locations. Since the clock rate of each ground station is independently timed, a preamble is needed for each TDM burst to establish clock synchronization in the receivers. After clock synchronization is established, a burst is synchronously demodulated and detected. The receive clock is then resynchronized to the preamble of the next TDM burst to detect data in that time slot.

8.4.2 Xerox Telecommunications Network

Although XTEN was intended for many of the same applications as SBS, it had a different philosophy for accessing the satellite. In the initial SBS concept, each user would lease its own earth station. Thus only heavy users of telecommunications services could justify the usage cost. As shown in Figure 8.9, however, XTEN would provide access to the earth stations by way of terrestial microwave links. Hence smaller

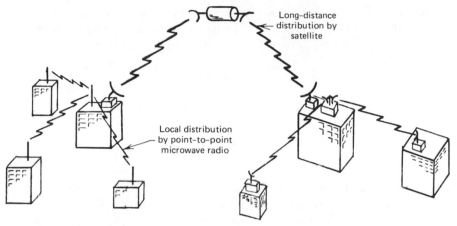

Figure 8.9   Xerox Telecommunications Network (XTEN).

organizations could economically utilize XTEN facilities by effectively
sharing an earth station. More recently, SBS has indicated they will develop facilities for shared earth station access also.

The following are the main applications that XTEN was intended to
address:

1   Facsimile document distribution.
2   Data communication.
3   Teleconferencing.
4   Store-and-forward message switching.

The thrust of the system was for document distribution (electronic
mail). Xerox anticipated that the cost for distribution would be comparable to anticipated first class mail rates in the 1980s. With XTEN
deliveries would be automated and occur in a matter of seconds.

Facsimile usage is currently growing at a 35% yearly rate [10]. If
XTEN or some similar network becomes available, facsimile growth will
become even greater. (Not coincidently, Xerox is a leading supplier of
facsimile equipment.)

As in data communications, facsimile communications is plagued by
a diversity of equipment types with different communications protocols and scanning formats. XTEN would help to alleviate the incompatibilities by establishing a standard XTEN facsimile code. All facsimile
machines connected to the network would have their individual formats
converted to and from the network standard. Thus, just as a packet-
switching network provides protocol and code translation for data communications, XTEN would provide translation of facsimile formats.
Furthermore, scanning and resolution differences would be resolved.

Hence a 96 line per inch machine would be capable of communicating with a 200 line per inch machine. Similar capabilities are also planned for the Compak/Faxpak network being developed by International Telephone and Telegraph [23].

8.5 INTEGRATED VOICE AND DATA NETWORKS

The preceding sections of this chapter described several different examples of digital networks. Except for Satellite Business Systems, which includes voice services, all of the networks are designed specifically for data traffic.* In view of the trend to use digital technology for public and private voice networks, it is clear that certain economical advantages are to be derived by designing digital transmission and switching facilities to provide both types of services. Although it will be some time before existing analog networks are digitized to the point that subscribers have ready access to digital voice circuits, new private networks can now be developed using digital facilities exclusively. This section discusses network-level considerations for new networks to efficiently accomodate all types of communications services.

8.5.1 Motivation for Integration

Obviously, the fundamental motivation for integrating all services into common facilities arises from the savings obtained by sharing the transmission links. There are two aspects of sharing transmission links that are important. The first of these is the reduction of total transmission capacity required to provide a given grade of service. When separate networks are used, each network includes spare transmission capacity to accommodate traffic overloads and failures. When the networks are combined so the various services share the transmission facilities on a dynamic basic, spare capacity is shared by all traffic types, and fewer total circuits are required.

The amount of savings to be obtained by sharing circuits in a dynamic manner is strongly dependent on circuit utilizations of the individual networks. If the circuits are lightly utilized, significant reductions in total transmission capacity are possible. If the transmission links of each network arc highly utilized, however, a combined operation may provide only marginal savings in total transmission requirements.†

---

*At this point, and in the rest of this chapter, data traffic is not restricted to computer communications, but includes all traffic of a digital nature, such as messages and digital facsimile.

†Chapter 9 provides the basic analytical techniques used to determine network transmission capacity requirements as a function of traffic intensity.

A second, and usually more important, reasons for combining special purpose networks is the savings to be realized by using common maintenance, testing, and repair procedures. Thus significant savings can be obtained even if individual networks are not integrated in an operational sense but use only common types of facilities. As an example, the primary metropolitan transmission facilities of the Digital Data System are T1 lines using the same routes as digital voice routes. DDS facilities have been installed only where significant amounts of T-carrier voice services existed already. When installed, the DDS facilities are operationally independent of the voice network but share the overhead costs of administration and maintenance. Similarly, DDS long-haul facilities (DUV) share the radio spectrum and physical microwave equipment.

The various transmission links of any communications network generally differ in their levels of utilization. Transmission links near the periphery of a network are normally lightly utilized, while interior links carrying heavily concentrated traffic are utilized quite efficiently. From this discussion, it can be seen that integrating interior portions of efficiently designed networks provides little savings in total equipment costs. However, overwhelming savings can be derived if the networks are integrated at their peripheries, where transmission link utilization is normally very low. Local distribution often represents the major cost of a network and is the most inefficiently utilized facility. Thus the motivation for integrating voice and data networks in an operational sense arises predominantly from the savings derived by sharing local distribution. An integrated network allows each user the ability to specify which service is needed from one access to the next by way of a single circuit.

8.5.2 Approaches to Integration

There are three basic network structures and operational procedures considered for integrated services:

1 Circuit switching.
2 Packet switching.
3 Hybrid circuit and packet switching.

In the following paragraphs these techniques are discussed in terms of their usefulness for a general purpose network with an arbitrary size and complexity (number of nodes).

Circuit Switching
Although the existence of an all-digital circuit-switched network for voice traffic provides higher-quality data communications at lower costs

than is available over analog facilities, conventional circuit switching does not service many forms of data traffic efficiently. A circuit-switched network designed to set up connections quickly (fast circuit switching) eliminates inefficiencies associated with circuit set-up times. Fast circuit switching, however, does not overcome inefficient use of an assigned circuit by interactive data traffic. Furthermore, single-packet messages such as a datagrams are serviced more effectively by packet switching. Thus if significant numbers of datagrams or interactive data communications are serviced, conventional circuit switching is an inefficient mode of operation.

Packet Switching

As already discussed, packet switching is an effective means of servicing most types of data traffic. The main exception would be long continuous file transfers or high-resolution facsimile, which are more efficiently serviced by a circuit-switching network. Packet switching has also been proposed as a technique for transmitting voice traffic [24]. In fact, the ARPANET has already conducted experiments for packet-switched voice [25]. The basic attraction for packetized voice arises from the characteristic that a voice circuit is only used in one direction at a time. Thus a full-duplex voice circuit is actually utilized for less than 50% of the duration of a call. If speech inactivity can be monitored, then no packets need to be transmitted when a user is not talking. The destination node merely fills in for unreceived packets with inaudible pauses.

Actually, the basic concept of assigning transmission channels only to active voice circuits is not new. This basic technique is used by AT&T to increase the capacity of analog circuits in transoceanic cables [26, 27]. The technique, referred to as time assignment speech interpolation (TASI), allows the number of active conversations to exceed the number of channels by switching a user onto a channel only when he is actively talking. If too many users on one end of a cable talk simultaneously, the speech of some of the users is clipped. With sufficiently large group sizes, however, the number of users can be doubled while maintaining an acceptably low probability of clipping.

The Bell System is planning to expand its usage of TASI by applying it to domestic voice circuits beginning in 1983. Recent use of this technique for digital voice signals is referred to as digital speech interpolation (DSI). DSI is incorporated into SBS for improved voice transmission efficiency.

In view of the flow control considerations discussed in Chapter 7, a large packet-switched network designed to accomodate voice traffic most assuredly requires a virtual circuit mode of operation to ensure a certain minimum throughput during networkwide traffic overloads. Thus the following discussion assumes that if packet switching is to be used for voice traffic, the network operates with virtual circuits for voice and multipacket data traffic.

Transmitting speech through a packet-switching network requires strict attention to two aspects of end-to-end transmission delay. First, the typical or average amount of delay must be limited so that the network does not introduce noticeable pauses into an interactive conversation. In a small network involving only a few tandem nodes, the delay is not noticeable. However, in large networks the delay through each individual node must be constrained to minimize the overall delay.

The average delay through a packet-switching node can be shortened by using smaller packets or by limiting the amount of traffic flowing through individual nodes. Both approaches reduce the utilization efficiency of the transmission links.

A second aspect of transmission delay that must be controlled is its variability. Speech is fundamentally a real time process requiring timely reconstruction of the encoded voice signals. The effects of variability in the arrival times can be overcome by initially delaying the packets in a buffer at the destination and smoothing the arrival times by removing packets at a regular rate (i.e., using an elastic store to remove arrival time jitter). Long network delays imply greater variability in arrival times and, hence, even longer elastic stores.

When voice packets are delayed so much that they arrive after their proper decoding time, they are effectively lost and produce a corresponding gap in the reconstructed speech. In one study of the effect of lost packets on speech quality [28], it is reported that 5% of the packets can be missing and still produce intelligible speech. This study, however, was made with military applications in mind and should not be construed as applicable to conventional telephone quality standards.

Both the delay and the delay variability of a packet-switching network increase rapidly as the network begins to get overloaded. In fact, a network that is designed to efficiently utilize its transmission links at a particular traffic level becomes completely overloaded when the traffic intensity increases by only a small percentage above the design conditions. For example, Figure 8.10 shows the relationship between average packet delay as a function of the output channel utilization. If the output utilization is 80%, the average queueing delay is 2.5 packet transmission times. If the channel loading increases to 90%, the queueing delay doubles to 5 packet transmission times. This result and others relating packet-switching delay to channel utilization are discussed in Chapter 9.

The main implication of the critical dependence of network delay on traffic loading is that a packet-switched voice network must use particularly tight flow control procedures to maintain acceptable delay performance. The delay performance of a packet-switched data network is not as critical since increased delay variability merely implies a

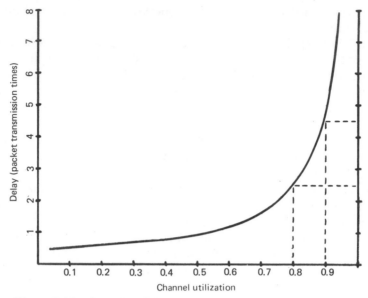

Figure 8.10   Queueing delay versus output channel utilization.

drop in individual user throughput. In voice applications a critical threshold for delay variability exists beyond which the reconstructed speech becomes objectionable or, unintelligible. Notice that delays are controlled more tightly by using a single path through the network— more justification for the use of virtual circuits.

Many of the analyses dealing with packet-switching combinations of voice and data traffic have considered the performance of a single link. While these analyses are appropriate for single-hop systems, as through a satellite, they do not include flow control considerations of a multi-node network. These analyses are therefore inadequate for predicting the performance of a packet-switching network with numerous tandem nodes. In the latter case, flow control procedures must be considered that inherently reduce the link utilization below the levels achieved with single-link systems.

One simplifying feature exploited in a voice packet-switched network is that voice packets do not need error control. As discussed in Chapter 3, digitized voice is virtually unaffected by the error rates obtainable on a reasonable quality digital transmission link. Thus the ARQ operations can be suspended for voice packets and the packets shortened by deleting the redundancy check bits. Considering the delay implications of retransmission, voice packets essentially require the simplified but different mode of operation anyway. Eliminating the need to check errors in voice packets also allows immediate transmission of packets while they are being received. Thus store-and-forward delays can be reduced by overlapping reception and transmission. If the transmission links are highly utilized, however, the desired outgoing link is normally busy, so that an arriving packet usually enters a queue anyway.

In light of the preceding discussion, packet-switching is not recommended as a general purpose mode of operation for integrating voice and data traffic. The reasons are summarized as follows:

1 Some types of traffic, such as file transfers or high-resolution facsimile, are inherently serviced more efficiently by another mode of operation (circuit switching).

2 Flow control and delay variation considerations require virtual circuit modes of operation at the very least.

3 Network size or transmission link efficiency is inherently limited by maximum end-to-end delay objectives.

4 Voice packets must be treated differently from data packets since voice packet error control is unnecessary and undesirable.

5 Voice activity sensing and accurate temporal reconstruction of the analog signal are not routine tasks. The complexity of these operations is not always justified.

6 The fundamental advantage attributed to packet-switching voice— improved link utilization provided by activity dependent transmission—is also obtainable within the confines of a circuit switched network.

Since neither circuit switching nor packet switching provides an optimum mode of operation for a general purpose network, the recommended approach is a hybrid network involving aspects of *both* circuit switching and packet switching. A hybrid approach is recommended primarily because a general purpose network must support many different and unanticipated applications with widely varying traffic statistics. It is unreasonable to assume that any single mode of operation can be efficient for all types of traffic.

Hybrid Circuit and Packet Switching

A number of hybrid networking techniques have been proposed in recent years [29, 30, 31]. In essence, these techniques establish two subnetworks: one for circuit switching and one for packet switching. All approaches provide some means of dynamically allocating the network resources to match the relative intensities of the various types of traffic. One such approach, referred to as a slotted envelope network (SENET) [30], utilizes a synchronous master frame format into which both circuit-switched and packet-switched data are inserted. As shown in Figure 8.11, each circuit-switched connection is allocated a fixed number of bits in every frame (envelope). Thus, each circuit-switched

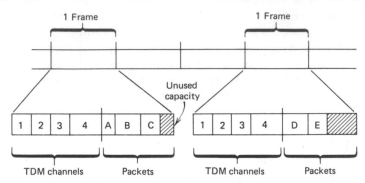

Figure 8.11   SENET frame structure.

connection is allocated a fixed capacity in a synchronous time division multiplex format. Packet-switched data are transmitted in the second half of each frame.

To provide dynamic allocation of the transmission capacity, the partition between synchronous TDM data and the packet data is adjusted whenever a circuit is set up or released. If a large packet queue exists at some node, that node can inhibit new circuits until the queue is lowered. However, a mere increase in the transmission capacity from an overloaded node does not ensure that packet transmission can be increased by the same rate. If the queues in the neighboring nodes are long (a likely reason for congestion in the first place), the packets have no where to go despite the increased channel capacity. Thus the flow control aspects of the circuit-switching subnetwork and the packet-switching subnetwork are complicated and interrelated.

A number of similar hybrid networks have been proposed that modify the basic SENET multiplex structure [31, 32]. Typically, these variations incorporate digital speech interpolation (DSI) into the circuit-switched portion of the frame and allow packet data to utilize the unused circuit-switched capacity. In essence, active voice information is given priority, and data packets utilize whatever capacity is left over. Thus the partition between circuit data and packet data is adjusted on a frame-to-frame basis. This approach provides greater link utilization with greater operational complexity.

If the number of circuit-switched connections in TASI-like systems is limited to the respective capacity of the links, no voice information is ever blocked, as in conventional TASI. In this case the average packet transmission capacity is somewhat greater than one-half the bit rate of the transmission links. If there is insufficient data traffic to fill out the frames, the voice circuits can be "oversubscribed" to improve the average utilization as in conventional TASI systems.

The SENET approach to combining circuit and packet switching defines the mode of operation of a single link and implies that networks are established by interconnecting nodes with similarly operating links. The following section describes a different approach to defining a hybrid network that begins at the network level and defines lower-level operations in a hierarchical fashion. This point of view is particularly useful because it provides a framework within which much flexibility is available for defining and controlling the subnetworks. Operations of one subnetwork are independent of other subnetworks. The approach is flexible enough that any particular networking structure (e.g. pure circuit switching, packet switching, or SENET) can be obtained as a special case implementation.

8.5.3 Dynamic Virtual Networking

As an alternative to the aforementioned approaches to integrating voice and data, this section describes a hybrid network using a hierarchical design to provide both circuit switching and packet switching services. The first level, or backbone of the network, is implemented as an all-digital circuit-switched network. Packet switching (or any other type of network such as a message-switched network) is implemented in a second level—using the facilities of the backbone network but remaining operationally independent of the circuit-switched network. This basic approach to integrated services networking has been described by Collins and Pedersen [33] as "adaptive trunking."

Figure 8.12 depicts the conceptual structure and operation of a hierarchical network. The backbone network is implemented with digital end offices (EOs), digital tandem switches (TSs), and digital TDM transmission links in between. Network terminals are connected to other terminals in a conventional circuit-switched manner. Superimposed upon the backbone network is a subnetwork to support specialized forms of communication. The nodes of the subnetwork are labeled as special service nodes (SSNs). These nodes are interconnected by leased lines, dial-up lines, or combinations of both.

As a specific example of a second-level network, consider the SSNs as packet-switching nodes. Data terminals are connected to the packet-switching nodes by way of local dial-up connections or leased lines. The leased lines can be shared as digital multipoint lines or loops. A message entering an SSN from a data terminal is packetized and transferred through the assigned transmission channels to intermediate SSNs in a store-and-forward manner and on to the destination SSN, where the message is reconstructed and delivered to the intended terminal or host computer.

Notice that packets in the subnetwork pass through the circuit-switching nodes in held connections. Thus the circuit-switched network is transparent to the packet-switched network. Furthermore, the

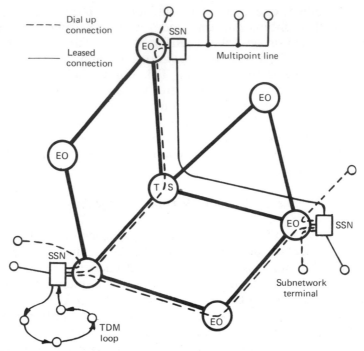

Figure 8.12 Circuit-switched network with dynamic virtual subnetwork.

internal operations of the second-level network are totally independent of the circuit-switched operations. The only time the two networks interact is when the second-level network requests or releases connections between SSNs. Thus the second-level network can change its transmission capacity in response to varying traffic conditions (packet queue length).

Because dial-up connections between SSNs are direct connections in a functional sense, the process of adding or releasing circuits between SSNs is referred to as "adaptive trunking." On a large scale, adding and releasing circuits between SSNs amounts to dynamically reconfiguring the subnetwork. Hence this mode of operation is also referred to as "dynamic virtual networking."

The most important aspect of a virtual networking approach to integrating various types of communications services is that the internal operations of virtual networks are independent of the backbone network and each other. Hence routing or flow control procedures in one network can be modified without affecting routing and flow control algorithms in another network. As traffic conditions and applications change, the operation and configuration of a subnetwork can be adjusted to improve its performance, while the backbone network continues to operate with well established control procedures for circuit-switched networks. In a functional sense, there are actually two or more networks. In a physical sense, however, the networks share the costs of maintenance, spare capacity, and local distribution.

Separate Logical Configurations

The community of interest structure of a data network is normally quite different from a voice network. Thus the traffic patterns and optimum configuration for each network are typically different. If the separate classes of traffic are not implemented with separate modules (both hardware and software), each node of the network supports both types of traffic. In contrast, notice that not all circuit-switching nodes in Figure 8.12 have an SSN attached to it. If there is insufficient need for special services, a node need not be implemented with the unnecessary capabilities. If the need for special services arises, an SSN can be added easily since, as far as the circuit-switching node is concerned, an SSN is just another circuit-switched terminal.

Another important feature of virtual configurability is that a packet network, for example, can be designed to match traffic patterns independently of the physical structure of the backbone network. If a significant amount of interactive data traffic flows between any two distant nodes, a circuit connecting the two appropriate SSNs can be established to carry packets directly between them. Thus, as shown in Figure 8.13, traffic between the upper and lower SSNs flows on a

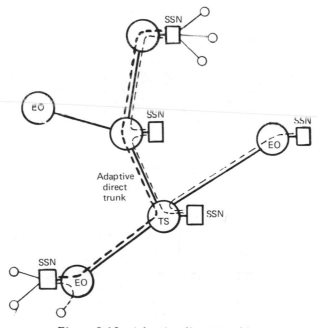

Figure 8.13 Adaptive direct trunking.

direct high-usage packet trunk, avoiding the delays and unnecessary congestion of tandem store-and-forward switching nodes. High-usage special service trunks do not have to be held, but can be set up and released as traffic patterns warrant. Adaptive trunking is a new tool for managing a communications network.

Multiple Virtual Networks

The virtual network concept is unlimited in the number of subnetworks that can be provided. Separate networks can exist for each type of application, or for distinct groups of users with identical applications. In either case, all subnetworks are functionally independent of each other.

Message switching and low-speed circuit switching are two more examples of useful subnetworks. The basic concept of a low-speed circuit-switched subnetwork is shown in Figure 8.14. In essence, each TDM channel of the backbone network, utilized by the low-speed net, is subdivided into channels. The backbone network is oblivious to the subdivision. As indicated, access to the low-speed network, which might

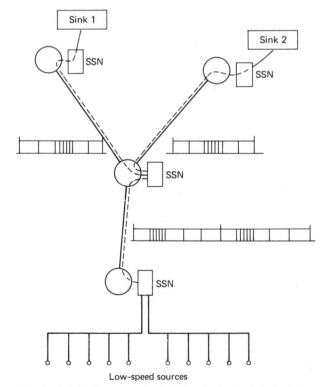

Figure 8.14   Low-speed circuit-switched subnetwork.

be used for continuous data collection, can be provided by leased multi-point lines or local dial-up connections.

Digital Speech Interpolation

Owing to long holding times and only modest inactivity factors, voice traffic is most appropriately transmitted by way of the backbone circuit-switched network. On any particular route, however, the backbone transmission links can be implemented with DSI techniques to improve the circuit utilization. Thus the complexity of a DSI operation is added only where the cost is warranted (just as in present networks).

When the number of active sources exceeds the number of channels in a DSI system, the usual operation is to block the overload and cause clipping of some speech syllables. More recently, however, a TASI system [34] has been developed by STC Communications of Broomfield, Colorado that delays blocked syllables for a short period of time in anticipation that a channel will soon become available. The reconstructed speech from this process is of better quality than when conventional syllable clipping occurs. Thus greater levels of line utilization are possible when delay capabilities are incorporated into a DSI or TASI system.

A packet-switched voice network inherently delays blocked packets rather than discard them. As discussed previously, the average amount of delay that can be tolerated at any one node gets smaller as the size of the network increases. Thus the degree of improvement in line utilization obtained by using delay is directly dependent on the number of tandem nodes in a packet network or the number of tandem DSI links in a circuit.

DSI operations inherently require the concentrating node to monitor each incoming signal to determine which channels are active. This process is complicated enough for voice signals with a wide dynamic range of signal levels, but it is particularly troublesome for channels carrying mixtures of traffic types.

Circuit-Switch Design

To support virtual subnetworks in the most flexible and efficient manner possible, the backbone circuit-switched network should be designed to include the following features:

1 Fast connect switching. By providing fast setup and release of network facilities, the circuit-switched network allows virtual sub-networks the opportunity to respond more quickly and more often to fluctuations in traffic patterns. As part of the fast release process, busy tones should be returned from the originating switch node instead of the destination node as is customary in current networks.

2 Expandable control facilities. In keeping with the desire to provide fast connect switching, call processing resources and control channel bandwidths must be expandable to support increasing percentages of short holding time connections.

3 Low switch blocking. As discussed in Chapter 5, a modern digital switch can be designed to be virtually nonblocking. This feature is particularly useful for supporting virtual subnetworks which may want to hold some connections for extended periods of time. Non-existent or low-blocking probabilities can eliminate the need for leased circuits altogether. Assuming that proper billing procedures are established, the equivalent of a leased circuit can be established as a held connection through a switching machine. Analog networks do not support this mode of operation because of blocking probability considerations in the switches and greater noise levels of dial-up connections. A modern digital switch can avoid both of these problems.

8.6 THE INTEGRATED DIGITAL NETWORK

The preceding section discusses basic considerations and approaches to integration of service offerings from a digital network. In this section, the word integration is used in a different context. Here, integration refers to a system design and implementation concept that maximizes the synergism of combined transmission and switching systems. Traditionally, the transmission and switching systems of telephone networks have been designed and administered by functionally independent organizations. In the operating companies, the two equipment classes have been referred to as "outside plant" and "inside plant." These equipments necessarily meet standardized interfaces, but other than that, the systems have been developed independently of each other.

With the development of digital time division multiplex equipment and digital time division switching, it is apparent that the traditional separation of functions is no longer desirable. A time division multiplex terminal has many of the basic properties of a time division switch. Furthermore, the interstage links of a time division switch are multiplexed in the same manner as a TDM trunk. Exploitation of these similarities is accomplished when a digital TDM trunk is connected directly to the first stage of a TDM switch. Thus the traditional appearance of first-level multiplexing and demultiplexing equipment within a switching office is eliminated. Instead of the 4 khz baseband interface, the standard interfaces become TDM group signals like the DS-1. An all-digital telephone network, in which the TDM transmission links are directly interfaced to TDM switches, has come to be referred to as the integrated digital network (IDN).

Virtually all of the telephone networks around the world are heading toward the implementation of an integrated digital network, at least at some levels within the network hierarchy. At first, integration of transmission and switching functions will occur only in localized areas and only within certain levels of the switching hierarchy. Three basic levels are considered:

1   Exchange area.
2   Local distribution and switching.
3   Toll network.

8.6.1   Exchange Area

At the present time, the major thrust of digitization in North America and other countries is in T-carrier systems for short-haul interoffice trunks. By the early 1980s more than half of the toll-connecting trunks will be digital. By 1990 practically all will be digital [35]. Helping to stimulate digital growth in the United States is the rapid introduction of No. 4ESS digital toll switches. DS-1 signals from T-carrier systems are interfaced directly to the digital toll switch—thereby providing *integration* of transmission and switching.

In the past, channel banks were a cost penalty for interfacing digital trunks to analog switches. However, analog lines entering No. 4ESS switching offices are necessarily connected to channel banks as their interface to the digital switch. Thus a channel bank now becomes a cost associated with the analog line. Since exchange area trunks are generally fairly short, the channel bank costs are a significant impetus to convert analog trunks to digital and to integrate them with a digital switch. Of course, the circuit quality is also improved when tandem analog-to-digital, digital-to-analog (A/D-D/A) conversions are eliminated.

Since tandem switching functions are also being implemented digitally, virtually all of the exchange area on the network side of the end offices will become digital in the near future. Until the long-distance network becomes digital, however, the exchange areas will be islands of transmission and switching. Unfortunately digital features of these islands will be inaccessable until local digitization occurs (except through special services like DDS).

8.6.2   Local Distribution and Switching

At the present time digital end offices are being installed in the U.S. telephone network for small, mostly rural, switching applications. The No. 5ESS digital end office, slated to begin service for the Bell System in 1981, is also initially intended for small offices. Large metropolitan offices of the Bell System and major independents have recently been equipped (or soon will be equipped) with analog switches of the No. 1ESS, No. 1EAX, ESC1 variety. Thus extensive digitization of end-office switching in the United States is a long way off.

Other countries around the world appear to be more poised for extensive introduction of digital end offices—primarily because they have not used the most recent analog switching technology as extensively, and can "skip" a technological generation by converting directly from step-by-step to digital. France, for example, actually introduced digital switching into end offices before installing digital toll and tandem offices [36]. The British have established a plan to not only integrate local transmission and switching, but the whole network as well [37]. Smaller countries are more likely to convert their higher-level transmission facilities to digital at an earlier date since they have fewer long-distance circuits and therefore have fewer circuits that are most economically implemented with bandwidth efficient analog transmission.

As discussed in earlier chapters, a number of the earlier digital subscriber carrier systems in the United States used delta modulation, and therefore are incompatible with PCM encoding used in higher levels of the network. Hence even if significant amounts of these subscriber carrier systems were to come into use, they would not provide the same stimulation for end-office digitization as has occurred in the exchange area and the lower levels of the toll network. Newer subscriber carrier systems, however, are being implemented with $\mu 255$ PCM encoding in anticipation of being connected through digital end offices to exchange and toll connecting trunks. A notable example is the DMS-1 carrier equipment, the lowest level of Bell Canada's family of integrated digital systems [38]. In the United States, the Bell System SLC-96 subscriber carrier system [39] will be directly interfaced to digital end offices also.

A major stimulus for local digitization will occur when digital fiber optics are used for local distribution. At the present time fiber optic systems are being developed primarily for interoffice trunks where the wide bandwidth is readily utilized. However, fiber optics technology also holds promise for high-density local distribution of voice and wide bandwidth signals such as Picturephone $^{®}$, facsimile, and perhaps even cable television.

It is somewhat ironic that the areas of the United States that could use local digital facilities most advantageously—the metropolitan areas—are the same areas currently being equipped with SPC analog switches. These switches will not be replaced by digital switches until the conversion is economically justifiable. Either the new equipment must provide demonstrable savings in maintenance, or it must provide a sufficient number of new revenue-producing services. As an alternative to replacing an analog switch with a digital one, the Bell System is currently developing the capability of switched digital service through a No. 1ESS to digital trunks [40]. Thus, new digital services can be supplied as an overbuild on the analog network.

A digital switch, in itself, can not provide new services of a digital nature when the local transmission facilities are predominantly analog. After a digital end office is installed, however, the prove-in distance for a digital carrier system is greatly reduced so that the use of remote digitization becomes more widespread. At first the digitization occurs at remote channel banks that provide the standard analog interface (BORSCHT in Chapter 5). Because the point of digitization is much closer to the customer however, it becomes much easier to extend digital channels to the customer on a special request basis.

Naturally, the first requests for digital connectivity will come from data processing concerns and their dial-up customers. Digital channels for the average residence will not occur until applications, undefined at present, arise. Some applications might be home computer networks, game networks, data retrieval, message switching, electronic mail, home shopping, and so on. Whether these, or any other applications, can evolve without the availability of a digital telephone network for relatively low-cost, high-quality data transmission is problematical. In a society becoming more and more informationally oriented, one must believe that the needs will arise, and that the demand for digital circuits will become explosive. Sometime in the distant future, data-oriented revenues for the telephone companies may exceed voice-oriented revenues.

8.6.3   Toll Network

The cost of a transmission system is divided between terminal facilities and transmission facilities. With long distances, transmission costs are obviously the more significant. Thus, long-haul systems are optimized by getting the most voice channels in an allotted bandwidth, irrespective of terminal (multiplexing) costs. However, digital multiplexing terminals are less expensive than analog terminals so that short-haul systems, or long-haul systems with many drop and insert points, are most economically implemented with digital techniques. At the present time, both wire-line and short-haul radio systems are being dominated by digital technology. The terminal economics are accentuated by digital switches that eliminate the need for lower-level digital multiplexers.

As mentioned in Chapter 1, the backbone of the toll network in North America is provided by FM-FDM radio systems of the TD-2 and TH-3 variety. Owing to bandwidth expansion factors of analog frequency modulation, existing digital radio technology can compete with existing analog radios in terms of voice circuits in a given bandwidth (at present, 1800 channels for TD2 and 1344 channels for digital). In an extensive economic study of the Canadian toll network [41], it was determined that the entire network should be digital and that the largest cost savings would accrue if existing analog equipment was converted to digital as rapidly as possible. Many other countries around the world also have plans to convert their long-distance telephone networks to all digital. European countries, in particular, can utilize digital technology most advantageously because of the more limited transmission distances involved. (Terminal and multiplexing costs dominate shorter-distance systems.)

Bell System engineers in the United States have recently developed a new analog single-sideband radio. This radio, the AR-6A, achieves 6000 voice circuits in a 30 MHz bandwidth. The AR-6A will ensure that long-haul routes with heavy traffic volumes remain analog in nature. Because of the greater sensitivity of the single-sideband channels to interference, however, this radio can not be used in as congested an environment as can FM or digital radios. Single-sideband transmission packs more circuits into a given bandwidth, but digital transmission, in particular, allows greater packing of radio channels into a geographic area.

Despite the fact that existing FM radio systems will remain in use for some time and new analog systems such as the AR-6A will be introduced in the future, digital transmission seems destined for satellite systems. Thus even though much of the terrestial network may remain analog, long-distance digital facilities will become available from satellites. Hence end-to-end digital transmission will be available in future toll connections or private networks.

8.6.4 Voice Coding in an Integrated Network

The dominant coding technique in use around the world is 64 kbps instantaneously companded PCM: $\mu255$ PCM in North America and Japan, and A-law elsewhere. As mentioned previously, PCM was originally chosen because it allowed sharing of compandors and coder/decoder circuitry. Furthermore, 64 kbps quality was selected for a network that would carry performance sensitive voiceband data and produce multiple analog-to-digital, digital-to-analog conversions. In an all-digital network, 64 kbps provides excess quality. Furthermore, more recent developments in voice coding techniques and integrated circuit technology imply that much lower bit rate voice coding techniques are possible. As indicated in Chapter 3, an adaptive differential PCM system can provide adequate quality in an all-digital network with one-half the bit rate of instantaneously companded PCM systems.

Considering the momentum established already by PCM channel banks, it seems that the public networks are locked into PCM coding.

Complete digitization of the public telephone network, if and when it occurs, is obviously going to occur in an evolutionary manner. At no time will obsoleting existing coders and decoders seem justified.

In the early stages of the present evolution, it was possible to make a transition from one coding format to another (7 bit $\mu100$ to 8 bit $\mu255$). This change was possible because the interfaces to the individual digital systems were always analog. Thus one T1 system could use one form of coding while another T1 system used a different form. In fact, many different forms of coding, including $\mu100$ PCM and various types of delta modulation, are in use at present—all with analog interfaces. The advent of digital switching, however, has virtually eliminated the possibility of a graceful transition to another coding technique. New equipment connected to the digital switches must be simultaneously compatible with the massive amount of equipment already connectable to the digital switches. Thus, one drawback to the integration of transmission and switching is that it precludes subsystem by subsystem conversion of coding algorithms.

The savings to be derived by significant reductions in the voice signal bit rate cannot be ignored completely. Within individual voice transmission or storage systems more efficient techniques are bound to be used even though they require conversion to and from the 64 kbps standards. When corporations lease digital tie-lines between PBXs, they will want bit-rate-efficient coders for the lines. The cost of even the most sophisticated vocoders and DSI techniques can be justified because they provide greater numbers of voice circuits per line. At the present time four vocoders operating at 2.4 kbps can be justified for long analog lines that have been conditioned to provide a 9.6 kpbs data rate. A single 64 kbps channel would carry 26 such subchannels—without additional line conditioning costs. These low bit rate vocoders do not provide standard telephone quality voice transmission, and therefore are not likely to be used in the public network.

One technique that might possibly be used to effect a widespread conversion of coding algorithms in an all-digital network is to use traveling class marks. If more than one type of coding algorithm is in use, the class mark indicates the need to insert the appropriate conversion equipment in a connection. In this manner, one coding scheme can be phased out gradually while another is phased in.

Presumably, the motivation for a coding conversion would be to reduce the voice channel bit rate. Initially, however, the conversions would not provide widespread bandwidth savings since existing multiplexers and switches are designed for 64 kbps channels. Only after a significant number of coders have been converted would it be justifiable to modify the multiplexing and switching equipment. At that time the class marks would also indicate how much bandwidth is required when setting up the connection. In some instances the low-rate signals would pass through low-rate channels, but where no low-rate channels existed, 64 kbps channels would be used.

REFERENCES

1 R. E. Kahn, "Resource-Sharing Computer Communications Networks," *Proceedings of IEEE*, November 1972, pp 1397–1407.

2 L. Kleinrock, "ARPANET Lessons," *International Communications Conference*, 1976, pp 20.1–10.6.

3 L. Kleinrock, "Principles and Lessons in Packet Communications," *Proceedings of IEEE*, November 1978, pp 1320–1329.

4 W. W. Clipsham, F. E. Glave, and M. L. Narraway, "Datapac Network Overview," *International Communications Conference*, 1976, pp 131–135.

5 A. Danet, R. Despres, A. LaRest, G. Pichon, and S. Ritzenthaler, "The French Public Packet Switching Service: The Transpac Network," *Proceedings Third ICCC*, Toronto, 1977, pp 251–260.

6 H. Opderbeck and R. Hovey, "Telenet-Network Features and Interface Protocols," *NTG Conference on Data Networks*, Baden-Daden, West Germany, 1976.

7 J. Kopf, "TYMNET as a Multiplexed Packet Network," *National Computer Conference*, 1977, pp 609–613.

8 C. R. Moster and L. R. Pamm, "The Digital Data System Launches a New Era in Data Communications," *Bell Labs Record*, December 1975, pp 421–426.

9 "SBS: Right on Target," *Computer Decisions*, November 1979, pp 50–54.

10 "FCC Reviews Xerox Network Proposal," *Datacomm Advisor*, December 1978, pp 1–7.

11 E. R. Berlekamp, *Algebraic Coding Theory*, McGraw-Hill, New York, 1968.

12 H. Rudin and H. Müller, "More on Routing and Flow Control," *National Telecommunications Conference*, 1979, pp 34.5.1–34.5.9.

13 F. E. Heart, R. E. Kahn, S. M. Ornstein, W. R. Crowther, and D. C. Walden, "The Interface Message Processor for the ARPA Computer Network," *Spring Joint Computer Conference*, 1970, pp 551–556.

14 J. Rinde, "Routing and Control in a Centrally Directed Network," *National Computer Conference*, 1977, pp 603–608.

15 S. C. K. Young and C. I. McGibbon, "The Control System of the Datapac Network," *International Conference on Communications*, 1976, pp 137–141.

16 T. S. Yum and M. Schwartz, "Comparison of Adaptive Routing Algorithms for Computer Communications Networks," *National Telecommunications Conference*, 1978, pp 4.1.1–4.1.5.

17 B. Cosell, A. Nemeth, and D. Walden, "X.25 Link Access Procedure," *Computer Communication Review of ACM*, October 1977, pp 15–35.

18 "Eight Special Features Keystone Bell's ACS," *Datacomm Advisor*, September 1978, pp 5–12.

19 "Digital Data System Channel Interface Specifications," Bell System Technical Reference PUB 41021, March 1973.

20 "Multistation Dataphone® Digital Service," Bell System Technical Reference PUB 41022, September 1974.

21 Members of Technical Staff, Bell Laboratories, *Telecommunications Transmission Engineering*, Vol. 3, Western Electric, 1977.

22 "SBS Communications Service Interfaces," SBS Document #3201-0004, Satellite Business Systems, McLean Virginia, January 5, 1978.

23 D. Minoli and G. Louit, "Local Network Design Strategies for FAXPAK," *International Communications Conference*, 1980, pp 39.1–39.1.6.

24 J. Gitman and H. Frank, "Economic Analysis of Integrated Voice and Data Networks: A Case Study," *Proceedings of IEEE*, November 1978, pp 1549–1570.

25 S. L. Casner, E. R. Mader, and E. R. Cole, "Some Initial Measurements of ARPANET Packet Voice Transmission," *National Telecommunications Conference*, 1978, pp 12.2.1–12.2.5.

26 K. Bullington and M. Fraser, "Engineering Aspects of TASI," *Bell System Technical Journal*, March 1959.

27 P. T. Brady, "A Technique for Investigating On-Off Patterns of speech," *Bell System Technical Journal*, January 1965, pp 1–21.

28 J. W. Forgie, "Speech Transmission in Packet-Switched Store-and-Forward Networks," *National Computer Conference*, 1975, pp 137–142.

29 M. J. Ross, A. C. Tabbot, and J. A. Waite, "Design Approaches and Performance Criteria for Integrated Voice/Data Switching," *Proceedings of IEEE*, September 1977, pp 1283–1295.

30 G. J. Coviello and P. A. Vena, "Integration of Circuit/Packet Switching By a SENET (Slotted Envelope Network) Concept," *National Telecommunications Conference*, 1975, pp 42–12 to 42-17.

31 E. Arthurs and B. Stuck, "A Theoretical Performance Analysis of an Integrated Voice-Data Virtual Circuit Packet Switch," *International Conference on Communications*, 1979, pp 24.2.1–24.2.3.

32 H. Miyahara and T. Hasegawa, "Performance Evaluation of Modified Multiplexing Technique with Two Types of Packet for Circuit and Packet Switched Traffic," *International Conference on Communications*, 1979, pp 20.5.1–20.5.5.

33 A. A. Collins and R. D. Pedersen, *Telecommunications: A Time for Innovation*, Merle Collins Foundation, Dallas, Texas, 1973.

34 J. M. Elder and J. F. O'Neill, "A Speech Interpolation System for Private Networks," *National Telecommunications Conference*, 1978, pp 14.6.1–14.6.5.

35 M. R. Aaron, "Digital Communications—The Silent (R)evolution?" *IEEE Communications*, January 1979, pp 16–26.

36 J. LeGuillou, J. Pernin, and A. J. Schwartz, "Electronic Systems and Equipment for Evolving Loop Plant," *IEEE Transactions on Communications*, July 1980, pp 967–975.

37 A. G. Orbell, "Preparations for Evolution Towards an Integrated Services Digital Network," *International Conference on Communications*, 1979, pp 29.1.1–29.1.6.

38 "Special Issue on DMS-10," *Telesis*, Vol. 5, No. 10, August 1978.

39 S. Brolin, Y. Cho, W. Michaud, and D. Williamson, "Inside the New Digital Subscriber Loop System," *Bell Labs Record*, April 1980, pp 110–116.

40 Irwin Dorros, "ISDN," *IEEE Communications Magazine*, March 1981, pp 16–19.

41 J. Hopkins and B. Richardson, "Towards an Integrated Digital Network for the TransCanada Telephone System," *Telesis*, April 1979, pp 2–7.

CHAPTER NINE

TRAFFIC ANALYSIS

Except for station sets and their associated loops, a telephone network is composed of a variety of common equipment such as digit receivers, call processors, interstage switching links, and interoffice trunks. The amount of common equipment designed into a network is determined under an assumption that not all users of the network need service at one time. The exact amount of common equipment required is unpredictable because of the random nature of the service requests. Networks conceivably could be designed with enough common equipment to instantly service all requests except for occurrences of very rare or unanticipated peaks. However, this solution is uneconomical because much of the common equipment is unused during normal network loads. The basic goal of traffic analysis is to provide a method for determining the cost-effectiveness of various sizes and configurations of networks.

Traffic in a communications network refers to the aggregate of all user requests being serviced by the network. As far as the network is concerned, the service requests arrive randomly and usually require unpredictable service times. The first step of traffic analysis is the characterization of traffic arrivals and service times in a probabilistic framework. Then the effectiveness of a network can be evaluated in terms of how much traffic it carries under normal or average loads and how often the traffic volume exceeds the capacity of the network.

The techniques of traffic analysis can be divided into two general categories: loss systems and delay systems. The appropriate analysis category for a particular system depends on the system's treatment of overload traffic. In a loss system overload traffic is rejected without being serviced. In a delay system overload traffic is held in a queue until the facilities become available to service it. Conventional circuit switching operates as a loss system since excess traffic is blocked and not serviced without a retry on the part of the user. In some instances "lost" calls actually represent a loss of revenue to the carriers by virtue of their not being completed.

Store-and-forward message or packet switching obviously possesses the basic characteristics of a delay system. Sometimes, however, a packet-switching operation can also contain certain aspects of a loss

413

system. Limited queue sizes and virtual circuits both imply loss operations during traffic overloads. Circuit-switching networks also incorporate certain operations of a delay nature in addition to the loss operation
of the circuits themselves. For example, access to a digit receiver, operator, or a call processor is normally controlled by a queuing process.

The basic measure of performance for a loss system is the probability
of rejection (blocking probability). A delay system, on the other hand,
is measured in terms of service delays. Sometimes the average delay is
desired, while at other times the probability of the delay exceeding
some specified value is of more interest.

Some of the analyses presented in this chapter are similar to those
presented in Chapter 5 for the blocking probabilities of a switch. Chapter 5 is concerned mostly with matching loss: the probability of not
being able to set up a connection through a switch under normal or
average traffic volumes. This chapter, however, is mostly concerned
with the probability that the number of active sources exceeds some
specified value. Typically, the specified value is the number of trunk
circuits in a route.

9.1   TRAFFIC CHARACTERIZATION

Owing to the random nature of network traffic, the following analyses
involve certain fundamentals of probability theory and stochastic processes. In this treatment only the most basic assumptions and results of
traffic analysis are presented. The intent is to provide an indication of
how to apply results of traffic analysis—not to delve deeply into analytical formulations. However, a few basic derivations are presented, to
acquaint the user with assumptions in the models so they can be appropriately applied.

In the realm of applied mathematics, where these subjects are treated
more formally, blocking probability analyses are referred to as congestion theory and delay analyses are referred to as queuing theory. These
topics are also commonly referred to as traffic flow analysis. In a
circuit-switched network, the "flow" of messages is not so much of a
concern as are the holding times of common equipment. A circuit-
switched network establishes an end-to-end circuit involving various
network facilities (transmission links and switching stages) that are held
for the duration of a call. From a network point of view, it is the holding of these resources that is important, not the flow of information
within individual circuits.

On the other hand, message-switching and packet-switching networks
are directly concerned with the actual flow of information, since in
these systems, traffic on the transmission links is directly related to the
activity of the sources.

As mentioned in Chapter 7, circuit switching does involve certain aspects of traffic flow in the process of setting up a connection. Connect requests flow from the sources to the destinations acquiring, holding, and releasing certain resources in the process. As was discussed, controlling the flow of connect-requests during network overloads is a vital function of network management.

The unpredictable nature of communications traffic arises as a result of two underlying random processes: call arrivals and holding times. An arrival from any particular user is generally assumed to occur purely by chance and be totally independent of arrivals from other users. Thus the number of arrivals during any particular time interval is indeterminant. In most cases holding times are also distributed randomly. In some applications this element of randomness can be removed by assuming constant holding times (e.g., fixed length packets). In either case the traffic load presented to a network is fundamentally dependent on both the frequency of arrivals and the average holding time for each arrival. Figure 9.1 depicts a representative situation in which both the arrivals and the holding times of 20 different sources are unpredictable. The bottom of the figure depicts activity of each individual source while the top displays the instantaneous total of all activity. If we assume that the 20 sources are to be connected to a trunk group, the activity curve displays the number of circuits in use at any particular time. Notice that the maximum number of circuits in use at any one time is 16 and that the average utilization is a little under 11 circuits. In general terms, the trunks are referred to as servers, and a trunk group is a server group.

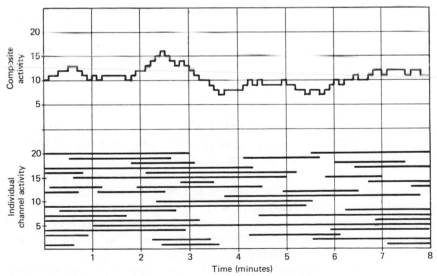

Figure 9.1   Activity profile of network traffic (all calls carried).

Traffic Measurements

One measure of network capacity is the volume of traffic carried over a period of time. Traffic volume is essentially the sum of all holding times carried during the interval. The traffic volume represented in Figure 9.1 is the area under the activity curve (approximately 84 call minutes).

A more useful measure of traffic is the traffic intensity (also called traffic flow). Traffic intensity is obtained by dividing the traffic volume by the length of time during which it is measured. Thus traffic intensity represents the average activity during a period of time (10.5 in Figure 9.1). Although traffic intensity is fundamentally dimensionless (time divided by time), it is usually expressed in units of erlangs, in honor of the Danish pioneer traffic theorist A. K. Erlang, or in terms of hundred (century) call seconds per hour (CCS). The relationship between erlangs and CCS units can be derived by observing that there are 3600 seconds in an hour:

$$1 \text{ erlang} = 36 \text{ CCS}$$

The maximum capacity of a single server (channel) is 1 erlang, which is to say that the server is always busy. Thus the maximum capacity in erlangs of a group of servers is merely equal to the number of servers. Because traffic in a loss system experiences infinite blocking probabilities when the traffic intensity is equal to the number of servers, the average activity is necessarily less than the number of servers. Similarly, delay systems operate at less than full capacity, on average, because infinite delays occur when the average traffic load approaches the number of servers.

Two important parameters used to characterize traffic are the average arrival rate λ and the average holding time t_m. If the traffic intensity A is expressed in erlangs, then

$$A = \lambda t_m \tag{9.1}$$

where λ and t_m are expressed in like units of time (e.g., calls per second and seconds per call, respectively).

Notice that traffic intensity is only a measure of average utilization during a time period and does not reflect the relationship between arrivals and holding times. That is, many short calls can produce the same traffic intensity as a few long ones. In many of the analyses that follow the results are dependent only on the traffic intensity. In some cases, however, the results are also dependent on the individual arrival patterns and holding time distributions.

Public telephone networks are typically analyzed in terms of the average activity during the busiest hour of a day. The use of busy hour traffic measurements to design and analyze telephone networks represents a compromise between designing for the overall average utilization (which includes virtually unused nighttime hours) and designing for short duration peaks that may occur by chance or as a result of TV commercial breaks, radio call-in contests, and so on.

Busy hour traffic measurements indicate that an individual residential telephone is typically in use between 5 and 10% of the busy hour. Thus each telephone represents a traffic load of between 0.05 and 0.10 erlangs. The average holding time is between 3 and 4 minutes, indicating that a typical telephone is involved in one or two phone calls during the busy hour.

Business telephones usually produce loading patterns different from residential phones. In the first place, a business phone is generally utilized more heavily. Second, the busy hour of business traffic is often different from the busy hour of residential traffic. Figure 9.2 shows a typical hourly variation for both sources of traffic. The trunks of a telephone network are sometimes designed to take advantage of variations in calling patterns from different offices. Toll connecting trunks from residential areas are often busiest during evening hours, and trunks from business areas are obviously busiest during midmorning or midafternoon. Traffic engineering depends not only on overall traffic volume, but also on time-volume traffic patterns within the network.

A certain amount of care must be exercised when determining the

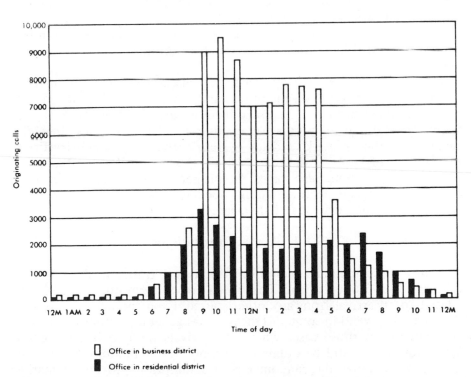

Figure 9.2   Traffic volume dependence on time of day.

total traffic load of a system from the loadings of individual lines or trunks. For example, since two telephones are involved in each connection, the total load on a switching system is exactly one-half the total of all traffic on the lines connected to the switch. In addition, it may be important to include certain setup and release times into the average holding times of some common equipment. A 10 second setup time is not particularly significant for a 4 minute voice call, but can actually dominate holding time of equipment used for short data messages. Common equipment setup times also become more significant in the presence of voice traffic overloads. A greater percentage of the overall load is represented by call attempts since they increase at a faster rate than completions.

An important distinction to be made when discussing traffic in a communications network is the difference between the offered traffic and the carried traffic. The offered traffic is the total traffic that would be carried by a network capable of sevicing all requests as they arise. Since economics generally precludes designing a network to immediately carry the maximum offered traffic, a small percentage of offered traffic typically experiences network blocking or delay. When the blocked calls are rejected by the network, the mode of operation is referred to as blocked calls cleared or lost calls cleared. In essence, blocked calls are assumed to disappear and never return. This assumption is most appropriate for trunk groups with alternate routes. In this case a blocked call is normally serviced by another trunk group and does not, in fact, return.

The carried traffic of a loss system is always less than the offered traffic. A delay system, on the other hand, does not reject blocked calls but holds them until the necessary facilities are available. With the assumption that the long-term average of offered traffic is less than the capacity of the network, a delay system carries all offered traffic. If the number of requests that can be waiting for service is limited, however, a delay system also takes on properties of a loss system. For example, if the queue for holding blocked arrivals is finite, requests arriving when the queue is full are cleared.

9.1.1   Arrival Distributions

The most fundamental assumption of classical traffic analysis is that call arrivals are independent. That is, an arrival from one source is unrelated to an arrival from any other source. Even though this assumption may be invalid in some instances, it has general usefulness for most applications. In those cases where call arrivals tend to be correlated, useful results can still be obtained by modifying a random arrival analysis. In this manner the random arrival assumption provides a mathematical formulation that can be adjusted to produce approximate solutions to problems that are otherwise mathematically intractable.

Negative Exponential Interarrival Times

Designate the average call arrival rate from a large group of independent sources (subscriber lines) as λ. Use the following assumptions:

1 Only one arrival can occur in any sufficiently small interval.
2 The probability of an arrival in any sufficiently small interval is directly proportional to the length of the interval. (The probability of an arrival is $\lambda \, \Delta t$ where Δt is the interval length.)
3 The probability of an arrival in any particular interval is independent of what has occurred in other intervals.

It is straightforward [1] to show that the probability distribution of interarrival times is

$$P_0(\lambda t) = e^{-\lambda t} \tag{9.2}$$

Equation 9.2 defines the probability that no arrivals occur in a randomly selected interval t. This is identical to the probability that t seconds elapse from one arrival to the next.

EXAMPLE 9.1

Assuming each of 10,000 subscriber lines originate one call per hour, how often do two calls arrive with less than 0.01 seconds between them?

Solution. The average arrival rate

$$\lambda = (10,000)\left(\frac{1}{3600}\right)$$

$$= 2.78 \text{ arrivals per second}$$

From Equation 9.2, the probability of no arrival in a 0.01 second interval is

$$P_0(0.0278) = e^{-0.0278}$$

$$= 0.973$$

Thus 2.7% of the arrivals occur within 0.01 seconds of the previous arrival. Since the arrival rate is 2.78 arrivals per second, the rate of occurrence of interarrival times less than 0.01 seconds is

$$(2.78) \, (0.027) = 0.075 \text{ times a second}$$

The first two assumptions made in deriving the negative exponential arrival distribution can be intuitively justified for most applications. The third assumption, however, implies certain aspects of the sources that can not always be supported. First of all, certain events, such as television commercial breaks, might stimulate the sources to place their calls at nearly the same time. In this case the negative exponential distribution may still hold, but for a much higher calling rate during the commercial.

A more subtle implication of the independent arrival assumption involves the number of sources, and not just their calling patterns. When the probability of an arrival in any small time interval is independent of other arrivals, it implies that the number of sources available to generate requests is constant. If a number of arrivals occur immediately before any subinterval in question, some of the sources become busy and can not generate requests. The effect of busy sources is to reduce the average arrival rate. Thus the interarrival times are always somewhat larger than what Equation 9.2 predicts them to be. The only time the arrival rate is truly independent of source activity is when an infinite number of sources exist.

If the number of sources is large and their average activity is relatively low, busy sources do not appreciably reduce the arrival rate. For example, consider an end office that services 10,000 subscribers with 0.1 erlangs of activity each. Normally, there are 1000 active links and 9000 subscribers available to generate new arrivals. If the number of active subscribers increases by an unlikely 50% to 1500 active lines, the number of idle subscribers reduces to 8500, a change of only 5.6%. Thus the arrival rate is relatively constant over a wide range of source activity. Whenever the arrival rate is fairly constant for the entire range of normal source activity, an infinite source assumption is justified.

Actually, some effects of finite sources have already been discussed in Chapter 5 when analyzing blocking probabilities of a switch. It is pointed out that Lee graph analyses overestimate the blocking probability because, if some number of interstage links in a group are known to be busy, the remaining links in the group are less likely to be busy. A Jacobaeus analysis produces a more rigorous and accurate solution to the blocking probability—particularly when space expansion is used. Accurate analyses of interarrival times for finite sources are also possible. These are included in the blocking analyses to follow.

Poisson Arrival Distribution

Equation 9.2 merely provides a means of determining the distribution of interarrival times. It does not, by itself, provide the generally more desirable information of how many arrivals can be expected to occur in some arbitrary time interval. Using the same assumptions presented, however, the probability of j arrivals in an interval t can be determined [1] as

$$P_j(\lambda t) = \frac{(\lambda t)^j}{j!} e^{-\lambda t} \tag{9.3}$$

Equation 9.3 is the well known Poisson probability law. Notice that when $j = 0$, the probability of a no arrivals in an interval t is $P_0(t)$, as obtained in Equation 9.2.

Again, Equation 9.3 assumes arrivals are independent and occur at a given average rate λ, irrespective of the number of arrivals occurring just prior to an interval in question. Thus the Poisson probability distribution should only be used for arrivals from a large number of independent sources.

Equation 9.3 defines the probability of experiencing exactly j arrivals in t seconds. Usually there is more interest in determining the probability of j or more arrivals in t seconds:

$$P_{>j}(\lambda t) = \sum_{i=j}^{\infty} P_i(\lambda t)$$

$$= 1 - \sum_{i=0}^{j-1} P_i(\lambda t) \qquad (9.4)$$

$$= 1 - P_{<j}(\lambda t)$$

where $P_i(\lambda t)$ is defined in Equation 9.3.

EXAMPLE 9.2

Given a message-switching node that normally experiences four arrivals per minute, what is the probability that eight or more arrivals occur in an arbitrarily chosen 30 second interval?

Solution. The average number of arrivals in a 30 second interval is

$$\lambda t = (4) \left(\frac{30}{60}\right) = 2$$

The probability of eight or more arrivals (when the average is 2) is

$$P_{\geq 8}(2) = \sum_{i=8}^{\infty} P_i(2)$$

$$= 1 - \sum_{i=0}^{7} P_i(2)$$

$$= 1 - e^{-2} \left(1 + \frac{2^1}{1!} + \frac{2^2}{2!} + \frac{2^3}{3!} + \cdots \frac{2^7}{7!}\right)$$

$$= 0.0011$$

EXAMPLE 9.3

What is the probability that a 1000 bit data block experiences exactly
four errors while being transmitted over a transmission link with a bit
error rate (BER) = 10^{-5}?

Solution. Assuming independent errors (a questionable assumption on
many transmission links), we can obtain the probability of exactly four
errors directly from the Poisson distribution. The average number of
errors (arrivals) $\lambda t = (10^3) \, 10^{-5} = 0.01$. Thus

$$\text{prob (4 errors)} = P_4 \, (0.01)$$

$$= \frac{(0.01)^4}{4!} \, e^{-0.01}$$

$$= 4.125 \, (10)^{-10}$$

An alternative solution can be obtained from the binomial probability
law:

$$\text{Prob (4 errors)} = \binom{1000}{4} p^4 \, (1-p)^{996} \qquad \text{where } p = 10^{-5}$$

$$= 4.101 \, (10)^{-10}$$

As can be seen, the two solutions of Example 9.3 are nearly identical.
The closeness of the two answers reflects the fact that the Poisson probability distribution is often derived as a limiting case of a binomial
probability distribution. Because it is easier to calculate, a Poisson distribution is often used as an approximation to a binomial distribution.

9.1.2 Holding Time Distributions

The second factor of traffic intensity as specified in Equation 9.1 is the
average holding time t_m. In some cases the average of the holding times
is all that needs to be known about holding times to determine blocking
probabilities in a loss system or delays in a delay system. In other cases
it is necessary to know the probability distribution of the holding times
to obtain the desired results. This section describes the two most commonly assumed holding time distributions: constant holding times and
exponential holding times.

Constant Holding Times
Although constant holding times cannot be assumed for conventional
voice conversations, it is a reasonable assumption for such activities as
per-call call processing requirements, interoffice address signaling, op-

erator assistance, and recorded message playback. Furthermore, constant holding times are obviously valid for transmission times in fixed length packet networks.

When constant holding time messages are in effect, it is straightforward to use Equation 9.3 to determine the probability distribution of active channels. Assume, for the time being, that all requests are serviced, then the probability of j channels being busy at any particular time is merely the probability that j arrivals occurred in the time interval of length t_m immediately preceding the instant in question. Since the average number of active circuits over all time is the traffic intensity $A = \lambda t_m$, the probability of j circuits being busy is dependent only on the traffic intensity:

$$P_j(\lambda t_m) = P_j(A)$$

$$= \frac{A^j}{j!} e^{-A}$$

(9.5)

where λ = the arrival rate
   t_m = the constant holding time
   A = the traffic intensity in erlangs

Exponential Holding Times

The most commonly assumed holding time distribution for conventional telephone conversations is the exponential holding time distribution:

$$P(>t) = e^{-t/t_m}$$

(9.6)

where t_m is the average holding time. Equation 9.6 specifies the probability that a holding time exceeds the value t. This relationship can be derived from a few simple assumptions concerning the nature of the call termination process. Its basic justification, however, lies in the fact that observations of actual voice conversations exhibit a remarkably close correspondence to an exponential distribution.

The exponential distribution possesses the curious property that the probability of a termination is independent of how long a call has been in process. That is, no matter how long a call has been in existence, the probability of it lasting another t seconds is defined by Equation 9.6. In this sense exponential holding times represent the most random process possible. Not even knowledge of how long a call has been in progress provides any information as to when the call will terminate.

Combining a Poisson arrival process with an exponential holding time process to obtain the probability distribution of active circuits is more complicated than it was for constant holding times because calls can last indefinitely. The final result, however, proves to be dependent on only the average holding time. Thus Equation 9.5 is valid for exponential holding times as well as for constant holding times (or any holding time distribution). Equation 9.5 is therefore repeated for emphasis: The probability of j circuits being busy at any particular instant, assuming a Poisson arrival process and that all requests are serviced immediately is

$$P_j(A) = \frac{A^j}{j!} e^{-A} \qquad (9.7)$$

where A is the traffic intensity in erlangs. This result is true for any distribution of holding times.

EXAMPLE 9.4

Assume that a trunk group has enough channels to immediately carry all of the traffic offered to it by a Poisson process with an arrival rate of one call per minute. Assume that the average holding time is 2 minutes. What percentage of the total traffic is carried by the first five circuits, and how much traffic is carried by all remaining circuits? (Assume that the traffic is always packed into the lowest-numbered circuits.)

Solution. The traffic intensity (offered load) of the system is $A =$ (1) (2) = 2 erlangs. The traffic intensity carried by i active circuits is exactly i erlangs. Hence the traffic carried by the first five circuits can be determined as follows:

$$A_5 = 1P_1(2) + 2P_2(2) + 3P_3(2) + 4P_4(2) + 5P_5(2)$$

$$= e^{-2} \left[2 + \frac{2(2)^2}{2!} + \frac{3(2)^3}{3!} + \frac{4(2)^4}{4!} + \frac{5(2)^5}{5!} \right]$$

$$= 1.89 \text{ erlangs}$$

All of the remaining circuits carry

$$2 - 1.89 = 0.11 \text{ erlangs.}$$

The result of Example 9.4 demonstrates the principle of diminishing returns as the capacity of a system is increased to carry greater and greater percentages of the offered traffic. The first five circuits in Example 9.4 carry 94.5% of the traffic while all remaining circuits carry only 5.5% of the traffic. If there are 100 sources, 95 extra circuits are needed to carry the 5.5%.

9.2 LOSS SYSTEMS

Example 9.4 provides an indication of the blocking probabilities that arise when the number of servers (circuits) is less than the maximum possible traffic load (number of sources). The example demonstrates that 94.5% of the traffic is carried by only five circuits. The implication is that the blocking probability, if only five circuits are available to carry the traffic, is 5.5%. Actually, Example 9.4 is carefully worded to indicate that all of the offered traffic is carried but that only the traffic carried by the first five circuits is of interest. There is a subtle but important distinction between the probability that six or more circuits are busy (as can be obtained from Equation 9.7) and the blocking probability that arises when only five circuits exist.

The basic reason for the discrepancy is indicated in Figure 9.3, which depicts the same traffic pattern arising from 20 sources as is shown previously in Figure 9.1. Figure 9.3, however, assumes that only 13 circuits are available to carry the traffic. Thus the three arrivals at $t = 2.2, 2.3$, and 2.4 minutes are blocked and assumed to have left the system. The total amount of traffic volume lost is indicated by the shaded area, which is the difference between all traffic being serviced as it arrives and traffic being carried by a blocked calls cleared system with 13 circuits. The most important feature to notice in Figure 9.3 is that the call arriving at $t = 2.8$ is not blocked, even though the original profile indicates that it arrives when all 13 circuits are busy. The reason it is not blocked is that the previously blocked calls left the system and therefore reduced the congestion for subsequent arrivals. Hence the percentage of time that the original traffic profile is at or above 13 is not the same as the blocking probability when only 13 circuits are available.

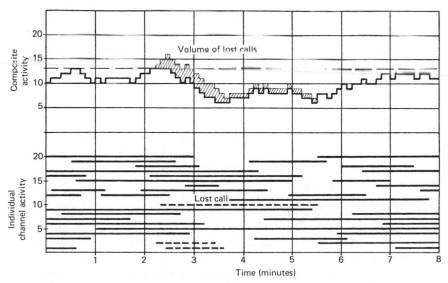

Figure 9.3 Activity profile of blocked calls cleared (13 channels).

9.2.1 Lost Calls Cleared

The first person to account fully and accurately for the effect of cleared calls in the calculation of blocking probabilities was A. K. Erlang in 1917. In this section we discuss Erlang's most often used result: his formulation of the blocking probability for a lost calls cleared system with Poisson arrivals. Recall that the Poisson arrival assumption implies infinite sources. This result is variously referred to as Erlang's formula of the first kind $E_{1,N}(A)$, the Erlang-B formula, or Erlang's loss formula.

A fundamental aspect of Erlang's formulation, and a key contribution to modern stochastic process theory, is the concept of statistical equilibrium. Basically, statistical equilibrium implies that the probability of a system's being in a particular state (number of busy circuits in a trunk group) is independent of the time at which the system is examined. In order for a system to be in statistical equilibrium, a long time must pass (several average holding times) from when the system is in a known state until it is again examined. For example, when a trunk group first begins to accept traffic it has no busy circuits. For a short time thereafter, the system is most likely to have only a few busy circuits. As time passes, however, the system reaches equilibrium. At this point the most likely state of the system is to have $A = \lambda t_m$ busy circuits.

When in equilibrium, a system is as likely to have an arrival as it is to have a termination. If the number of active circuits happens to increase above the average A, departures become more likely than arrivals. Similarly, if the number of active circuits happens to drop below A, an arrival is more likely than a departure. Thus if a system is perturbed by chance from its average state, it tends to return.

Although Erlang's elegant formulation is not particularly complicated, it is not presented here because we are mostly interested in application of the results. The interested reader is invited to see Reference 2 or 3 for a derivation of the result:

$$B = E_{1,N}(A) = \frac{A^N}{N! \, \Sigma_{i=0}^{N} \left(\dfrac{A^i}{i!} \right)} \tag{9.8}$$

where N is the number of servers (channels) and $A = \lambda t_m$ is the offered traffic intensity in erlangs.

Equation 9.8 specifies the probability of blocking for a system with random arrivals from an infinite source and arbitrary holding time distributions. The blocking probability of Equation 9.8 is plotted in Figure 9.4 as a function of offered traffic intensity for various numbers of channels. An often more useful presentation of Erlang's results is provided in Figure 9.5, which presents the output channel utilization

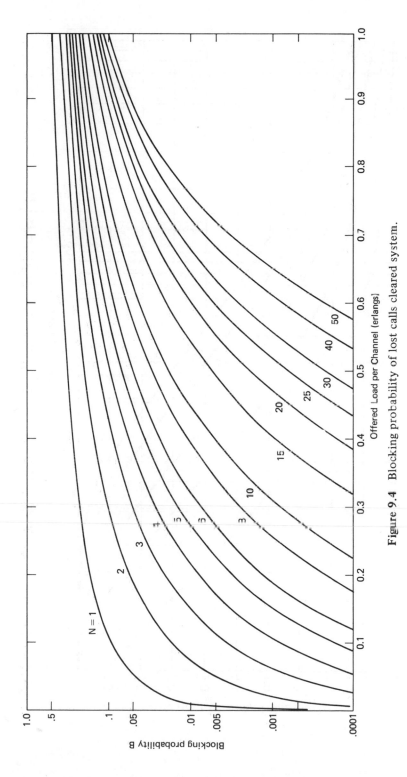

Figure 9.4 Blocking probability of lost calls cleared system.

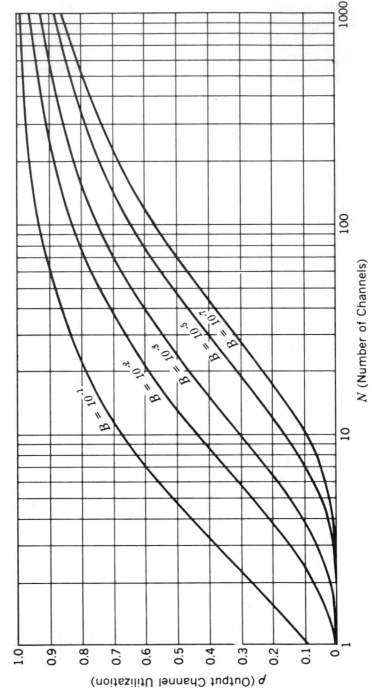

Figure 9.5 Output channel utilization of lost calls cleared system.

for various blocking probabilities and numbers of servers. The output utilization ρ represents the traffic carried by each circuit:

$$\rho = \frac{(1 - B)A}{N} \tag{9.9}$$

where A = the offered traffic
              B = the blocking probability
     $(1 - B)A$ = the carried traffic

Blocking probabilities are also provided in tabular form in Appendix D.

EXAMPLE 9.5

A T1 line is to be used as a high-usage trunk group between two end offices. How much traffic can the trunk group carry if the blocking probability is to be .1? What is the offered traffic intensity? (The blocking probability between switching offices is normally designed to be less than .01. Because alternate routes are available, however, direct trunks can be blocked more than 10%.)

Solution. From Figure 9.5 it can be seen that the output circuit utilization for B = 0.1 and N = 24 is 0.8. Thus the carried traffic intensity is $(0.8)(24)$ = 19.2 erlangs. Since the blocking probability is .1, the maximum level of offered traffic is

$$A = \frac{19.2}{1 - 0.1}$$

$$= 21.3 \text{ erlangs}$$

EXAMPLE 9.6

Four clusters of data terminals are to be connected to a computer by way of leased circuits as shown in Figure 9.6. In Figure 9.6a the traffic from the clusters uses separate groups of shared circuits. In Figure 9.6b the traffic from all clusters is concentrated onto one common group of circuits. Determine the total number of circuits required in both cases when the maximum desired blocking probability is 5%. Assume that 22 terminals are in each cluster and each terminal is active 10% of the time. (Use a blocked calls cleared analysis.)

Solution. The offered traffic from each cluster is $22(0.1)$ = 2.2 erlangs. Since the average number of active circuits is much smaller than the number of sources, an infinite source analysis can be used. Using Table D.1, the number of circuits required for B = 5% at a loading of 2.2 erlangs is 5. Thus the configuration of Figure 9.6a, requires a total of 20 circuits.

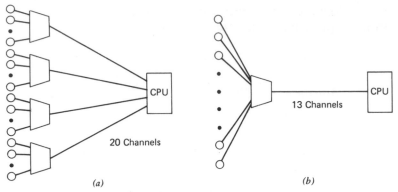

Figure 9.6 Data terminal network of Example 9.6. (*a*) Four separate groups. (*b*) All traffic concentrated into one group.

> The total offered traffic to the concentrator of the configuration of Figure 9.6*b* is 4(2.2) = 8.8 erlangs. From D.1, 13 circuits are required to support the given traffic load.

Example 9.6 demonstrates that consolidation of small traffic groups into one large traffic group can provide significant savings in total circuit requirements. Large groups are more efficient than multiple small groups because it is unlikely that the small groups will become overloaded at the same time (assuming independent arrivals). In effect, excess traffic in one group can use idle circuits in another group. Thus those circuits that are needed to accommodate traffic peaks, but are normally idle, are utilized more efficiently when the traffic is combined into one group. This feature is one of the motivations mentioned in Chapter 8 for integrating voice and data traffic into a common network. The total savings in transmission costs is most significant when the individual traffic intensities are low. Hence it is the peripheral area of a network that benefits the most by concentrating the traffic.

The greater circuit efficiency obtained by combining traffic into large groups is often referred to as the advantage of large group sizes.

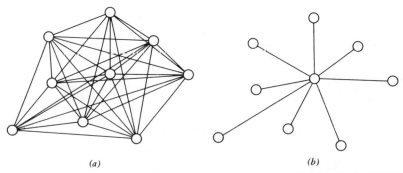

Figure 9.7 Use of tandem switching to concentrate traffic. (*a*) Mesh. (*b*) Star.

This efficiency of circuit utilization is the basic motivation for hierarchical switching structures. Instead of interconnecting a large number of nodes with rather small trunk groups between each pair, it is more economical to combine all traffic from individual nodes into one large trunk group and route the traffic through a tandem switching node. Figure 9.7 contrasts a mesh versus a star network with a centralized switching node at the center. Obviously, the cost of the tandem switch becomes justified when the savings in total circuit miles is large enough.

EXAMPLE 9.7

What happens to the blocking probabilities in Figure 9.6a and b discussed in Example 9.6 when the traffic intensity increases by 50%?

Solution. If the traffic intensity of each group increases from 2.2 to 3.3 erlangs, the blocking probability of the configuration of Figure 9.6a increases from 5% to almost 14%.

In the configuration of Figure 9.6b, a 50% increase in the traffic intensity causes a 400% increase in the blocking probability (from 5% to 20%).

Example 9.7 demonstrates a couple of important considerations in network design. As indicated, blocking probabilities are very sensitive to increases in traffic intensities, particularly when the channels are heavily utilized. Thus because large trunk groups utilize their channels more efficiently they are more vulnerable to traffic increases than are a number of smaller groups designed to provide the same grade of service. Furthermore, failures of equal percentages of transmission capacity affect the performance of a large group more than the performance of several small groups. In both cases the vulnerability of the large groups arises because large groups operate with less spare capacity than do multiple small groups.

A second aspect of blocking analyses demonstrated in Example 9.7 is that the calculated results are highly dependent on the accuracy of the traffic intensities. Accurate values of traffic intensities are not always available. Furthermore, even when accurate traffic measurements are obtainable, they do not provide an absolute indication of how much growth to expect. Thus only limited confidence can be attached to calculations of blocking probabilities in an absolute sense. The main value of these analyses is that they provide an objective means of comparing various network sizes and configurations. The most cost-effective design for a given grade of service is the one that should be chosen, even if the traffic statistics are hypothetical. If a network is liable to experience wildly varying traffic patterns or rapid growth, these factors must be considered when comparing design alternatives.

A network with a somewhat larger initial cost may be more desirable if it can absorb or grow to accomodate unanticipated traffic volumes more easily.

9.2.2 Lost Calls Returning

In the lost calls cleared analyses just presented, it is assumed that unserviceable requests leave the system and never return. As mentioned, this assumption is most appropriate for trunk groups whose blocked requests overflow to another route and are usually serviced elsewhere. However, lost calls cleared analyses are also used in instances where blocked calls do not get serviced elsewhere. In many of these cases, blocked calls tend to return to the system in the form of re-tries. Some examples are: subscriber concentrator systems, corporate tie lines and PBX trunks, calls to busy telephone numbers, and access to WATS lines (if DDD alternatives are not used). This section derives blocking probability relationships for lost calls cleared systems with random re-tries.

The following analysis involves three fundamental assumptions regarding the nature of the returning calls:

1 All blocked calls return to the system and eventually get serviced, even if multiple re-tries are required.

2 The elapsed times between call blocking occurrences and the generation of re-tries are random and statistically independent of each other. (This assumption allows the analysis to avoid complications arising when re-tries are correlated to each other and tend to cause recurring traffic peaks at a particular waiting time interval.)

3 The typical waiting time before re-tries occur is somewhat longer than the average holding time of a connection. This assumption essentially states that the system is allowed to reach statistical equilibrium before a re-try occurs. Obviously , if re-tries occur too soon, they are very likely to encounter congestion since the system has not had a chance to "relax." In the limit, if all re-tries are immediate and continuous the network operation becomes similar to a delay system discussed in latter sections of this Chapter. In this case, however, the system does not queue requests—the sources do so by continually "redialing."

When considered in their entirety, these assumptions characterize re-tries as being statistically indistinguishable from first-attempt traffic.* Hence blocked calls merely add to the first-attempt call arrival rate.

---

*First-attempt traffic is also referred to as demand traffic: the service demands assuming all arrivals are serviced immediately. The offered traffic is the demand traffic plus the re-tries.

Consider a system with a first-attempt call arrival rate of λ. If a percentage B of the calls are blocked, B times λ re-tries will occur in the future. Of these retries, however, a percentage B will be blocked again. Continuing in this manner, the total arrival rate λ', after the system has reached statistical equilibrium, can be determined as the infinite series:

$$\lambda' = \lambda + B\lambda + B^2\lambda + B^3\lambda + \cdots$$

$$= \frac{\lambda}{(1 - B)} \qquad\qquad (9.10)$$

where B is the blocking probability from a lost calls cleared analysis with traffic intensity $A' = \lambda' t_m$.

Equation 9.10 relates the average arrival rate λ', including the re-tries, to the first-attempt arrival rate and the blocking probability in terms of λ'. Thus this relationship does not provide a direct means of determining λ' or B since each is expressed in terms of the other. However, the desired result can be obtained by iterating the lost calls cleared analysis of Equation 9.8. First, determine an estimate of B using λ and then calculate λ'. Next, use λ' to obtain a new value of B and an updated value of λ'. Continue in this manner until values of λ' and B are obtained.

EXAMPLE 9.8

What is the blocking probability of a PBX to central office trunk group with 10 circuits and an offered traffic load of 7 erlangs? What is the blocking probability if the number of circuits is increased to 13? Assume random re-tries for all blocked calls.

Solution. It can be assumed that the 7 erlangs of traffic arise from a large number of PBX stations. Thus an infinite source analysis is justified. The blocking probability for $A = 7$ erlangs and $N = 10$ servers is about 8%. Thus the total offered load, including re-tries, is approximately 7.6 erlangs. With $N = 10$ and $A = 7.6$, the blocking probability is 10%. Two more iterations effectively produce convergence at $A = 8$ erlangs and $B = 12\%$. If the number of circuits in the trunk group is increased to 13, the blocking probability of a lost calls cleared system is 1.5%. Thus a first approximation to the returning traffic intensity is $7/0.985 = 7.1$ erlangs. Hence the blocking probability including all returning traffic increases only slightly above the 1.5%.

Example 9.8 demonstrates that the effect of returning traffic is insignificant when operating at low blocking probabilities. At high blocking probabilities, however, it is necessary to incorporate the effects of the returning traffic into the analysis. This relationship between lost calls cleared and lost calls returning is shown in Figure 9.8.

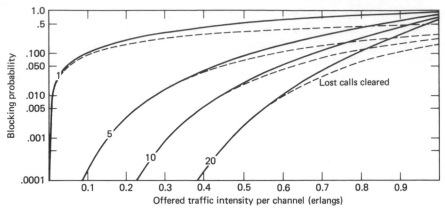

Figure 9.8   Blocking probability of lost calls returning.

When measurements are made to determine the blocking probability of an outgoing trunk group, the measurements can not distinguish between first-attempt calls (demand traffic) and re-tries. Thus if a significant number of re-tries are contained in the measurements, this fact should be incorporated into an analysis of how many circuits must be added to reduce the blocking of an overloaded trunk group. The apparent offered load will decrease as the number of servers increases because the number of re-tries decreases. Thus fewer additional circuits are needed than if no re-tries are contained in the measurements.

9.2.3 Lost Calls Held

In a lost calls held system, blocked calls are held by the system and serviced when the necessary facilities become available. Lost calls held systems are distinctly different from the delay systems discussed later in one important respect: the total elapsed time of a call in the system, including waiting time and service time, is independent of the waiting time. In essence, each arrival requires service for a continuous period of time and terminates its request independently of its being serviced or not. Figure 9.9 demonstrates the basic operation of a lost calls held system. Notice that most blocked calls eventually get some service, but only for a portion of the time that the respective sources are busy.

Although a switched telephone network does not operate in a lost calls held manner, some systems do. Lost calls held systems generally arise in real time applications in which the sources are continuously in need of service, whether or not the facilities are available. When operating under conditions of heavy traffic, a lost calls held system typically provides service for only a portion of the time a particular source is active.

Even though conventional circuit switching does not operate accord-

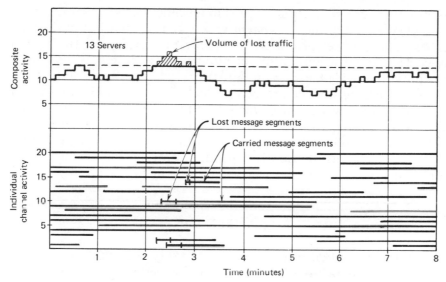

Figure 9.9   Activity profile of lost calls held.

ing to the theoretical model of lost calls held, Bell System traffic engineers have used it to calculate blocking probabilities for trunk groups [4]. A lost calls held analysis always produces a larger value for blocking than does Erlang's loss formula. Thus the lost calls held analysis produces a conservative design that helps account for re-tries and day-to-day variations in the busy hour calling intensities. In contrast, CCITT recommendations [5] stipulate Erlang-B formulas should be used in determining blocking probabilities.

One example of a system that closely fits the lost calls held model is time assignment speech interpolation (TASI). A TASI system concentrates some number of voice sources onto a smaller number of transmission channels. A source receives service (is connected to a channel) only when it is active. If a source becomes active when all channels are busy, it is blocked and speech clipping occurs. Each speech segment starts and stops independently of whether it is serviced or not. In contrast to conventional TASI systems, a system developed by STC communications adds a small amount of delay to speech segments, when necessary, to minimize the clipping. In this case, a lost calls held analysis is not rigorously justified because the total time a speech segment is "in the system" increases as the delay for service increases. However, if the average delay is a small percentage of the holding time, or if the coding rate of delayed speech is reduced to allow the transmission channel time to "catch up," a lost calls held analysis is still justified.

Lost calls held systems are easily analyzed to determine the probability of the total number of calls in the system at any one time. Since the duration of a source's activity is independent of whether it is being serviced, the number in the system at any time is identical to the number of active sources in a system capable of carrying all traffic as it arises. Thus the distribution of the number in the system is the Poisson distribution provided earlier in Equation 9.3. The probability that i sources requesting service are being blocked is simply the probability that $i + N$ sources are active when N is the number of servers. Recall that the Poisson distribution essentially determines the desired probability as the probability that $i + N$ arrivals occurred in the preceding t_m seconds. The distribution is dependent only on the product of the average arrival rate λ and the average holding time t_m.

EXAMPLE 9.9

What is the probability of clipping in a TASI system with 10 sources and five channels? With 100 sources and 50 channels? Assume that the activity factor of each talker is 0.4.

Solution. For the first case, the clipping probability can be determined as the probability that five or more sources are busy in a Poisson process with an average of $A = (0.4)(10) = 4$ busy servers. Using Equation 9.3.

$$\text{prob (clipping)} = \sum_{j=5}^{\infty} P_j(4)$$

$$= 1 - \sum_{j=0}^{4} P_j(4)$$

$$= 1 - e^{-4} \left(1 + 4 + \frac{4^2}{2!} + \frac{4^3}{3!} + \frac{4^4}{4!} \right)$$

$$= 0.37$$

With 100 sources, the average number of busy circuits $A = (0.4)(100) = 40$. A speech segment is clipped if 50 or more talkers are active at once. Thus the clipping probability can be determined as

$$\text{prob (clipping)} = 1 - \sum_{j=0}^{49} P_j(40)$$

$$= 0.04$$

Example 9.9 demonstrates that TASI systems are much more effective for large group sizes than for small ones. The 37% clipping factor occurring with five channels produces unacceptable voice quality. On the other hand, the 4% clipping probability for 50 channels can be tolerated when the line costs are high enough.

In reality, the values for blocking probabilities obtained in Example 9.9 are higher than actual because an infinite source assumption was used. A more accurate solution is obtained in a later section using a finite source analysis.

9.2.4   Lost Calls Cleared–Finite Sources

As mentioned previously, a fundamental assumption in the derivation of the Poisson arrival distribution, and consequently Erlang's loss formula, is that call arrivals occur independently of the number of active callers. Obviously, this assumption can be justified only when the number of sources is much larger than the number of servers. This section presents some fundamental relationships for determining blocking probabilities of lost calls cleared systems when the number of sources is not much larger than the number of servers. The blocking probabilities in these cases are always less than those for infinite source systems since the arrival rate decreases as the number of busy sources increases.

When considering finite source systems, traffic theorists introduce another parameter of interest called time congestion. Time congestion is the percentage of time that all servers in a group are busy. It is identical to the probability that all servers are busy at randomly selected times. However, time congestion is not necessarily identical to blocking probability (which is sometimes referred to as call congestion). Time congestion merely specifies the probability that all servers are busy. Before blocking can occur, there must be an arrival.

In an infinite source system, time congestion and call congestion are identical because the percentage of arrivals encountering all servers busy is exactly equal to the time congestion. (The fact that all servers are busy has no bearing on whether or not an arrival occurs.) In a finite source system, however, the percentage of arrivals encountering congestion is smaller because fewer arrivals occur during periods when all servers are busy. Thus in a finite source system, call congestion (blocking probability) is always less than the time congestion. As an extreme example, consider equal numbers of sources and servers. The time congestion is the probability that all servers are busy. The blocking probability is obviously zero.

The same basic techniques introduced by Erlang when he determined the loss formula for infinite sources can be used to derive loss formulas for finite sources [3]. Using these techniques, we find the probability of n servers being busy in a system with M sources and N servers is

$$P_n = \frac{\binom{M}{n}(\lambda' t_m)^n}{\sum_{i=0}^{N} \binom{M}{i}(\lambda' t_m)^i} \tag{9.11}$$

where λ' is the calling rate per idle source and t_m is the average holding time. Equation 9.11 is known as the truncated Bernoullian distribution and also as the Engset distribution.

Setting $n = N$ in Equation 9.11 produces an expression for the time congestion:

$$P_N = \frac{\binom{M}{N}(\lambda' t_m)^N}{\sum_{i=0}^{N} \binom{M}{i}(\lambda' t_m)^i} \qquad (9.12)$$

Using the fact that the arrival rate when N servers are busy is $(M - N)/M$ times the arrival rate when no servers are busy, we can determine the blocking probability for lost calls cleared with a finite source as follows:

$$B = \frac{\binom{M-1}{N}(\lambda' t_m)^N}{\sum_{i=0}^{N} \binom{M-1}{i}(\lambda' t_m)^i} \qquad (9.13)$$

which is identical to P_N for $M - 1$ sources.

Equations 9.11, 9.12, and 9.13 are easily evaluated in terms of the parameters λ' and t_m. However, λ' and t_m do not, by themselves. Specify the average activity of a source. If the average activity of a source, assuming no traffic is cleared, is designated as ρ, the value of $\lambda' t_m$ can be determined as

$$\lambda' t_m = \frac{\rho}{1 - \rho(1 - B)} \qquad (9.14)$$

where B is the blocking probability defined by Equation 9.13.

The difficulty with using the unblocked source activity factor ρ to characterize a source's offered load is now apparent. The value of $\lambda' t_m$ depends on B which, in turn, depends on $\lambda' t_m$. Thus some form of iteration is needed to determine B when the sources are characterized by ρ (an easily measured parameter) instead of λ'. If the total

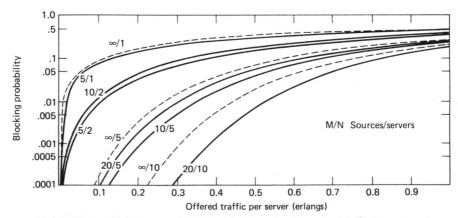

Figure 9.10 Blocking probability of lost calls cleared with finite sources.

offered load is considered to be $M\rho$, the carried traffic is

$$A_{\text{carried}} = M\rho(1 - B) \qquad (9.15)$$

A table of traffic capacities for finite sources is provided in Appendix D where the offered load $A = M\rho$ is listed for various combinations of M, N, and B. Some of the results are plotted in Figure 9.10 where they can be compared to blocking probabilities of infinite source systems. As expected, infinite source analyses (Erlang-B) are acceptable when the number of sources M is large.

EXAMPLE 9.10

A group of callers generate requests at a rate of five calls per hour per telephone (including incoming and outgoing calls). Assuming the average holding time is 4 minutes, what is the average calling rate of each idle source? How many callers can be supported by a 12 channel concentrator/multiplexer if the maximum acceptable blocking probability is 1%?

Solution. Since each caller typically is active for 20 minutes of every hour, the calling rate for idle sources $\lambda' = 5/40 = 0.125$ calls per minute. The offered load for M sources, assuming all traffic is carried, is $0.33M$. Thus Table D.2 must be searched to find the largest M such that $0.33M$ is less than or equal to the maximum offered load for $B = 1\%$ and $N = 12$. Using interpolation for $M = 21$ reveals that 12 servers can support 7.11 erlangs at $B = 1\%$. Since $21(0.33) = 6.93$ is the offered load, 21 sources is an acceptable solution. If 22 sources are used, the offered load of 7.26 erlangs is higher than the 7.04 erlangs obtainable from interpolation in Table D.2 as the maximum offered load for $B = 1\%$.

It is worthwhile comparing the result of Example 9.10 to a result obtained from an infinite source analysis (Erlang-B). For a blocking probability of 1%, Table D.1 reveals that the maximum offered load for 12 servers is 5.88 erlangs. Thus the maximum number of sources can be determined as $5.88/0.333 = 17.64$. Hence in this case an infinite source analysis produces a result that is conservative by 15%.

9.2.5 Lost Calls Held—Finite Sources

A lost calls held system with finite sources is analyzed in the same basic manner as a lost calls held systems with infinite sources. At all times the number of calls "in the system" is defined to be identical to the number of calls that would be serviced by a strictly nonblocking server group. Thus Equation 9.11 is used to determine the probability that exactly n calls are in the system when all M sources are active at once:

$$P_n = \frac{\binom{M}{n}(\lambda' t_m)^n}{\sum_{i=0}^{M} \binom{M}{i}(\lambda' t_m)}$$

$$= \frac{\binom{M}{n}(\lambda' t_m)^n}{(1 + \lambda' t_m)^M} \tag{9.16}$$

Since no calls are cleared, the source activity (offered load per source) can be determined as

$$= \frac{\lambda' t_m}{1 + \lambda' t_m} \tag{9.17}$$

Combining Equations 9.16 and 9.17 produces a more useful expression for the probability that n calls are "in the system":

$$P_n = \binom{M}{n}^n (1 - \lambda)^{M-n} \tag{9.18}$$

If there are N servers, the time congestion is merely the probability that N or more servers are busy:

$$P_N = \sum_{n=N}^{M} P_n \tag{9.19}$$

The blocking probability, in a lost calls held sense, is the probability of an arrival encountering N or more calls in the system:

$$B_h = \frac{\sum_{n=N}^{N} P_n \cdot \text{prob (arrival} \mid n \text{ sources are busy)}}{(\text{average arrival rate})}$$

$$= \sum_{n=N}^{M-1} \binom{M-1}{n} \rho^n (1 - \rho)^{M-1-n} \tag{9.20}$$

where ρ = the offered load per source
       M = the number of sources
       N = the number of servers

EXAMPLE 9.11

Determine the probability of clipping for the TASI systems described in Example 9.9. In this case, however, use a lost calls held analysis for finite sources.

Solution. In this example we are concerned only with the probability that a speech utterance is clipped for some period of time until a channel becomes available. Thus Equation 9.20 provides the desired answer using $\rho = 0.4$ for the offered load per source. In the first case, with 10 sources and five channels,

$$B_h = \sum_{n=5}^{9} \binom{9}{n}(0.4)^n (0.6)^{9-n}$$

$$= 0.27$$

In the second case with 100 sources and 50 servers,

$$B_h = \sum_{n=50}^{99} \binom{99}{n}(0.4)^n (0.6)^{99-n}$$

$$= 0.023$$

The results of Example 9.11 show again that a TASI system requires large group sizes to provide low clipping probabilities. When compared to the results of Example 9.9, these results also indicate that an infinite source analysis overestimates the clipping probability in both cases (0.37 versus 0.27, and 0.04 versus 0.023). Notice that the percentage error in the infinite source analysis is almost identical for the 10 source system and the 100 source system. Hence the validity of choosing an infinite source model is more dependent on the ratio of sources to servers than it is on the number of sources.

Example 9.11 merely determines the probability that a speech segment encounters congestion and is subsequently clipped. A complete analysis of a TASI system must consider the time duration of clips in addition to their frequency of occurrence. In essence, the desired information is represented by the amount of traffic volume in the clipped segments. Weinstein in Reference 6 refers to the clipped segments as "fractional speech loss" or simply "cutout fraction." This is not the same as the lost traffic, since a conventional lost calls held analysis considers any arrival that encounters congestion as being completely "lost"—even if it eventually receives some service. The cutout fraction is determined as the ratio of untransmitted traffic intensity to offered traffic intensity:

$$B_{cf} = \frac{1}{A} \sum_{n=N+1}^{M} (n - N) P_n \qquad (9.21)$$

where M = the number of sources

$\quad A = M_\rho$ is the offered load

$\quad N$ = the number of servers

$\quad P_n$ = the probability of n calls "in the system" (Equation 9.18)

EXAMPLE 9.12

Determine the average duration of a clip in the two TASI systems of
Example 9.11. Assume the average duration of a speech segment is
300 ms. (The average length of a speech segment is dependent on the
volume threshold, which also influences the activity factor ρ.)

Solution. In the first case,

$$B_{cf} = \frac{1}{4} \sum_{n=6}^{10} (n-5) \binom{10}{n} (0.4)^n (0.6)^{10-n}$$

$$= 0.059$$

Thus, on average, 5.9% or 17.7 ms of every 300 ms speech segment is
clipped. Since 27% of the segments experience clipping (Example 9.11),
the average duration of a clip, for clipped segments, is $0.059/0.27 =$
22%, or 66 ms—an obviously intolerable amount.
In the second case for 100 sources and 50 channels,

$$B_{cf} = \frac{1}{40} \sum_{n=51}^{100} (n-50) \binom{100}{n} (0.4)^n (0.6)^{100-n}$$

$$= 0.001$$

Thus in this case only 0.1% of all speech is clipped, which implies that
when clipping occurs $(300)(0.001)/0.023 = 13$ ms of the segment is
lost.

Example 9.12 shows that not only do large group sizes greatly reduce
the clipping probability of TASI systems, but also they reduce the duration of the clips. The relationship of clipping probabilities and clipping
durations (fractional speech loss) to group size and source activity is
provided in Figure 9.11. As shown, the clipping probability is extremely sensitive to source activity (offered load). For a discussion of
the effects of clipping on speech quality see References 7 and 8.

9.3   NETWORK BLOCKING PROBABILITIES

In the preceding sections basic techniques of congestion theory are presented to determine blocking probabilities of individual trunk groups.
In this section techniques of calculating end-to-end blocking probabilities of a network with more than one route between endpoints is considered. In conjunction with calculating the end-to-end blocking probabilities, it is necessary to consider the interaction of traffic on various

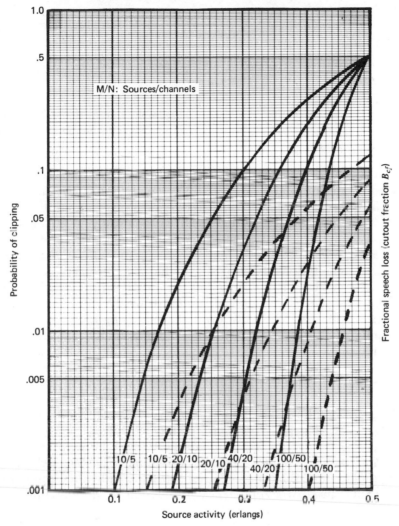

Figure 9.11 Clipping probability and clipping duration of TASI.

routes of a network. Foremost among these considerations is the effect of overflow traffic from one route onto another. The following sections discuss simplified analyses only. More sophisticated techniques for more complex networks can be obtained in References 9, 10, and 11.

9.3.1 End-To-End Blocking Probabilities

Generally, a connection through a large network involves a series of transmission links, each one of which is selected from a set of alternatives. Thus an end-to-end blocking probability analysis usually involves

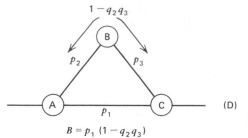

Figure 9.12 Probability graph for end-to-end blocking analysis.

a composite of series and parallel probabilities. The simplest procedure is identical to the blocking probability (matching loss) analyses presented in Chapter 5 for switching networks. For example, Figure 9.12 depicts a representative set of alternative connections through a network and the resulting composite blocking probability.

The blocking probability equation in Figure 9.12 contains several simplifying assumptions. First of all, the blocking probability (matching loss) of the switches is not included. In a digital time division switch, matching loss can be low enough that it is easily eliminated from the analysis. In other switches, however, the matching loss may not be insignificant. When necessary, switch blocking is included in the analysis by considering it a source of blocking in series with the associated trunk groups.

When more than one route passes through the same switch, as in node c of Figure 9.12, proper treatment of correlation between matching losses is an additional complication. A conservative approach considers the matching loss to be completely correlated. In this case the matching loss is in series with the common link. On the other hand, an optimistic analysis assumes that the matching losses are independent, which implies that they are in series with the individual links. Figure 9.13 depicts these two approaches for including the matching loss of

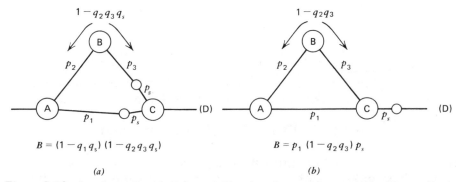

Figure 9.13 Incorporating switch matching loss into end-to-end blocking analysis. (*a*) Independent switch blocking. (*b*) Correlated switch blocking.

switch C into the end-to-end blocking probability equation of Figure 9.12. In this case, the link from C to D is the common link.

A second simplifying assumption used in deriving the blocking probability equation in Figure 9.12 involves assuming independence for the blocking probabilities of the trunk groups. Thus the composite blocking of two parallel routes is merely the product of the respective probabilities (Equation 5.4). Similarly, independence implies that the blocking probability of two paths—in series—is 1 minus the product of the respective availabilities (Equation 5.5). In actual practice individual blocking probabilities are never completely independent. This is particularly true when a large amount of traffic on one route results as overflow from another route. Whenever the first route is busy, it is likely that more than the average amount of overflow is being diverted to the second route. Thus an alternate route is more likely to be busy when a primary route is busy.

In a large public network, trunks to tandem or toll switches normally carry traffic to many destinations. Thus no one direct route contributes an overwhelming amount of overflow traffic to a particular trunk group. In this case independent blocking probabilities on alternate routes are justified. In some instances of the public network, and often in private networks, overflow traffic from one route dominates the traffic on tandem routes. In these cases failure to account for the correlation in blocking probabilities can lead to overly optimistic results.

EXAMPLE 9.13

Two trunk groups are to be used as direct routes between two switching offices. The first group has 12 channels and the second group has 6 channels. Assume 10.8 erlangs of traffic is offered to the 12 channel group, and overflows are offered to the 6 channel group when the first group is busy. What is the blocking probability of the first group, and how much traffic overflows to the second group? Using the overflow traffic volume as an offered load, determine the blocking probability of the second trunk group. What is the probability that both trunk groups are busy? Compare this answer to the blocking probability of one 18 channel trunk group.

Solution. Using a lost calls cleared analysis with an infinite source, we determine that blocking of the first group is 15% ($A = 10.8$, $N = 12$). Therefore the overflow traffic is $(10.8)(0.15) = 1.62$ erlangs. The blocking probability (assuming random arrivals???) of the second group is 0.5% ($A = 1.62$, $N = 6$). The probability that both trunk groups are busy simultaneously can be determined (assuming independence ???) as:

$$B = (0.15)(0.005)$$

$$= 0.00075$$

In contrast, the correct blocking probability of an 18 channel trunk group with an offered load of 10.8 erlangs is

$$B = 0.013$$

The question marks in the solution of Example 9.13 point to two sources of error in the determination of the first blocking probability value. The first error is the assumption of independence of blocking in the two trunk groups. The second error results from the use of an analysis predicated on purely random (Poisson) arrivals for overflow traffic into the second trunk group. Resolution of this error is discussed in the next section.

Separating the 18 channels of Example 9.13 into two groups is obviously an artifice. This example is useful in that it demonstrates an extreme case of correlation between blocking probabilities of two trunk groups. When correlation exists, the composite blocking probability of a direct route and an alternate route should be determined as follows:

$$B = (B_1)(B_2|1) \qquad\qquad (9.22)$$

where B_1 = the blocking probability of group 1
  $B_2|1$ = the blocking probability of group 2 given that group 1 is busy

In the artificial case of dividing a trunk group into two subgroups, the conditional blocking probability can be determined as

$$B_2|1 = B(N|N_1)$$

$$= \text{prob}\ (N\ \text{servers are busy when}\ N_1\ \text{are known to be busy})$$

$$= \frac{P_N}{\sum_{n=N_1}^{N} P_n} \qquad\qquad (9.23)$$

$$= \frac{(A^N/N!)}{\sum_{n=N_1}^{N} (A^n/n!)}$$

where P_n is the probability that exactly n of N servers are busy (Equation 9.3).

Evaluating Equation 9.23 for $A = 10.8$, $N_1 = 12$, and $N = 18$ reveals that the appropriate conditional probability $B_2|1$ for Example 9.13 is 0.033. Thus the composite blocking probability per Equation 9.22 is: $B = (0.15)(0.033) = 0.005$. The remaining inaccuracy (0.005 versus 0.013 actually) is due to nonrandom characteristics of overflow traffic.

Equation 9.23 is only valid for the contrived case of an alternate route carrying overflow traffic from only one primary route. It can be used, however, as a worst case solution to situations where overflow from one route tends to dominate the traffic on an alternate route.

The correlations between the blocking probabilities of individual routes arise because congestion on one route produces overflows that tend to cause congestion on other routes. External events stimulating networkwide overloads also cause the blocking probabilities to be correlated. Thus a third assumption in the end-to-end blocking probability equation of Figure 9.12 is that traffic throughout the network is independent. If fluctuations in the traffic volume on individual links tend to be correlated (presumably because of external events such as television commercials, etc.), significant degradation in overall performance results.

9.3.2 Overflow Traffic

The second source of error in Example 9.13 occurred because an Erlang-B analysis used the average volume of overflow traffic from the first group to determine the blocking probability of the second trunk group. An Erlang-B analysis assumes traffic arrivals are purely random, that is, they are modeled by a Poisson distribution. However, a Poisson arrival distribution is an erroneous assumption for the traffic offered to the second trunk group. Even though arrivals to the first group may be random, the overflow process tends to select groups of these arrivals and pass them on to the second trunk group. Thus instead of being random the arrivals to the second group occur in bursts. This overflow effect is illustrated in Figure 9.14, which portrays a typical random arrival pattern to one trunk group and the overflow pattern to a second group. If a significant amount of the traffic flowing onto a trunk group results as overflow from other trunk groups, overly optimistic values of blocking probability arise when all of the traffic is assumed to be purely random.

The most common technique of dealing with overflow traffic is to

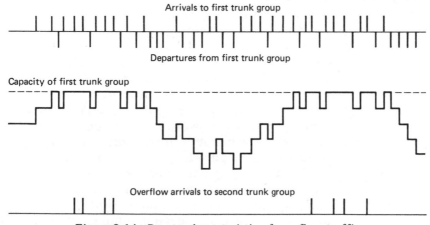

Figure 9.14 Bursty characteristic of overflow traffic.

relate the overflow traffic volume to an equivalent amount of random traffic in a blocking probability sense. For example, if the 1.62 erlangs of overflow traffic in Example 9.12 is equated to 2.04 erlangs of random traffic, a blocking probability of 1.3% is obtained for the second trunk group. (This is the correct probability of blocking for the second group since both groups are busy if and only if the second group is busy.)

This method of treating overflow traffic is referred to as the equivalent random theory [12]. Tables of traffic capacity are available [13] that incorporate the overflow effects directly into the maximum offered loads. The Neal-Wilkinson tables used by Bell System traffic engineers comprise one such set of tables. The Neal-Wilkinson tables, however, also incorporate the effects of day-to-day variations in the traffic load. (Forty erlangs on one day and 30 erlangs on another is not the same as 35 erlangs on both days.) These tables are also used for trunk groups that neither generate nor receive overflow traffic. The fact that cleared traffic does not get serviced by an alternate route implies that re-tries are likely. The effect of the re-tries, however, is effectively incorporated into the value of B by equivalent randomness.

9.4 DELAY SYSTEMS

The second category of teletraffic analysis concerns systems that delay nonserviceable requests until the necessary facilities become available. These systems are variously referred to as delay systems, waiting-call systems, and queuing systems. Call arrivals occurring when all servers are busy are placed in a queue and held until service commences. The queue might consist of storage facilities in a physical sense, such as blocks of memory in a message-switching node, or the queue might consist only of a list of sources waiting for service. In the latter case, storage of the messages is the responsibility of the sources themselves.

Using the more general term, queueing theory, we can apply the following analyses to a wide variety of applications outside of telecommunications. Some of the more common applications are data processing, supermarket check-out counters, aircraft landings, inventory control, and various forms of service bureaus. These and many other applications are considered in the relatively new field of operations research. The foundations of queuing theory, however, rest on fundamental techniques developed by early telecommunications traffic researchers. In fact, Erlang is credited with the first solution to the most basic type of delay system. Examples of delay system analysis applications in telecommunications are message switching, packet switching, statistical time division multiplexing, multipoint data communications, automatic call distribution, digit receiver access, signaling equipment usage, and call processing. Furthermore, many of the newer, stored-program control PBXs have features allowing queued access to corporate tie lines or WATS lines. Thus some systems formerly operating as loss systems now operate as delay systems.

In general, a delay operation allows for greater utilization of servers (transmission facilities) than does a loss system. Basically, the improved utilization is achieved because peaks in the arrival process are "smoothed" by the queue. Even though arrivals to the system are random, the servers see a somewhat regular arrival pattern. The effect of the queuing process on overload traffic is illustrated in Figure 9.15. This figure displays the same traffic patterns presented earlier in Figures 9.1, 9.3, and 9.9. In this case, however, overload traffic is delayed until call terminations produce available channels.

In most of the following analyses it is assumed that all traffic offered to the system eventually gets serviced. One implication of this assumption is that the offered traffic intensity A is less than the number of servers N. Even when A is less than N, there are two cases in which the carried traffic might be less than the offered traffic. First, some sources might tire of waiting in a long queue and abandon the request. Second, the capacity for storing requests may be finite. Hence requests may occasionally be rejected by the system.

A second assumption in the following analyses is that infinite sources exist. In a delay system, there may be a finite number of sources in a physical sense, but an infinite number of sources in an operational sense because each source may have an arbitrary number of requests out-

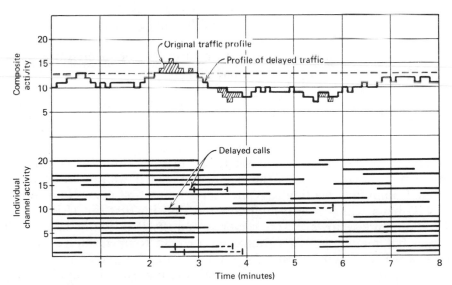

Figure 9.15 Activity profile of blocked calls delayed (13 servers).

standing (e.g., a packet-switching node). There are instances in which a finite source analysis is necessary, but not in the applications considered here.

An additional implication of servicing all offered traffic arises when infinite sources exist. This implication is the need for infinite queuing capabilities. Even though the offered traffic intensity is less than the number of servers, no statistical limit exists on the number of arrivals occurring in a short period of time. Thus the queue of a purely lossless system must be arbitrarily long. In a practical sense, only finite queues can be realized, so either a statistical chance of blocking is always present, or all sources can be busy and not offer additional traffic.

When analyzing delay systems it is convenient to separate the total time that a request is in the system into the waiting time and the holding time. In delay systems analysis the holding time is more commonly referred to as the service time. In contrast to loss systems, delay system performance is generally dependent on the distribution of service times, and not just the mean value t_m. Two service time distributions are considered here: constant service times and exponential service times. Respectively, these distributions represent the most deterministic and the most random service times possible. Thus a system that operates with some other distribution of service times performs somewhere between the performance produced by these two distributions.

The basic purpose of the following analyses is to determine the probability distribution of waiting times. From the distribution, the average waiting time is easily determined. Sometimes only the average waiting time is of interest. More generally, however, the probability that the waiting time exceeds some specified value is of interest. In either case, the waiting times are dependent on the following factors:

1 The intensity and the probabilistic nature of the offered traffic.
2 The distribution of service times.
3 The number of servers.
4 The number of sources.
5 The service discipline of the queue.

The service discipline of the queue can involve a number of factors. The first of these concerns the manner in which waiting calls are selected. Commonly, waiting calls are selected on a first come, first served (FCFS) basis which is also referred to as first in, first out (FIFO) service. Sometimes, however, the server system itself does not maintain a queue, but merely polls its sources in a round robin fashion to determine which ones are waiting for service. Thus the queue may be serviced in sequential order of the waiting sources. In some applications waiting requests may even be selected at random. Furthermore, additional service variations arise if any of these schemes are augmented with a priority discipline that allows some calls to move ahead of others in the queue.

A second aspect of the service discipline that must be considered is the length of the queue. If the maximum queue size is smaller than the effective number of sources, blocking can occur in a lost calls sense. The result is that two characteristics of the grade of service must be considered: the delay probability and the blocking probability. A common example of a system with both delay and loss characteristics is an automatic call distributor with more access circuits than attendants (operators or reservationists). Normally, incoming calls are queued for service. Under heavy loads, however, blocking occurs before the ACD is even reached. Reference 14 contains an analysis of a delay system with finite queues and finite servers.

In order to simplify the characterization of particular systems, queuing theorists have adopted a concise notation for classifying various types of delay systems. This notation, which was introduced by D. G. Kendall, uses letter abbreviations to identify alternatives in each of the categories listed. Although the discussions in this book do not rely on this notation, it is introduced and used occasionally so the reader can relate the following discussions to classical queuing theory models. The interpretation of each letter is specified in Figure 9.16.

The specification format presented in Figure 9.16 actually represents an extension of the format commonly used by most queuing theorists. Thus this format is sometimes abbreviated by eliminating the last one or two entries. When these entries are eliminated, infinite case specifica-

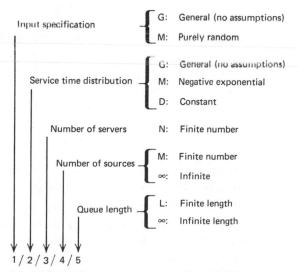

Figure 9.16 Queuing system notation.

tions are assumed. For example, a single-server system with random input and negative exponential service times is usually specified as: M/M/1. Both the number of sources and the permissible queue length are assumed infinite.

9.4.1 Exponential Service Times

The simplest delay system to analyze is a system with random arrivals and negative exponential service times: M/M/N. Recall that a random arrival distribution is one with negative exponential interarrival times. Thus in the shorthand notation of queuing theorists, the letter M always refers to negative exponential distributions (an M is used because a purely random distribution is memoryless).

In the M/M/1 system, and all other systems considered here, it is assumed that calls are serviced in the order of their arrival. The following analyses also assume that the probability of an arrival is independent of the number of requests already in the queue (infinite sources). From these assumptions, the probability that a call experiences congestion and is therefore delayed was derived by Erlang:

$$\text{prob}(\text{delay}) = p(>0) = \frac{NB}{N - A(1 - B)} \tag{9.24}$$

where N = the number of servers

   A = the offered load in erlangs

   B = the blocking probability for a lost calls cleared system (Equation 9.8)

The probability of delay $p(>0)$ is variously referred to as Erlang's second formula, $E_{2,N}(A)$ Erlang's delay formula; or the Erlang-C formula. For single-server systems $N = 1$, the probability of delay reduces to ρ, which is simply the output utilization or traffic carried by the server. Thus the probability of delay for a single-server system is also equal to the offered load λt_m (assuming $\lambda t_m < 1$).

The distribution of waiting times for random arrivals, random service times, and a FIFO service discipline is

$$p(>t) = p(>0)\, e^{-(N - A)\, t/t_m} \tag{9.25}$$

where $p(>0)$ = the probability of delay given in Equation 9.24,

   t_m = the average service time of the negative exponential service time distribution

Equation 9.25 defines the probability that a call arriving at a randomly chosen instant is delayed for more than t/t_m service times. Figure 9.17 presents the relationship of Equation 9.25 by displaying the traffic capacities of various numbers of servers as a function of acceptable delay times. Given a delay time objective t/t_m, Figure 9.17a displays the max-

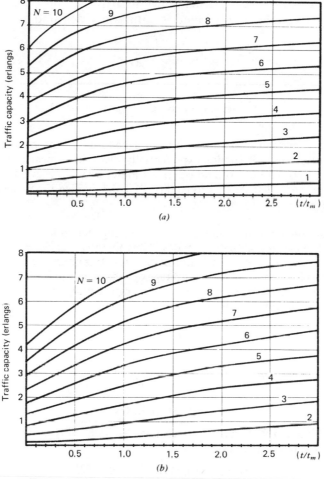

Figure 9.17 Traffic capacity of multiple-server delay systems with exponential service times. (a) Probability of exceeding t, $p(>t) = 10\%$. (b) Probability of exceeding t, $p(>t) = 1\%$.

imum traffic intensity if the delay objective is to be exceeded for only 10% of the arrivals. Similarly, Figure 9.17b displays the maximum traffic intensity if the delay objective is exceeded for only 1% of the arrivals. Notice that at $p(>t) = 0.01$, the server systems do not approach their maximum capacity (number of servers) unless the acceptable delay is several times larger than t_m.

By integrating Equation 9.25 over all time, the average waiting time for all arrivals can be determined as

$$\bar{t} = \frac{p(>0)\, t_m}{N - A} \tag{9.26}$$

Notice that $\bar{t}$ is the expected delay for all arrivals. The average delay of only those arrivals that get delayed is commonly denoted as

$$t_w = \frac{t_m}{N - A} \tag{9.27}$$

EXAMPLE 9.14

A message-switching network is to be designed for 95% utilization of its transmission links. Assuming exponentially distributed message lengths and an arrival rate of 10 messages per minute, what is the average waiting time, and what is the probability that the waiting time exceeds 5 minutes?

Solution. Assume that the message-switching network uses a single channel between each pair of nodes. Thus there is a single server and a single queue for each transmission link. Since p is given to be 0.95 and $\lambda = 10$ arrivals per minute, the average service time can be determined as $t_m = 0.95/10 = 0.095$ minutes. The average waiting time (not including the service time) is easily determined as

$$\bar{t} = \frac{p(>0)\, t_m}{N - A}$$

$$= \frac{(0.95)\,(0.095)}{1 - 0.95}$$

$$= 1.805 \text{ minutes}$$

Using Equation 9.25, we can determine the probability of the waiting time exceeding 5 minutes as

$$p(>5) = (0.95)\, e^{-(1 - .95)\, 5/.095}$$

$$= 0.068$$

Thus 6.8% of the messages experience queuing delays of more than 5 minutes.

EXAMPLE 9.15

Determine the number of digit receivers required to support 1000 telephones with an average calling rate of two calls per hour. Assume the dialing time is exponentially distributed with an average service time of 6 seconds. The grade of service objective is to return dial tone within 1 second of the off-hook signal for 99% of the call attempts. Compare the answer obtained from a delay system analysis to an answer obtained from a loss system analysis at $B = 1\%$. If the blocking probability is less than 1%, fewer than 1% of the calls are delayed.

Solution. The calling rate λ and the offered traffic intensity A are easily determined as 0.555 calls per second and 3.33 erlangs, respectively. Since the number of servers N cannot be solved for directly from the equations, Figure 9.17b is used to obtain a value of eight servers for $t/t_m = \frac{1}{6}$.

Examination of Table B.1 reveals that 99% of the call attempts can be serviced immediately if there are nine digit receivers. Thus in this case the ability to delay service provides a savings of only one server.

Example 9.15 demonstrates that a blocking probability analysis produces approximately the same results as a delay system analysis when the maximum acceptable delay is a small percentage of the average service time. The two results are almost identical because, if a digit receiver is not immediately available, there is only a small probability that one will become available within a short time period. (With an average service time of 6 seconds, the expected time for one of eight digit receivers to be released is $\frac{6}{8} = 0.75$ seconds. Hence the delay operation in this case allows a savings of one digit receiver.)

Because a digit receiver must be available within a relatively short time period after a request is generated, digit receiver group sizing is often determined strictly from a blocking probability analysis. The fact that digit receiver access is actually operated as a delay system implies the grade of service is always better than that calculated.

9.4.2 Constant Service Times

This section considers delay systems with random arrivals, constant service times, and a single server (M/D/1). Again, FIFO service disciplines and infinite sources are assumed. The case for multiple servers has been solved [3] but is too involved to include here. Graphs of multiple-server systems with constant service times are available in Reference 15.

The average waiting time for a singler server with constant service times is determined as

$$\bar{t} = \frac{\rho t_m}{2(1 - \rho)} \tag{9.28}$$

where $\rho = A$ is the server utilization. Notice that Equation 9.28 produces an average waiting time that is exactly one-half of that for a single-server system with exponential service times. Exponential service times cause greater average delays because there are two random processes involved in creating the delay. In both types of systems, delays occur when a large burst of arrivals exceeds the capacity of the servers. With exponential service times, however, long delays also arise because of excessive service times of just a few arrivals. (Recall that this aspect of conventional message switching systems is one of the motivations for breaking messages up into packets in a packet-switching network.)

If the activity profile of a constant service time system (M/D/1) is compared with the activity profile of an exponential service time system (M/M/1), the M/D/1 system is seen to be active for shorter and more frequent periods of time. That is, the M/M/1 system has a higher variance in the duration of its busy periods. The average activity of both systems is, of course, equal to the server utilization ρ. Hence the probability of delay for a single-server system with constant service times is identical to that for exponential service times: $p(>0) = \lambda t_m$.

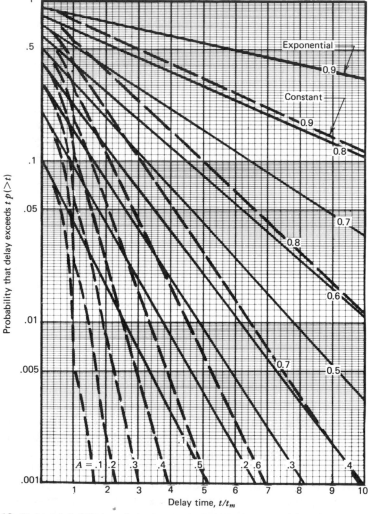

Figure 9.18. Delay probabilities of single-server systems (exponential and constant service times).

The probability of congestion for larger N is relatively close to that for exponential service times. Thus Equation 9.25 can be used as a close approximation for $p(>0)$ for multiple-server systems with arbitrary service time distributions.

For single-server systems with constant holding times, the probability of delay greater than an arbitrary value t is

$$p(>t) = p(>(k + r) t_m)$$

$$= 1 - (1 - p) \sum_{i=0}^{k} \frac{p^i (i - t/t_m)^i \, e^{-p(i - t/t_m)}}{i!} \qquad (9.29)$$

where k = the largest integral quotient of t/t_m
$\quad\quad r$ = the remainder of t/t_m
$\quad\quad p = \lambda t_m$ is the server utilization

Comparisons of the waiting time distributions for single-server systems with exponential and constant service times are shown in Figure 9.18. For each pair of curves, the upper one is for exponential service times and the lower one is for constant service times. Since all other service time distributions produce delay probabilities between these extremes, Figure 9.18 provides a direct indication of the range of possible delays.

EXAMPLE 9.16

A packet-switching node operates with fixed length packets of 300 bits on 9600 bps lines. If the link utilization is to be 90%, what is the average delay through a node? What percentage of packets encounter more than 0.35 seconds of delay? What is the average delay if the offered load increases by 10%?

Solution. Message lengths of 300 bits and a data rate of 9600 bps imply that the fixed length service time is $300/9600 = 0.031$ seconds. From Equation 9.28, the average waiting time is

$$t = \frac{(0.9)\,(0.031)}{2(1 - 0.9)} = 0.140 \text{ seconds}$$

The total average delay through the node, excluding processing, is obtained by adding the average waiting time to the service time:

$$\text{average delay} = 0.140 + 0.031 = 0.171 \text{ seconds}$$

Since the service time is 0.031 seconds, 0.35 seconds of delay occur when the waiting time is $0.35 - 0.031 = 0.319$. This corresponds to $0.319/0.031 = 10$ service times. From Figure 9.18, the probability of delay for $t/t_m = 10$ is approximately .12. Thus 12% of the packets experience delays of greater than 0.35 seconds. An increase of 10% in the traffic intensity implies that the new offered load is 0.99 erlangs. From Equation 9.30, the average waiting time becomes

$$t = \frac{(0.9)(0.031)}{2(1 - 0.99)} = 1.40 \text{ seconds}$$

Thus when the offered load increases by only 10%, the average delay through the node increases more than eightfold to 1.43 seconds.

Example 9.16 demonstrates the same characteristic for heavily utilized delay systems that was demonstrated for loss systems: the performance is very sensitive to increases in traffic intensity. Thus, as discussed in Chapter 7, flow control is a critical aspect of a packet-switching operation—particularly when there are real time delivery objectives.

9.4.3 Finite Queues

All of the delay system analyses presented so far have assumed that an arbitrarily large number of delayed requests could be placed in a queue. In many applications this assumption is invalid. Examples of systems that sometimes have significantly limited queue sizes are store-and-forward switching nodes, automatic call distributors, and various types of computer input/output devices. These systems treat arrivals in three different ways, depending on the number "in the system" at the time of an arrival:

1 Immediate service if one or more of N servers is idle.
2 Delayed service if all servers are busy and less than L requests are waiting.
3 Blocked or no service if the queue of length L is full.

In finite queue systems the arrivals getting blocked are those that would otherwise experience long delays in a pure delay system. Thus an indication of the blocking probability of a combined delay and loss system can be determined from the probability that arrivals in pure delay systems experience delays in excess of some specified value. However, there are two basic inaccuracies in such an analysis. First, the effect of blocked or lost calls cleared is to reduce congestion for a period of time and thereby to reduce the delay probabilities for subsequent arrivals. Second, delay times do not necessarily indicate how many calls are "in the system." Normally, queue lengths and blocking probabilities are determined in terms of the number of waiting requests, not the amount of work or total service time represented by the requests. With constant service times, there is no ambiguity between the size of a queue and its implied delay. With exponential service times, however, a given queue size can represent a wide range of delay times.

A message-switching node is an example of a system in which the queue length is most appropriately determined by implied service time and not by the number of pending requests. That is, the maximum queue length may be determined by the amount of store-and-forward memory available for variable length messages, and not by some fixed number of messages.

Whenever a message-switching or packet-switching node rejects an arriving message, the source of the message must become aware of the rejection in some manner. First, the receiving node can remain passive and ignore the arrival. The source then determines that the transmission was unsuccessful by timing-out for an acknowledgement. This approach is attractive because time-outs are generally needed anyway, and the overloaded node or network does not need to respond actively to the arrival. The main drawback to the passive approach is that the source node does not learn why the transmission was unsuccessful and may continue to transmit undeliverable messages—thereby continuing to congest the network.

Obviously, a second technique is to respond to all blocked requests with a short control message that indicates why and where the rejection occurred and what the appropriate response should be.

For a system with random input, exponential service times, N servers, an infinite source, and a maximum queue length of L (M/M/N/∞/L), the probability of j calls in the system is

$$P_j(A) = P_0(A) \frac{A^j}{j!} \quad (0 \leqslant j \leqslant N)$$

$$= \frac{P_0(A) A^j}{N! \, N^{j-N}} \quad (N \leqslant j \leqslant N+L)$$

(9.30)

$$P_0(A) = \left\{ \sum_{j=0}^{N+L} \left(\frac{A^j}{j!} \right) \right\}^{-1}$$

$A = \lambda t_m$ = offered load in erlangs
N = the number of servers
L = maximum number in the queue

The time congestion, or probability of delay, can be determined from Equation 9.30 as

$$p(>0) = \sum_{j=N}^{N+L} P_j(A)$$

$$= P_N(A) \frac{1 - \rho^{L+1}}{1 - \rho}$$

(9.31)

where $\rho = A/N$ is the offered load per server.

The loss, or blocking probability, is determined as

$$B = P_{N+L}(A) = \frac{P_0(A)\,A^{N+L}}{N!\,N^L} \tag{9.32}$$

It is worth noting that if there is no queue ($L = 0$), these equations reduce to those of the Erlang loss equation (9.8). If L is infinite, Equation 9.31 reduces to Erlang's delay formula, Equation 9.24. Thus these equations represent a general formulation that produces the pure loss and pure delay formulas as special cases.

The waiting time distribution [3] is

$$p(>t) = P_N(A) \sum_{j=0}^{L-1} \frac{p^j}{j!} \int_{Nt/t_m}^{\infty} x^j\, e^{-x}\, dx \tag{9.33}$$

from which the average delay can be determined as:

$$\bar{t} = \frac{[p(>0) - L P_{N+L}(A)]\, t_m}{N - A} \tag{9.34}$$

Again, Equation 9.34 is identical to Equation 9.26 for an infinite queue ($L = \infty$).

The blocking probability of a single-server system ($N = 1$) is plotted in Figure 9.19. When using Figure 9.19, keep in mind that the blocking probability (Equation 9.32) is determined by the number of waiting calls and not by the associated service time. Furthermore, since the curves of Figure 9.19 are based on exponential service times, they overestimate the blocking probabilities of constant holding time systems (e.g. fixed length packet networks). However, if fixed length packets arise primarily from longer, exponentially distributed messages, the arrivals are no longer independent, and the use of Figure 9.19 (or Equation 9.32) as a conservative analysis is more appropriate.

9.4.4 Tandem Queues

All of the equations provided in previous sections for delay system analysis have dealt with the performance of a single queue. In many applications a service request undergoes several stages of processing— each one of which involves queuing. Thus it is often desirable to analyze the performance of a system with a number of queues in series. Figure 9.20 depicts a series of queues that receive, as inputs, locally generated requests and outputs from other queues. Two principal examples of applications with tandem queues are data processing systems and store-and-forward switching networks.

Researchers in queuing theory have not been generally successful in deriving formulas for the performance of tandem queues. Often,

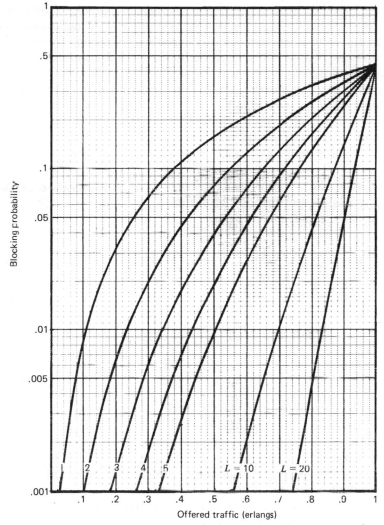

Figure 9.19 Blocking probability of single server loss/delay system (exponential service times).

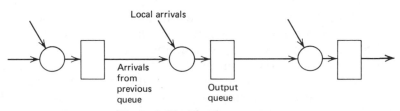

Figure 9.20 Tandem queues.

simulation is used to analyze a complex arrangement of interdependent queues arising in systems like store-and-forward networks. Simulation has the advantage that special aspects of a network's operation—like routing and flow control—can be included in the simulation model. The main disadvantages of simulation are expense and, often, less visibility into the dependence of system performance on various design parameters.

One tandem queuing problem that has been solved [16] is for random inputs and random (negative exponential) holding times for all queues. The solution of this system is based on the following theorem:

In a delay system with purely random arrivals and negative exponential holding times, the instants at which calls terminate is also a negative exponential distribution.

The significance of this theorem is that outputs from an M/M/R system have statistical properties that are identical to its inputs. Thus a queuing process in one stage does not affect the arrival process in a subsequent stage, and all queues can be analyzed independently. Specifically, if a delay system with N servers has exponentially distributed interarrival times with average $1/\lambda$, and if the average service time is t_m, calls leave each of the servers according to exponentially distributed intercompletion times with average $1/\lambda N$.

Although independent analysis of tandem queues can be rigorously justified only for purely random arrivals and service times, independence is often assumed in other cases. Before using such assumptions, however, the systems in question should be examined closely to determine if the state of one queue can influence the operation of another queue in the system.

As a specific example consider the connection setup process of a circuit-switched network. This process is basically a tandem queuing process because connect requests propagate from one switching node to another with each node processing a request before forwarding it to the next node. Depending on the manner in which signaling information is transmitted between nodes, congestion in one node can influence the operation of another node.

In the public network switching offices commonly exchange information by way of senders using in-channel multifrequency or dial-pulse signaling. The senders are used on a switched basis so they can be shared across a number of circuits. When the stored program control in one office wants to transfer a connect request to another office, a sender is assigned to the appropriate output trunk. However, before the information can be transferred, the destination office must assign a sender (also used for receiving) to the corresponding input trunk. In the meantime the sender in the first office is allocated and unavailable to service other requests. Hence if congestion exists in the second office, the first office is affected. In essence, the holding time of the sender is as much, or more, a function of the response time at the destination node as a function of the underlying message length. One advantage of common channel interoffice signaling is that signaling information goes directly from one call processor to another and does not involve selection of common equipment like senders. (It is also an advantage to the traffic analyst since the independence assumption is easier to justify.)

Another example of interacting queues arises in packet-switching networks that do not preallocate storage space for all packets. Since the amount of available storage space is finite, the queue is finite and blocking can occur. When blocking does occur, a blocked packet returns (in a functional sense) to the node that sent it. Thus congestion in a destination node prevents the transmitting node from releasing internal storage space assigned to the packet and causes the packet to reenter the same or a different output queue for retransmission. Notice that in this case queue interaction arises because the maximum queue length of the destination node is finite instead of infinite, as assumed in the model for independent queues.

REFERENCES

1 W. Feller, *An Introduction to Probability Theory and its Applications*, John Wiley & Sons, New York, 1968.

2 A. A. Collins and R. D. Pederson, *Telecommunications A Time For Innovation*, Merle Collins Foundation, Dallas, Texas, 1973.

3 R. Syski, *Introduction to Congestion Theory in Telephone Systems*, Oliver and Boyd, London, 1965.

4 Members of Technical Staff, Bell Telephone Laboratories, *Engineering and Operations in the Bell System*, Western Electric, 1977.

5 "Determination of the Number of Circuits in Automatic and Semi-Automatic Operation," CCITT Recommendation E. 520, *Orange Book*, Vol. II.2, 1976, pp 211.

6 C. J. Weinstein, "Fractional Speech Loss and Talker Activity Model for Packet-Switched Speech," *IEEE Transactions on Communications Technology*, August 1978, pp 1253–1256.

7 H. Miedema and M. Schachtman, "TASI Quality-Effect of Speech Detection and Interpolation," *Bell System Technical Journal*, July 1962, pp 1455–1473.

8 G. Szarvas and H. Suyderhoud, "Voice-Activated-Switch Performance Criteria," *Comsat Technical Review*, Spring 1980, pp 151–177.

9 S. S. Katz, "Improved Network Administration Process Utilizing End-to-End Service Considerations," *International Teletraffic Conference*, 1979.

10 R. Dayem, "Alternate Routing in High Blocking Communications Networks," *National Telecommunications Conference*, 1979, pp 28.4.1–28.4.6.

11 P. R. Boorstyn and H. Frank, "Large-Scale Network Topological Optimization," *IEEE Transactions on Communications*, January 1977, pp 29–47.

12 R. I. Wilkinson, "Theories for Toll Traffic Engineering in U.S.A.," *Bell System Technical Journal*, March 1956.

13 "Calculation of the Number of Circuits in a Group Carrying Overflow Traffic," CCITT Recommendation E.521, *Orange Book*, Vol. 2, No. 2, pp. 218.

14 J. A. Morrison, "Analysis of Some Overflow Problems with Queueing," *Bell System Technical Journal*, October 1980, pp 1427–1462.

15 *Telephone Traffic Theory, Tables, and Charts*, Siemens Aktiengesellschaft, Munich, 1970.

16 L. Kleinrock, *Queueing Systems Volume 1: Theory*, John Wiley & Sons, New York, 1975.

PROBLEMS

9.1 A central office to PBX trunk group contains four circuits. If the average call duration is 3 minutes and the busy hour offered traffic intensity is 2 erlangs, determine each of the following:
(a) The busy hour calling rate.
(b) The probability that two arrivals occur less than 1 second apart.
(c) The blocking probability assuming a lost calls cleared operation.
(d) The amount of lost traffic.
(e) The proportion of time the fourth circuit is in use (assuming fixed order selection).

9.2 A T1 line is used to carry traffic from a remote concentrator to a central office. How many 10 CCS subscribers can the concentrator system support at 0.5% blocking. Compare an infinite source analysis to a finite source analysis.

9.3 Two switching offices experience 20 erlangs of average busy hour traffic load between them. Assume a single T1 line provides 24 direct trunks between the offices. How much busy hour traffic overflows to a tandem switch?

9.4 A PBX with 200 stations has five trunks to the public network. What is the blocking probability if each station is involved in three external calls per 8 hour working day with an average duration of 2 minutes per call? Assume blocked calls return with random re-tries. What is the offered load? What is the demand traffic?

9.5 How many dial-up I/O ports are needed for a computer center to support 40 users with a blocking probability limit of 5%?

Assume each user averages four calls per day with an average session duration of 30 minutes. If three users remain connected all day, what is the grade of service for the remaining 37 users?

9.6 A 24 channel trunk group is divided into two groups of 12 one-way trunks in each direction. (A one-way trunk is one that can only be seized at one end.) How many erlangs of traffic can this system support at 0.5% blocking? How many erlangs can be supported if all 24 trunks are two-way trunks? (That is, every trunk can be seized at either end.)

9.7 The following 10 a.m. to 11 a.m. busy hour statistics have been observed on a 32 channel interoffice trunk group. What is the overall blocking probability? What is the blocking probability for the same busy hour, if day-to-day fluctuations are averaged together?

Monday	Tuesday	Wednesday	Thursday	Friday
20E	19E	22E	19E	30E

9.8 Traffic measurements on a PBX to central office trunk group indicate that during the busiest hour of the day the trunks are 80% utilized. If there are eight trunks in the group, what is the blocking probability, assuming blocked calls do not return? How many trunks must be added to achieve a maximum blocking probability of 5%?

9.9 Repeat Problem 9.8 assuming blocked calls return with random re-tries.

9.10 A small community with 400 subscribers is to be serviced with a community dial office switch. Assume that the average subscriber originates 0.1 erlangs of traffic. Also assume that 20% of the originations are local (intracommunity) calls and that 80% are transit calls to the serving central office. How many erlangs of traffic are offered to the community dial office to central office trunk group? How many trunks are needed for 0.5% blocking of the transit traffic?

9.11 For the community of Problem 9.10 determine the number of concentrator channels required if local calls are not switched locally but are merely concentrated into pair-gain systems and switched at the central office.

9.12 Repeat Problems 9.10 and 9.11 if 80% of the originations are intracommunity calls and 20% are transit calls.

9.13 A group of eight remote farm houses are serviced by four lines. If each of the eight families utilizes their telephones for 10% of

the busy hour, compare the blocking probabilities of the following configurations:
(a)   Four party lines with two stations per line.
(b)   An 8 to 4 concentration system.

9.14   A PBX provides queuing and automatic call back for access to
       outgoing WATS lines. If there are 20 requests per hour for the
       WATS lines, and if the average call is 3 minutes in length, how
       many WATS lines are needed to provide delays of less than
       1 hour for 90% of the requests?

9.15   A call processor has 50% of its time available for servicing re-
       quests. If each request requires 50 ms of processing time, what
       arrival rate can be supported if only 1% of the service requests
       are delayed by more than 1 second? Assume that processor time
       is sliced into 500 ms time slots. (That is, 500 ms are allocated to
       call processing and then 500 ms to overhead functions etc.)

9.16   A group of 100 sources offers messages with exponentially dis-
       tributed lengths to a 1200 bps line. The average message length
       is 200 bits, including overhead, and each source generates one
       message every 20 seconds. Access to the line is controlled by
       message-switching concentration with an infinite queue. Deter-
       mine the following:
       (a)   The probability of entering the queue.
       (b)   The average queueing delay for all arrivals.
       (c)   The probability of being in the queue for more than 1 sec-
             ond.
       (d)   The utilization of the transmission link.

9.17   An airline company uses an automatic call distributor to service
       reservations and ticket purchases. Assume that the processing
       time of each inquiry is randomly distributed with a 40 second
       average. Also assume that if customers are put on hold for more
       than 2 minutes, they hang up and call another airline. If each of
       200 inquiries per hour produces $30 worth of sales, on average,
       what is the optimum number of reservationists? Assume each
       reservationist costs the company $20/hour (including overhead).

9.18   A radio station talk show solicits the listening public for com-
       ments on the ineptness of government (I assume this will be a
       topical subject for the life of this book). Assume that each caller
       talks for a random length of time with an average duration of
       1 minute. (Either the show is unpopular or the public has given
       up on the government.) How many incoming lines must the radio
       station have to keep the idle time below 5%?

APPENDIX A

DERIVATION OF EQUATIONS

A.1 QUANTIZING NOISE POWER: EQUATION 3.2

The probability density function of a noise sample η is assumed to be uniform:

$$p(\eta) = \frac{1}{q} \qquad \frac{-q}{2} \leqslant \eta \leqslant \frac{q}{2}$$

$$= 0 \qquad \text{otherwise}$$

The average or expected value of noise power is determined as

$$\text{quantization noise power} = \int_{-q/2}^{q/2} \left(\frac{1}{q}\right) \eta^2 \, dn$$

$$= \frac{q^2}{12}$$

A.2 NRZ LINE CODE: EQUATION 4.1

$$f(t) = 1 \qquad |t| \leqslant \frac{T}{2}$$

$$= 0 \qquad \text{otherwise}$$

$$F(j\omega) = \int_{-\infty}^{\infty} f(t) \, e^{-j\omega t} \, dt$$

$$= \int_{-T/2}^{T/2} e^{-j\omega t} \, dt$$

$$= \left(\frac{1}{j\omega}\right) (e^{j\omega T/2} - e^{-j\omega T/2})$$

$$= (T) \frac{\sin (\omega T/2)}{(\omega T/2)}$$

467

Note. $F(j\omega)$ is the spectrum of a single pulse. $|F(j\omega)|^2 \, (1/T)$ is the power spectral density of a random pulse train assuming positive and negative pulses are equally likely and occur independently.

A.3 DIGITAL BIPHASE: FIGURE 4.13

$$f(t) = 1 \qquad -\tfrac{1}{2}T \leqslant t < 0$$
$$= -1 \qquad 0 \leqslant t < \tfrac{1}{2}T$$
$$= 0 \qquad \text{otherwise}$$

$$F(j\omega) = \int_{-(1/2)T}^{0} e^{-j\omega t} \, dt - \int_{0}^{(1/2)T} e^{-j\omega t} \, dt$$

$$= \left(\frac{-1}{j\omega}\right)\{1 - e^{j(1/2)\omega T} - e^{-j(1/2)\omega T} + 1\}$$

$$= \left(\frac{j4}{\omega}\right) \sin^2 \left(\frac{\omega T}{4}\right)$$

A4. FRAME ACQUISITION TIME OF SINGLE-BIT FRAME CODE: EQUATION 4-9

Framing is established by successively examining one bit position after another until a sufficiently long framing pattern is detected. In this derivation it is assumed that the framing pattern alternates 1's and 0's. Furthermore, it is assumed that when beginning to test a particular bit position for framing, the value of the first appearance is saved and compared to the second appearance. Thus the minimum time to reject an invalid framing position is one frame time. If we denote by p the probability of a 1 and by $q = 1 - p$ the probability that a 0 is received first, the average number of frames required to receive a mismatch is

$A_0 = (1)$ (probability of mismatch at end of first frame)

$\quad + (2)$ (probability of mismatch at end of second frame)

$\quad + (3)$ (probability of mismatch at end of third frame) . . .

$\quad = (1)q + (2)(1 - q)p + (3)(1 - q)(1 - p)q$

$\quad\quad + (4)(1 - q)^2(1 - p)p + \cdots$

$\quad = (1)q + (2)p^2 + (3)pq^2 + (4)p^3q + (5)p^2q^3 + \cdots$

$\quad = (q + 2p^2 + pq^2)(1 + (2)pq + (3)p^2q^2 + (4)p^3q^3 + \cdots)$

$$= (1 - p + 2p^2 + p(1 - p)^2)(1 + pq + p^2q^2 + p^3q^3 + \cdots)^2$$

$$= \frac{(1 + p^3)}{(1 - pq)^2}$$

$$= \frac{(1 + p)}{(1 - pq)}$$

Similarly, if a 1 is received first, the expected number of frames before receiving a mismatch is

$$A_1 = \frac{(1 + q)}{(1 - pq)}$$

The overall average number of frames required to detect a mismatch is

$$A = qA_0 + pA_1$$

$$= \frac{(1 + 2pq)}{(1 - pq)}$$

If we assume a random starting point in a frame with N bits, the average number of bits that must be tested before the true framing bit is encountered is

$$\text{frame time} = (\tfrac{1}{2}N)(A \cdot N)$$

$$= \tfrac{1}{2}AN^2 \text{ bit times}$$

If 1's and 0's are equally likely ($p = q = \tfrac{1}{2}$), $A = 2$ so: frame time $= N^2$ bit times (Equation 4.9).

A.5 FRAME ACQUISITION TIME OF SINGLE-BIT FRAMING CODE USING A PARALLEL SEARCH: EQUATION 4.10

This framing algorithm assumes that all bit positions in a frame are simultaneously scanned for the framing pattern. If we assume that an alternating-bit frame code is used and that 1's and 0's in the information bits are equally likely, the probability that a particular information bit does not produce a framing violation in n frames is

$$p_n = (\tfrac{1}{2})^n$$

The probability that a framing violation has been received in n or less frames is $1 - p_n$. The probability that all $N - 1$ information bits in a frame produce a framing violation in n or less frames is

$$P_n = [1 - (\tfrac{1}{2})^n]^{N-1} \tag{4.10}$$

A.6 FRAME ACQUISITION TIME OF MULTIBIT FRAME CODE: EQUATION 4.12

N = length of frame including framing code

L = length of framing code

$p = (\frac{1}{2})^L$ = probability of matching frame code with random data

The expected number of frames examined before a particular frame position mismatches the frame code follows:

$$A = (0)(1 - p) + (1)p(1 - p) + (2)p^2(1 - p) + \cdots$$
$$= (1 - p)p(1 + 2p + 3p^2 + 4p^3 + \cdots)$$
$$= (1 - p)p(1 + p + p^2 + p^3 \cdots)^2$$
$$= (1 - p)p\left(\frac{1}{1 - p}\right)^2$$
$$= \frac{p}{1 - p}$$

The average number of bits that pass before the frame position is detected (assuming a random starting point and the test field is moved one bit position when a mismatch occurs) is

$$T = \left(\frac{N}{2}\right)(A)(N)$$
$$= \frac{N^2(1/2)^{L+1}}{1 - (1/2)^L} \tag{4.12}$$

Note. Equation 4.12 with $L = 1$ is not identical to Equation 4.9 because Equation 4.12 assumes a fixed frame code while Equation 4.9 assumes an alternating code.

A.7 PATHFINDING TIME: EQUATION 5.11

Assume that all paths through a switch are independently busy with probability p. Let the probability that a path is not busy be denoted by $q = 1 - p$. The probability p_i that exactly i paths are tested before an idle one is found is the probability that the first $i - 1$ are busy and the ith is not:

$$p_i = p^{(i-1)}q$$

The expected number of paths tested before an idle path is found is

$$N_p = (1)q + (2)pq + (3)p^2q + \cdots (k)p^{k-1}q + (k)p^k$$

where the last term represents the expectation that all possible paths k are unavailable. A closed form for A is determined as

$$N_p = (1 - p)\, 1 + 2p + 3p^2 + \cdots kp^{k-1} + kp^k$$

$$= 1 + p + p^2 + p^3 + \cdots p^{k-1}$$

$$= \frac{1}{1-p} - p^k \left(\frac{1}{1-p} \right) \qquad\qquad (5.11)$$

$$= \frac{1 - p^k}{1 - p}$$

ENCODING/DECODING ALGORITHMS FOR SEGMENTED PCM

B.1 EIGHT BIT μ255 CODE

The encoded representations of μ255 PCM code words use a sign-magnitude format wherein one bit identifies the sample polarity and the remaining bits specify the magnitude of the sample. The seven magnitude bits are conveniently partitioned into a 3 bit segment identifier S and a 4 bit quantizing step identifier Q. Thus the basic structure of an 8 bit μ255 PCM code word is shown in Figure B.1.

In the following descriptions of encoding and decoding algorithms, it is assumed, for convenience in using integer representations, that analog input signals are scaled to a maximum amplitude of 8159. Furthermore, all amplitudes and segment identifiers are assumed to be encoded using conventional binary representations. The actual encoders used in T1 transmission systems, however, complement the code words to increase the density of 1's in a transmitted bit stream.

B.1.1 Algorithm 1: Direct Encoding (Table B.1)

$$\text{Polarity bit } P = 0 \quad \text{for positive sample values}$$
$$= 1 \quad \text{for negative sample values}$$

Given a sample value with a magnitude x, the first step in the magnitude encoding process is to determine the segment identifier S. The

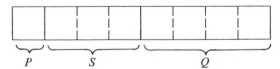

Figure B.1 Eight bit μ225 PCM code format.

major segments are identified by the segment endpoints: 31, 95, 223, 479, 991, 2015, 4063, and 8159. Thus S can be determined as the smallest endpoint that is greater than the sample value x. S is equal to the smallest a such that

$$x < 64 \cdot 2^a - 33 \qquad a = 0, 1, \ldots, 7$$

After the major segment containing the sample value has been determined, the particular quantization interval within the major segment must be identified. As a first step, a residue r is determined as the difference between the input amplitude and the lower endpoint of the segment:

$$r = x \qquad\qquad\qquad S = 0$$
$$r = x - (32 \cdot 2^S - 33) \qquad S = 1, 2, \ldots, 7$$

The value of Q can now be determined as the quantization interval containing the residue r. Q is equal to the smallest b such that

$$r < 2b + 1 \qquad\qquad\qquad S = 0$$
$$r < (2^{S+1})(b + 1) \qquad\qquad S = 1, 2, \ldots, 7$$

where $b = 0, 1, \ldots, 15$. Notice that this process identifies quantization intervals in segment $S = 0$ as having upper endpoints at $1, 3, 5, \ldots, 31$ while the other segments have quantization endpoints that are multiples of 4, 8, 16, 32, 64, 128, 256 for $S = 1, 2, 3, 4, 5, 6, 7$, respectively.

After S and Q have been determined, they are binary encoded into 3 and 4 bits respectively. The concatenation of S and Q produces a 7 bit word that can be conveniently represented as an integer between 0 and 127. In essence, this integer identifies one of the 128 quantization intervals of a compressed signal amplitude.

The decoding process involves assigning the designated polarity to an analog output sample at the midpoint of the nth quantization interval $n = 0, 1, \ldots, 127$. Using the values of S and Q directly, we can determine a discrete output sample value y_n as

$$y_n = (2Q + 33)(2^S) - 33$$

where n is the integer obtained by concatenating the binary representations of S and Q.

Example

An input sample of +242 produces the following code word:

$$0, 3, 1 = \boxed{0}\,\boxed{011}\,\boxed{0001}$$

The decoder output becomes:

$$y_{49} = (2 \cdot 1 + 33)(2^3) - 33$$

$$= 247$$

which is the midpoint of the forty-ninth quantization interval from 239 to 255.

B.1.2 Algorithm 2: Linear Code Conversion

The fundamental reason for using a μ-law companding characteristic with $\mu255$ is the ease with which the segmented approximation can be

TABLE B.1 PIECEWISE LINEAR APPROXIMATION TO $\mu255$ COMPANDING[a]

	Segment Code S								Quantization Code Q	
	000	001	010	011	100	101	110	111		
Quantization Endpoints	0	31	95	223	479	991	2015	4063	0000	0
	1	35	103	239	511	1055	2143	4319	0001	1
	3	39	111	255	543	1119	2271	4575	0010	2
	5	43	119	271	575	1183	2399	4831	0011	3
	7	47	127	287	607	1247	2527	5087	0100	4
	9	51	135	303	639	1311	2655	5343	0101	5
	11	55	143	319	671	1375	2783	5599	0110	6
	13	59	151	335	703	1439	2911	5855	0111	7
	15	63	159	351	735	1503	3039	6111	1000	8
	17	67	167	367	767	1567	3167	6367	1001	9
	19	71	175	383	799	1631	3295	6623	1010	10
	21	75	183	399	831	1695	3423	6879	1011	11
	23	79	191	415	863	1759	3551	7135	1100	12
	25	83	199	431	895	1823	3679	7391	1101	13
	27	87	207	447	927	1887	3807	7647	1110	14
	29	91	215	463	959	1951	3935	7903	1111	15
	31	95	223	479	991	2015	4063	8159		

[a](1) Sample values are referenced to a full-scale value of 8159. (2) Negative samples are encoded in sign-magnitude format with a polarity bit of 1. (3) In actual transmission the codes are inverted to increase the density of 1's when low signal amplitudes are encoded. (4) Analog output samples are decoded as the center of the encoded quantization interval. (5) Quantization error is the difference between the reconstructed output value and the original input sample value.

digitally converted to and from a uniform code. This section describes the basic algorithms that implement the conversions. The first algorithm provides the means of implementing a μ255 PCM encoder using a 13 bit uniform encoder followed by digital logic to provide the compression function. The second algorithm indicates how to implement the decoder function of first expanding a compressed code into a 13 bit linear code to be used in generating the output samples.

Just as in algorithm 1, the polarity bit P is determined as

$$P = 0 \qquad \text{for positive sample values}$$

$$P = 1 \qquad \text{for negative sample values}$$

The simplicity of converting from a linear code to a compressed code is most evident if the linear code is biased by adding the value 33 to the magnitude of all samples. Notice that this bias shifts the encoding range from (0 to 8159) to (33 to 8192). The addition process can be performed directly on the analog samples before encoding or with digital logic after encoding. In either case, the general form of all biased linear code patterns and the corresponding compressed codes are shown in the following table.

μ255 Encoding Table

Biased Linear Input Code	Compressed Code
0 0 0 0 0 0 0 1 w x y z a	0 0 0 w x y z
0 0 0 0 0 0 1 w x y z a b	0 0 1 w x y z
0 0 0 0 0 1 w x y z a b c	0 1 0 w x y z
0 0 0 0 1 w x y z a b c d	0 1 1 w x y z
0 0 0 1 w x y z a b c d e	1 0 0 w x y z
0 0 1 w x y z a b c d e f	1 0 1 w x y z
0 1 w x y z a b c d e f g	1 1 0 w x y z
1 w x y z a b c d e f g h	1 1 1 w x y z

From the foregoing table it can be seen that all biased linear codes have a leading 1 that indicates the value of the segment number S. Specifically, the value of S is equal to 7 minus the number of leading 0's before the 1. The value of Q is directly available as the four bits (w, x, y, z) immediately following the leading 1. All trailing bits (a through h) are merely ignored when generating a compressed code.

In reverse fashion the following table indicates how to generate a biased linear code from a compressed code. An unbiased output can be obtained by subtracting 33 from the biased code.

μ255 Decoding Table

Compressed Code	Biased Linear Output Code
0 0 0 w x y z	0 0 0 0 0 0 0 1 w x y z 1
0 0 1 w x y z	0 0 0 0 0 0 1 w x y z 1 0
0 1 0 w x y z	0 0 0 0 0 1 w x y z 1 0 0
0 1 1 w x y z	0 0 0 0 1 w x y z 1 0 0 0
1 0 0 w x y z	0 0 0 1 w x y z 1 0 0 0 0
1 0 1 w x y z	0 0 1 w x y z 1 0 0 0 0 0
1 1 0 w x y z	0 1 w x y z 1 0 0 0 0 0 0
1 1 1 w x y z	1 w x y z 1 0 0 0 0 0 0 0

Both of these tables indicate that 13 bits of the linear code are used to represent the magnitude of the signal. In Chapter 3 it is mentioned that a μ255 PCM coder has an amplitude range equivalent to 12 bits. The discrepancy occurs because the first quantization interval has length one while all others in the first segment are of length two. Thus the extra bit is needed only to specify the first quantization interval. Notice further that the least significant bit in the tables carries no information but is included only to facilitate the integer relationships. In particular, the least significant bit in the encoding table is completely ignored when determining a compressed code (assuming that the bias is added to the analog sample). Furthermore, the least significant bit in the output codes is completely specified by the segment number S. It is a 0 for segment zero and a 1 for all other segments.

Example

An input code word of +242 is biased to produce a value of 275. The binary representation of 275 is

$$0\ 0\ 0\ 0\ 1\ 0\ 0\ 0\ 1\ 0\ 0\ 1\ 1 \qquad \text{(biased linear code)}$$

Hence from the encoding table $S = 011$ and wxyz = 0001, and the compressed code is

$$\boxed{0\ |\ 011\ |\ 0001} \qquad \text{(compressed code)}$$

Using the decoding table, this compressed code produces the following biased linear output code:

$$0\ 0\ 0\ 0\ 1\ 0\ 0\ 0\ 1\ 1\ 0\ 0\ 0 \qquad \text{(biased linear output)}$$

The decimal representation of the foregoing code is 280, which corresponds to an unbaised output equal to +247.

B.2 EIGHT BIT A-LAW CODE

The following companding algorithms for segmented A-law codes use the same basic procedures as those presented for the $\mu255$ codes. One difference that does occur, however, involves the elimination of a bias in the linear code for conversion to and from a compressed code. Another difference occurs in the use of the integer 4096 as the maximum amplitude of a sample in an A-law representation. If desired, the scale factors for the two systems can be brought into close agreement by doubling the A-law scale to 8192.

As mentioned in Chapter 3, the segmented A-law code is usually referred to as a 13 segment code owing to the existence of seven positive and seven negative segments with the two segments near the origin being colinear. In the following descriptions, however, the first segment for each polarity is divided into two parts to produce eight positive and eight negative segments. This point of view permits a code format that is identical to the $\mu255$ code format. Thus, a compressed code word consists of a sign bit P followed by 3 bits of a segment identifier S and 4 bits of quantizer level Q.

TABLE B.2 SEGMENTED A-LAW ENCODING TABLE

	Segment Code								Quantization Code	
	000	001	010	011	100	101	110	111		
	0	32	64	128	256	512	1024	2048	0000	0
	2	34	68	136	272	544	1088	2176	0001	1
	4	36	72	144	288	576	1152	2304	0010	2
	6	38	76	152	304	608	1216	2432	0011	3
	8	40	80	160	320	640	1280	2560	0100	4
	10	42	84	168	336	672	1344	2688	0101	5
	12	44	88	176	352	704	1408	2816	0110	6
	14	46	92	184	368	736	1472	2944	0111	7
Quantization Endpoints	16	48	96	192	384	768	1536	3072	1000	8
	18	50	100	200	400	800	1600	3200	1001	9
	20	52	104	208	416	832	1664	3328	1010	10
	22	54	108	216	432	864	1728	3456	1011	11
	24	56	112	224	448	896	1792	3584	1100	12
	26	58	116	232	464	928	1856	3712	1101	13
	28	60	120	240	480	960	1920	3840	1110	14
	30	62	124	248	496	992	1984	3968	1111	15
	32	64	128	256	512	1024	2048	4096		

B.2.1   Algorithm 1: Direct Encoding

The segment endpoints of an A-law code are: 32, 64, 128, 256, 512, 1024, 2048, and 4096. Thus for a sample with magnitude x the major segment identifier S can be determined as the smallest a such that

$$x < 32 \cdot 2^a \qquad a = 0, 1, \ldots, 7$$

After S has been determined, the residue r can be obtained as

$$r = x \qquad\qquad\qquad S = 0$$
$$= x - 16 \cdot 2^S \qquad S = 1, 2, \ldots, 7$$

The value of Q can then be determined as the smallest b such that

$$r < 2(b + 1) \qquad S = 0$$
$$r < 2^S (b + 1) \qquad S = 1, 2, \ldots, 7$$

Just as in the case for $\mu 255$ coding, an A-law magnitude can be represented as an integer $n = 0, 1, \ldots, 127$ derived from the concatenation of 3 S-bits and 4 Q-bits. An output magnitude y_n can then be expressed as:

$$y_n = 2Q + 1 \qquad\qquad S = 0$$
$$= 2^S (Q + 16\tfrac{1}{2}) \qquad S = 1, 2, \ldots, 7$$

Example

An input sample of +121 produces the following code word:

$$\boxed{0\,|\,010\,|\,1110} = +46 \text{ in decimal}$$

The decoder output becomes:

$$y_{46} = 2^S (14 + 16\tfrac{1}{2})$$
$$= 122$$

which is the midpoint of the quantization interval from 120 to 124.

B.2.2   Algorithm 2: Linear Code Conversion

The following tables indicate how to convert a 12 bit linear code directly into a compressed A-law code. The algorithm is basically the same as for the $\mu 255$ conversion except that biasing the linear code is unnecessary and a first segment code does not have a leading 1. Thus the segment number S can be determined as 7 minus the number of leading zeros as before. The Q field data is obtained as the four bits (wxyz) immediately following the leading 1, except when $S = 0$, in which case the Q field is contained in the 4 bits following the 7 leading 0's.

A-Law Encoding Table

Linear Code	Compressed Code
0 0 0 0 0 0 0 w x y z a	0 0 0 w x y z
0 0 0 0 0 0 1 w x y z a	0 0 1 w x y z
0 0 0 0 0 1 w x y z a b	0 1 0 w x y z
0 0 0 0 1 w x y z a b c	0 1 1 w x y z
0 0 0 1 w x y z a b c d	1 0 0 w x y z
0 0 1 w x y z a b c d e	1 0 1 w x y z
0 1 w x y z a b c d e f	1 1 0 w x y z
1 w x y z a b c d e f g	1 1 1 w x y z

The following table provides the means of generating a linear code word directly from a compressed code word. The output value corresponds to the middle of the quantization interval designated by S and Q.

A-Law Decoding Table

Compressed Code	Linear Output Code
0 0 0 w x y z	0 0 0 0 0 0 0 w x y z 1
0 0 1 w x y z	0 0 0 0 0 0 1 w x y z 1
0 1 0 w x y z	0 0 0 0 0 1 w x y z 1 0
0 1 1 w x y z	0 0 0 0 1 w x y z 1 0 0
1 0 0 w x y z	0 0 0 1 w x y z 1 0 0 0
1 0 1 w x y z	0 0 1 w x y z 1 0 0 0 0
1 1 0 w x y z	0 1 w x y z 1 0 0 0 0 0
1 1 1 w x y z	1 w x y z 1 0 0 0 0 0 0

Each of these tables relates 12 bits of magnitude in a linear code to a compressed code with 7 bits of magnitude. Notice, however, that the least significant bit of the encoder is always ignored. Thus the encoder needs only 11 bits of resolution if all of its outputs are immediately compressed. If any signal processing (such as adding two signals together) is to take place before compression, however, the extra bit is useful in reducing the composite quantization error.

Example

The previously used sample value 121 is represented in binary form as: 000001111001. From the encoding table $S = 010$ and $Q = 1110$. Thus the compressed code word is

0	010	1110

Using the decoding table, the linear output can be determined as 000001111010, which is 122 decimal.

APPENDIX C

ANALYTIC
FUNDAMENTALS OF
DIGITAL TRANSMISSION

C.1 PULSE SPECTRA

This section presents the frequency spectra of common pulse wave-forms used for digital transmission. These are square pulses as generated at a source. Since the spectra of square pulses have infinite frequency content, the spectra presented here do not correspond to pulse responses at the output of a channel where the pulse shapes have been altered by band limiting filters. In the next section channel output pulse responses are described. Then the necessary combinations of input pulse shapes and filter designs to produce particular output pulses are considered.

The various pulse shapes and corresponding frequency spectra are presented in Figure C.1. In deriving the spectra, the following conditions and assumptions were made:

1 All pulses have equal energy.
2 All systems signal at rate $1/T$.
3 The waveforms shown are used to encode a 1.
4 Opposite polarities are used for a 0.
5 It is equally likely for 1's and 0's to occur, and occur at random (complete independence).

C.2 CHANNEL OUTPUT PULSE RESPONSES

Although a digital transmission system can be designed to produce a variety of output pulse responses, the most common is defined as

$$y_{rc}(t) = \frac{\sin (\pi t/T)}{(\pi t/T)} \frac{\cos (\alpha \pi t/T)}{1 - (2\alpha t/T)^2} \tag{C.1}$$

480

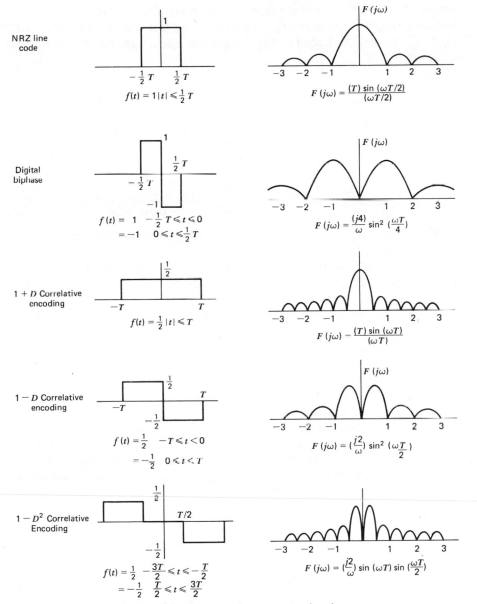

Figure C.1 Spectra of common pulse shapes.

NRZ line code

$$f(t) = 1 \, |t| \leqslant \tfrac{1}{2}T$$

$$F(j\omega) = \frac{(T)\sin(\omega T/2)}{(\omega T/2)}$$

Digital biphase

$$f(t) = 1 \quad -\tfrac{1}{2}T \leqslant t \leqslant 0$$
$$= -1 \quad 0 \leqslant t \leqslant \tfrac{1}{2}T$$

$$F(j\omega) = \frac{(j4)}{\omega}\sin^2\left(\frac{\omega T}{4}\right)$$

1 + D Correlative encoding

$$f(t) = \tfrac{1}{2} \, |t| \leqslant T$$

$$F(j\omega) = \frac{(T)\sin(\omega T)}{(\omega T)}$$

1 − D Correlative encoding

$$f(t) = \tfrac{1}{2} \quad -T \leqslant t < 0$$
$$= -\tfrac{1}{2} \quad 0 \leqslant t < T$$

$$F(j\omega) = \left(\frac{j2}{\omega}\right)\sin^2\left(\omega\frac{T}{2}\right)$$

1 − D² Correlative Encoding

$$f(t) = \tfrac{1}{2} \quad -\tfrac{3T}{2} \leqslant t \leqslant -\tfrac{T}{2}$$
$$= -\tfrac{1}{2} \quad \tfrac{T}{2} \leqslant t \leqslant \tfrac{3T}{2}$$

$$F(j\omega) = \left(\frac{j2}{\omega}\right)\sin(\omega T)\sin\left(\frac{\omega T}{2}\right)$$

481

where $1/T$ is the signaling rate and α is an excess bandwidth factor between 0 and 1. Equation C.1 represents the response of a raised cosine channel—so called because the frequency spectrum corresponding to $y_{rc}(t)$ in C.1 is

$$Y_{rc}(f) = 1 \qquad\qquad\qquad |f| \leqslant \frac{(1-\alpha)}{2T}$$

$$= \frac{1}{2}\left\{1 + \cos\left[\frac{\pi|f|T}{\alpha} - \frac{\pi(1-\alpha)}{2\alpha}\right]\right\} \quad \frac{(1-\alpha)}{2T} \leqslant |f| \leqslant \frac{(1+\alpha)}{2T}$$

$$= \cos^2\left[\frac{\pi|f|T}{2\alpha} - \frac{\pi(1-\alpha)}{4\alpha}\right]$$

$$= 0 \qquad\qquad\qquad\qquad \text{otherwise} \qquad\qquad \text{(C.2)}$$

The origin of the apellation "raised cosine" is apparent in the third line of Equation C.2.

The parameter α in Equations C.1 and C.2 is referred to as an excess bandwidth parameter. If $\alpha = 0$, the spectrum defined in Equation C.2 is exactly equal to the theoretical minimum bandwidth $1/2T$ for signaling rate $1/T$. As α increases from 0 to 1, the excess spectrum width increases to 100%. Raised cosine channel spectra are illustrated in Figure C.2 for several values of α.

Practical systems are typically designed for excess bandwidths of 30% or more for several reasons. First, "brick wall" filters needed to produce the infinite attenuation slopes implied by $\alpha = 0$ are physically unrealizable. Second, as shown in Figure C.3, the time domain pulse response for small values of α exhibits large amounts of ringing. Slight errors in the sample times cause significant degradations in performance due to intersymbol interference. Third, a slight deviation in the signaling rate from the design rate also produces significant intersymbol interference.

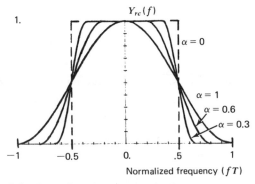

Figure C.2   Raised-cosine spectrums for various values of α.

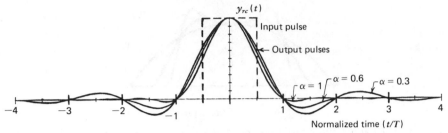

Figure C.3   Raised-cosine pulse responses for various values of α.

It must be emphasized that Equation C.2 defines the desired spectrum at the output of the channel (the input to the decision circuit). Thus the desired response results from a combination of the input pulse spectrum and the channel filter responses. Often, the input pulse spectrum arises from a square pulse of duration T:

$$x_s(t) = 1 \qquad |t| \leqslant \frac{T}{2}$$

$$= 0 \qquad \text{otherwise}$$

(C.3)

The frequency spectrum corresponding to $x_s(t)$ in Equation C.3 is the "sin $(x)/x$" spectrum also referred to as a "sinc" function:

$$X_s(f) = \frac{(T)\sin(\pi f T)}{(\pi f T)}$$

$$= T \operatorname{sinc}(\pi f T)$$

(C.4)

When the channel input spectrum is as defined in Equation C.4, the filter function of a channel to produce a raised cosine output is determined as

$$H(f) = \frac{Y_{rc}(f)}{X_s(f)}$$

(C.5)

The channel filter functions implied in Equation C.5 are shown in Figure C.4 for the same values of α shown in Figure C.2. Figure C.2 displays channel output spectra while Figure C.4 displays corresponding frequency domain transfer functions for the channel.

As mentioned in Chapter 3, the design of a smoothing filter in a digital voice decoder sometimes requires a modification like that defined in Equation C.5. In the case mentioned, the ideal "flat" filter for reconstructing speech from narrow impulselike samples should be modified by dividing the flat response by the spectrum of the finite width samples (Equation C.4). When the samples are narrower than the signaling interval, the sin $(x)/x$ response is essentially flat across the bandwidth of the filter. However, when the samples are made to last for the entire

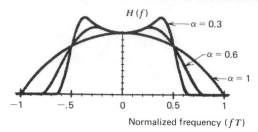

Figure C.4   Channel filter responses needed to produce raised-cosine outputs when excited by sin $(\omega T/2)/(\omega T/2)$ pulses.

duration of the sample interval, the filter response should be "peaked" to compensate for sin $(x)/x$ roll off.

As discussed in Chapter 6, a channel filter function is usually partitioned between the transmitter and the receiver. The so-called "optimum" filter partitioning occurs when the receive filter response is the square root of the desired output response $Y_{rc}(f)$, and the transmit filter response is whatever is necessary to transform the channel input spectrum into the complex conjugate of the receive filter response. (The output spectrum of the transmitter is also the square root of the desired channel response.) When the channel input is a pulse of duration T, the "optimum" filter functions are defined as

$$H_{RX}(f) = \{Y_{rc}(f)\}^{1/2} \tag{C.6}$$

$$H_{TX}(f) = \frac{\{Y_{rc}(f)\}^{1/2}}{X_s(f)} \tag{C.7}$$

Notice that the transmit spectrum resulting from $H_{TX}(f)$ in Equation C.7 is equal to the square root of the desired output response—no matter what the input pulse shape is. Hence, when optimum partitioning is used, the transmit spectrum and the receiver design are independent of the channel excitation.

The transmit and receive filter functions for square wave excitations (Equations C.6 and C.7) are shown in Figure C.5.* Notice that the transmit filter function has peaking at frequencies other than direct current. The implication for midband attenuation is one of the reasons why optimum partitioning may not be optimum in a system with source power limitations.

C.2.1   Optimum Filtering for Minimum Shift Keying

As discussed in Chapter 6, minimum shift keyed (MSK) modulation can be represented as quadrature channel modulation with baseband excita-

---

*This discussion assumes baseband transmission. The principles are easily extended to carrier based systems by translating the filter functions to the carrier frequency.

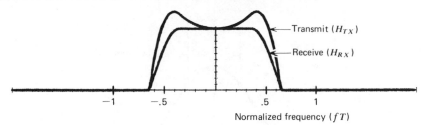

Figure C.5 "Optimum" transmit and receive filter functions for raised cosine response with $\alpha = 0.3$ and $\sin(\omega T/2)/(\omega T/2)$ excitation.

tion defined as cosine pulse shapes:

$$x_c(t) = \cos\left(\frac{\pi t}{T}\right) \quad |t| \leqslant \frac{T}{2} \tag{C.8}$$

$$= 0 \qquad\qquad \text{otherwise}$$

The transform of $x_c(t)$ is

$$X_c(f) = \left(\frac{\pi}{2T}\right)\frac{\cos \pi f T}{1 - (2fT)^2} \tag{C.9}$$

When a raised cosine output response and optimum filter partitioning is desired, the MSK filter functions are

$$H_{RX}(f) = \{Y_{rc}(f)\}^{1/2} \tag{C.10}$$

$$H'_{TX}(f) = \frac{\{Y_{rc}(f)\}^{1/2}}{X_c(f)} \tag{C.11}$$

The transmit spectrum and the receiver of an optimally partitioned MSK system is identical to counterparts in an optimally partitioned, offset-keyed 4-PSK system!

C.2.2 Partial Response Systems

As another digital transmission system design example, consider a $1 + D$ partial response system. The desired time response of the channel is defined in Equation C.12 and the associated frequency spectrum in Equation C.13.

$$y_c(t) = \frac{4 \cos(\pi t/T)}{\pi(1 - (2t/T)^2)} \tag{C.12}$$

$$Y_c(f) = \cos \pi f T \quad |f| \leqslant \frac{1}{2T} \tag{C.13}$$

$$= 0 \qquad\qquad \text{otherwise}$$

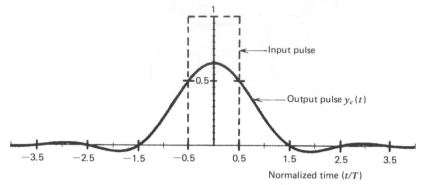

Figure C.6 Output pulse of $1 + D$ partial response channel.

The pulse response of a $1 + D$ partial response channel is shown in Figure C.6. Notice that a single pulse contributes equally to the response at two successive sample times but crosses zero at all other sample times.

Optimum filter partitioning is again achieved with a receive filter having an amplitude response equal to the square root of the desired output response $Y_c(f)$. If the channel is excited by square pulses of duration T, as defined in Equation C.3, the optimum filter functions are

$$H_{RX}(f) = \{Y_c(f)\}^{1/2} \qquad |f| \leqslant \frac{1}{2T} \tag{C.14}$$

$$H_{TX}(f) = \frac{\{Y_c(f)\}^{1/2}}{X_s(f)} \qquad |f| \leqslant \frac{1}{2T} \tag{C.15}$$

where $X_s(f)$ is defined in Equation C.4.

The optimum filter functions defined in Equations C.14 and C.15 are shown in Figure C.7 along with the desired output response of the channel. Notice that, unlike the full response (raised cosine) systems, optimum partitioning of a partial response system does not require peaking of the transmit filter.

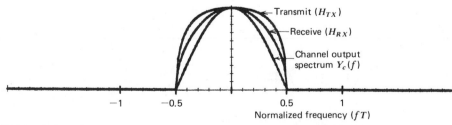

Figure C.7 Spectra of $1 + D$ PRS channel and "optimum" filter responses for $\sin(\omega T/2)/(\omega T/2)$ excitation.

C.3 ERROR RATE ANALYSES–BASEBAND SYSTEMS

C.3.1 Binary Transmission

Consider the receiver model of a digital transmission system shown in Figure C.8. The receiver consists of two parts: signal processing circuitry and a data detector (decision circuit). For the time being, assume that the output of the signal processing circuit produces a pulse of amplitude $+V$ when a 1 is transmitted and a pulse of amplitude $-V$ when a 0 is transmitted. Obviously, the detector merely examines the polarity of its input at the sample times defined by the sample clock. A decision error occurs if noise at the sample times has an amplitude greater than V and a polarity opposite to the transmitted pulse.

The most commonly analyzed type of noise is assumed to have a Gaussian or normal probability distribution. Thus the probability of error P_E can be determined as

$$P_E = \frac{1}{\sqrt{2\pi}\sigma} \int_V^\infty e^{-t^2/2\sigma^2}\, dt \tag{C.16}$$

where σ^2 is the rms noise power at the detector. Using the error function:

$$\text{erf } z = \frac{2}{\sqrt{\pi}} \int_0^z e^{-t^2}\, dt \tag{C.17}$$

Equation C.16 is sometimes rewritten as

$$P_E = \tfrac{1}{2}[1 - \text{erf } z] \tag{C.18}$$

where $z = V/\sqrt{2}\sigma$. The error probability P_E can also be expressed in terms of the complimentary error function:

$$P_E = \tfrac{1}{2}\,\text{erfc } z \tag{C.19}$$

where erfc $z = 1 - \text{erf } z$
$z = V/\sqrt{2}\sigma$

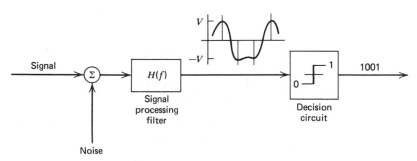

Figure C.8 Digital receiver model.

In lieu of evaluating the integral in Equation C.17 or C.19 (which has no closed form solution), the error functions can be approximated as

$$\text{erfc } z \doteq \frac{e^{-z^2}}{(z\sqrt{\pi})} \qquad (z \gg 1) \tag{C.20}$$

C.3.2 Multilevel Transmission

The error rate of a multilevel baseband system is easily determined by an appropriate reduction in the error distance. If the maximum amplitude is V, the error distance d between equally spaced levels at the detector is

$$d = \frac{V}{L - 1} \tag{C.21}$$

where L is the number of levels. Adjusting the error distance V of a binary system to that defined in C.21 provides the error rate of a multilevel system as

$$P_E = \left(\frac{1}{\log_2 L}\right)\left(\frac{L - 1}{L}\right) \text{erfc}\left\{\frac{V}{(L - 1)\sqrt{2}\sigma}\right\} \tag{C.22}$$

where the factor $(L - 1)/L$ reflects the fact that interior signal levels are vulnerable to both positive and negative noise, and the factor $1/\log_2 L$ arises because the multilevel system is assumed to be coded so symbol errors produce single bit errors ($\log_2 L$ is the number of bits per symbol).

Equation C.22 reveals that, with respect to peak signal-to-noise ratios at the detector, L-level transmission incurs a penalty of 20 $\log_{10}(L - 1)$ dB. If V is increased by a factor $(L - 1)$, the error rate of the L-level system is identical to the error rate of the binary system. (Except for the factors $1/\log_2 L$ and $(L - 1)/L$, which typically represent only a few tenths of a decibel.)

Equation C.22 relates error rate to the peak signal power V^2. To determine the error rate with respect to average power, the average power of an L-level system is determined by averaging the power associated with the various pulse amplitude levels:

$$(V^2)_{\text{avg}} = \frac{2}{L}\left\{\left(\frac{V}{L - 1}\right)^2 + \left(\frac{3V}{L - 1}\right)^2 + \cdots V^2\right\}$$

$$= \frac{2V^2}{L(L - 1)^2} \sum_{i=1}^{L/2} (2i - 1)^2 \tag{C.23}$$

where the levels

$$\frac{V}{L-1} \{\pm 1, \pm 3, \cdots \pm(L-1)\}$$

are assumed to be equally likely.

C.3.3   Energy-per-Bit to Noise Density Ratios

The foregoing error rate equations relate P_E to the signal energy at the sample times and the rms noise power at the detector. When comparing various digital modulation formats, multilevel systems in particular, it is more relevant to relate error performance to signal power and noise power at the input to the receiver (in front of the signal processing circuitry). As a first step in developing an error rate equation based on signal-to-noise ratios at the receiver input, the noise power at the detector is determined.

The variance of the noise σ^2 used in the previous equations is exactly equal to the rms power that would be measured at the detector in the absence of a signal. This noise power can be determined analytically as

$$\sigma^2 = \int_{-\infty}^{\infty} |H(f)(\tfrac{1}{2}N_0)|^2 \, df \qquad \text{(C.24)}$$

$$= N_0 \int_{0}^{\infty} |H(f)|^2 \, df$$

$$= N_0 \cdot B_N \qquad \text{(C.25)}$$

where N_0 is the one-sided noise power spectral density in watts per hertz and $B_N = \int_0^\infty |H(f)|^2 \, df$ is the noise-equivalent bandwidth or simply the noise bandwidth of the receiver filter function $H(f)$.

In Equation C.25, the noise source is assumed to be white. That is, a uniform spectral density exists across the entire band of interest. This noise may exist in the transmission medium itself, or it may occur in the "front end" amplifier of the receiver. If the rms noise power passing through a square (ideal) filter of bandwidth B is measured, a reading of $(N_0)(B)$ W would be obtained. Thus B_N represents the bandwidth of a perfectly square filter that passes the same amount of noise as the receiver filter $H(f)$. [$H(f)$ may be decidedly nonsquare in its amplitude response.]

The receiver function $H(f)$ necessarily provides a compromise between two conflicting objectives. First, it must minimize the amount of noise passed to the detector (i.e. minimize B_N). Second, the difference between sample values ($+V$ and $-V$) must be maximized. Obviously, the signal-to-noise ratio at the detector (V^2/σ^2) should be maximized to minimize the error probability. A classical result of digital communication theory states that V/σ is maximized when $H(f)$ is "matched" to the received signal.

When viewed in the time domain, a "matched filter" is implemented by correlating (multiplying) the received signal with each of the receivable (noise-free) pulse shapes. The outputs of the correlators are integrated across a signal interval to determine the overall average correlation during the interval. The integrator with the maximum output indicates the most likely symbol to have been transmitted.

In most systems all signals or pulses have the same shape but differ only in amplitude and polarity. Thus a single matched filter can be used. Detection merely involves comparing the matched filter output to appropriate decision levels. The output of a single correlator $h(t)$ and its integrator is expressed as

$$V = \int_0^T s(t)\,h(t)\,dt$$

$$= \int_0^T |s(t)|^2\,dt \qquad\qquad (C.26)$$

where $s(t)$ is the signal or pulse shape being measured. Notice that V is, in essence, a measure of the energy in the signal over the signaling interval T.

When analyzed in the frequency domain, a matched filter response $H(f)$ is the complex conjugate of the channel pulse spectrum $S(f)$. Thus the matched filter output $Y(f)$ can be expressed in the frequency domain as

$$Y(f) = H(f) \cdot S(f)$$
$$= S^*(f) \cdot S(f) \qquad\qquad (C.27)$$

Frequency domain representations are most convenient when the transmit signal is strictly bandlimited—implying that the duration of the pulse response is theoretically unlimited. In this case the energy in a pulse (e.g. raised cosine) is directly proportional to the detector voltage at the optimum sample time. Hence optimum detection is achieved by merely sampling the output of the receive filter at the proper time.

Using the relationship of Equation C.25 and the parameter E_s to represent symbol energy, we express the binary error rate Equation C.19 as

$$P_E = \tfrac{1}{2}\,\text{erfc}\,(z) \qquad\qquad (C.28)$$

where $z^2 = E_s/(N_0 B_N)$.

Notice that for a given system (fixed B_N), the error rate is dependent only on the ratio of the symbol energy E_s and the noise density N_0. This ratio is commonly referred to as a signal-to-noise ratio although it is not a signal-power to noise-power measurement. Equation C.28 is the preferred form of the error rate equation for comparing different modulation schemes. In a binary scheme, the symbol energy E_s is equal to the energy per bit E_b.

As an example of a specific system, consider a baseband raised cosine channel with optimum partitioning. The output spectrum of the transmitter is the square root of the raised cosine spectrum $Y_{rc}(f)$ defined in Equation C.2. The matched receiver filter also has a square root of a raised cosine response (Equation C.6). Hence the noise bandwidth of the receiver can be determined as

$$B_N = \int_0^\infty |H(f)|^2 \, df$$

$$= \int_0^{(1+\alpha)/2T} |Y_{rc}(f)| \, df \qquad\qquad \text{(C.29)}$$

$$= \frac{1}{2T} \qquad\qquad \text{(independent of } \alpha)$$

As defined in Equation C.1 and shown in Figure C.3, the (normalized) peak sample value at the detector of a raised cosine pulse is 1. Using unnormalized pulses of amplitude E_s, we can determine the error rate of a binary $(+E_s, -E_s)$ raised cosine channel as

$$P_E = \tfrac{1}{2} \, \text{erfc} \, (z) \qquad\qquad \text{(C.30)}$$

where $z^2 = (E_s/N_0)T$ and T is the duration of a signal interval.

Although Equation C.30 was derived for a raised cosine channel, it is more general in that it is applicable to any binary system using antipodal signaling. Thus Equation C.30 is plotted in Figure 4.23 as the best performance achievable by any digital transmission system detecting one pulse at a time. (Lower error rates are possible when redundant signals or error correcting codes are used.)

The error rate performance presented in Equation C.22 for multilevel systems is based on noise power at the detector. As long as the signal bandwidth is identical for all systems, Equation C.22 is valid since the noise bandwidth of the receiver is independent of the number of levels. When the signaling rate is held constant, however, the bit rate increases with the number of levels. To compare multilevel systems on the basis of a given data rate, the signaling interval T and, hence, the noise bandwidths must be adjusted accordingly.

If T is the signaling interval for a two-level system, the signaling interval T_L for an L-level system providing the same data rate is determined as

$$T_L = T \log_2 L \tag{C.31}$$

Using the noise bandwidth of a raised cosine filter in Equation C.29, we extend Equation C.22 to multilevel systems as

$$P_E = \left(\frac{1}{\log_2 L}\right) \left(\frac{L-1}{L}\right) \text{erfc} (z) \tag{C.32}$$

where $z = \dfrac{V/(L-1)}{(N_0/T_L)^{1/2}}$

Equation C.32 can be simplified and presented in a more customary form by using the relationship that energy per symbol $E_s = E_b \log_2 L = V^2 T_L$ where E_b is the energy per bit:

$$P_E = \left(\frac{1}{\log_2 L}\right) \left(\frac{L-1}{L}\right) \text{erfc} (z) \tag{C.33}$$

where $z = \dfrac{(\log_2 L)^{1/2}}{L-1} \left(\dfrac{E_b}{N_0}\right)^{1/2}$

Equation C.33 is plotted in Figure 4.26 for 2, 4, 8, and 16 levels. These curves represent the ideal relative performances of multilevel baseband systems at a constant data rate. The bandwidth requirements of the higher-level systems decrease in proportion to $\log_2 L$.

Equation C.33 and Figure 4.26 represent the performance of multilevel systems with respect to E_b/N_0 (common data rate but different bandwidths). The following relationship can be used to determine error rates with respect to signal-to-noise ratios at the decision circuit (different data rates but common bandwidth).

$$\text{SNR} = \frac{\text{signal power}}{\text{noise power}}$$

$$= \frac{(E_b)(\log_2 L)(1/T_L)}{(N_0)(1/2T_L)} \tag{C.34}$$

$$= (2)(\log_2 L)\left(\frac{E_b}{N_0}\right)$$

where $1/2T_L$ is the minimum "Nyquist" bandwidth of the signal.

The SNR obtained in Equation C.34 represents the ratio of signal power at the sample time to noise power at the detector. This ratio is sometimes referred to as a postdetection SNR because it is the signal-to-noise ratio at the output of the signal processing circuitry.

Some communications theorists use predetection signal-to-noise ratios in determining error rates. Since predetection SNRs are measured prior to bandlimiting the noise, a noise bandwidth must be hypothesized to establish a finite noise power. Commonly, a bit rate bandwidth $(1/T)$ or a Nyquist bandwidth $(1/2T)$ is specified. The latter specification produces SNRs identical to that in Equation C.34. Exceptions occur with double-sideband modulation using coherent demodulation (e.g. 2-PSK) where the predetection SNR is 3 dB higher than the postdetection SNR. (All signal power is coherent to the demodulator carrier reference, but only half of the noise is "coherent.")

C.3.4 Partial Response Systems

The error rate equation for a $1 + D$ partial response system is obtained by incorporating the following modifications to a full response system:

1 The error distance is exactly one-half the error distance of a corresponding full response system (see Figure C.6).
2 The noise bandwidth of the receiver (assuming square root partitioning) is obtained by integrating H_{RX} of Equation C.14:

$$B_N = \int_0^{1/2T} \cos \pi f T \, df$$

$$= 1/(\pi T)$$

(C.35)

Thus the noise bandwidth of a $1 + D$ partial response system is 2 dB lower than the noise bandwidth of a raised cosine (full response) system obtained in Equation C.29. Since the error distance is reduced by 6 dB, partial response filtering incurs a net penalty of 4 dB—with respect to unfiltered signal power at the source. In terms of channel powers, the partial response system incurs a smaller penalty owing to greater spectrum truncation in the transmit filter.

The difference in channel powers of the two systems can be obtained by integrating the respective channel power spectra. If square root partitioning is used, the channel power of the partial response system is 2 dB below the channel power of the corresponding full response systems. In this case, the difference in channel powers is exactly equal to the difference in noise bandwidths because the receiver filter responses are matched to the respective channel spectra. That is, integration of channel spectra has essentially been accomplished in Equations C.29 and C.35.

When spectrum truncation loss in a transmitter is considered, a square root partitioned partial response system is only 2 dB worse than that of a corresponding full response system. Of course, the partial response system requires less bandwidth than the full response system. For completeness, the error rate equation of a square root partitioned

$1 + D$ partial response system is

$$P_E = \left(\frac{1}{\log_2 L}\right) \left(\frac{L - 1}{L}\right) \text{erfc } z \qquad (C.36)$$

where $z = (\pi/4)(E_b/N_0)^{1/2} (\log_2 L)^{1/2}/(L - 1)$ and E_b is the energy per bit on the channel.

Equation C.36 is identical to Equation C.32 except for the factor $\pi/4$ resulting from the lower error distance, the lower noise bandwidth, and the lower channel power of the partial response system.

C.4 CARRIER TRANSMISSION SYSTEMS

C.4.1 Filter Design

Except for a few relatively uncommon frequency modulated systems, digitally modulated carrier systems can be designed and analyzed with baseband-equivalent channels. Carrier based filters are derived by translating the baseband filters to bandpass filters centered at the carrier frequency. The output pulse response of the channel is determined by the composite of the baseband-equivalent filters. Thus pulse shaping can be achieved by filtering the baseband signals or the modulated signals. Partitioning of the channel filter function is dependent on the application. In all cases the composite channel response is identical to that defined in Equation C.2 for raised cosine channels or Equation C.13 for partial response systems.

C.4.2 Error Rate Analysis

Error rate analyses of baseband systems can be directly applied to carrier systems—under one important condition: coherent modulation and demodulation must be used. For example, coherent demodulation of a 2-PSK signal $y(t) = \cos [\omega t + \phi(t)]$ involves implementing the following equations:

$$\phi(t) = \text{low pass } \{\cos [\omega t + \phi(t)] \cdot 2 \cos (\omega t)\}$$
$$= \text{low pass } \{[(\cos \phi(t) \cos \omega t - \sin \phi(t) \sin \omega t] \, 2 \cos \omega t\}$$
$$= \text{low pass } \{\cos \phi(t) (1 + \cos 2\omega t) - \sin \phi(t) \sin 2\omega t\}$$
$$= \cos \phi(t)$$
$$= +1 \text{ for } \phi(t) = 0 \text{ and } -1 \text{ for } \phi(t) = \pi$$
$$(C.37)$$

Notice that coherent demodulation involves multiplying the received signal by a local carrier that is exactly in phase with the respective

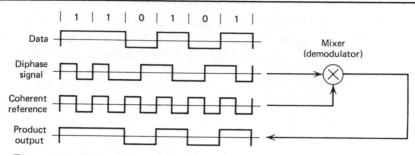

Figure C.9   Coherent demodulation (detection) of digital biphase signal.

incoming signal. Hence coherent demodulation is closely related to matched filter detection as presented in Equations C.26 and C.27. To complete optimum detection of a digitally modulated signal, the baseband equivalent of the receiver filter function must also be matched to the carrier envelope: cos $\phi(t)$. When coherent demodulation (also called coherent detection) is used, the error rate performance is identical to the analagous baseband system. Hence the error rate of a coherently detected 2-PSK system is provided in Equations C.19 or C.28.

The correspondence of coherent demodulation to matched filter detection is illustrated in Figure C.9. For convenience, digital biphase (diphase) is compared to an NRZ baseband signal. The basic principle also applies to sine wave carriers at any frequency. The important point to notice in Figure C.9 is that the output of the coherent demodulator (or equivalently, the matched filter) is identical to the NRZ signal. Furthermore, the noise power density coming out of the coherent demodulator is identical to the **noise power density** of the baseband system. (Positive weighted white noise is statistically no different than negative weighted white noise.)

Figure C.9 demonstrates that coherently demodulated carrier based signals produce the same signal-to-noise ratios at the detector as baseband systems—despite the fact that (double-sideband) carrier systems require twice as much bandwidth. Coherent demodulation leads to a receiver noise bandwidth equal to the baseband-equivalent noise bandwidth because only one-half of the noise power in the carrier signal bandwidth is passed by the coherent demodulator. (Noise in the carrier signal bandwidth that is out of phase with respect to the coherent reference is translated to twice the carrier frequency and therefore eliminated by the low-pass filter, sin $\omega t \cdot \cos \omega t = \frac{1}{2} \sin 2\omega t$.)

C.4.3   QAM Error Rates

The error rate equation of a coherently detected QAM system is identical to the error rate of the corresponding multilevel baseband system applied individually to each quadrature channel. Thus the error performance of a 16-QAM system is provided in Equation C.33 for $L = 4$ levels. In a QAM system coherent demodulation causes one-half of the noise power in the carrier signal bandwidth to show up at the I-channel detector and one-half shows up at the Q-channel detector. Of course, the total signal power is divided in two so the signal-to-noise ratio at the individual detectors is identical to the carrier signal-to-noise ratio (i.e. predetection SNR is equal to postdetection SNR). The error rate performance of QAM systems is plotted in Figure 6.20 in terms of E_b/N_0. To relate those results to signal-to-noise ratios, use Equation C.34 using L as the number of levels on each quadrature channel.

C.4.4 PSK Error Rates

The error rate of a multilevel PSK system is derived most easily by using quadrature channel representations for the signals. For example, Figure C.10 displays the regions of decision errors for a representative signal phase in an 8-PSK signal. The received signal is processed by two orthogonal phase detectors to produce quadrature signals $Y_I(t)$ and $Y_Q(t)$ in Equations 6.10 and 6.11, respectively.

If the transmitted phase is $\pi/8$ (corresponding to data values 011), a decision error results if noise causes $Y_Q(t)$ to go positive at the sample time. [$Y_Q(t)$ is positive downward to represent the Q (sine) channel as lagging the I (cosine) channel.] The normalized error distance is sin $(\pi/8)$. A decision error also results if noise causes $-Y_Q(t)$ to exceed $Y_I(t)$, indicating the phase is greater than $\pi/4$. This latter condition can be represented as a negative value for the transformed signal $Y_B(t) = 0.707\ Y_I(t) + 0.707\ Y_Q(t)$. [$Y_B(t)$ is a projection of the received signal onto a $-\pi/4$ basis vector (see Figure 6.13)]. Examination of Figure C.10 reveals that the error distance for this second type of error is also

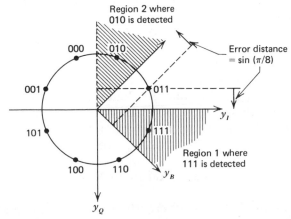

Figure C.10 Regions of decision error for 8-PSK signal at $\pi/8$ (011).

sin $(\pi/8)$. Since the noise variance $Y_B(t)$ is identical to the noise variance of $Y_I(t)$ (by virtue of the 0.707 multipliers), both types of errors are equally likely.

In general terms, the error distance of a PSK system with N phases is $V \cdot \sin(\pi/N)$ where V is the signal amplitude at a detector (i.e. the radius of a PSK signal constellation). A detection error occurs if noise of the proper polarity is present at the output of either of two phase detectors. A detection error, however, is assumed to produce only a single-bit error. The general expression for the theoretical error rate for PSK modulation is now determined by modifying Equation C.22 as

$$P_E = \left(\frac{1}{\log_2 N}\right) \text{erfc}(z)$$

where

$$z = \frac{\sin(\pi/N) \cdot V}{\sqrt{2}\sigma} \tag{C.38}$$

The signal amplitude V can be expressed as

$$V = \left[E_b \, (\log_2 N) \left(\frac{1}{T}\right)\right]^{1/2} \tag{C.39}$$

and the rms noise voltage σ as

$$\sigma = \left[N_0 \left(\frac{1}{2T}\right)\right]^{1/2} \tag{C.40}$$

for noise in a Nyquist bandwidth.

Combining Equations C.38, C.39, and C.40 relates PSK error rates to energy per bit to noise density on the channel:

$$P_E = \left(\frac{1}{\log_2 N}\right) \text{erf}(z) \tag{C.41}$$

where

$$z = \sin\left(\frac{\pi}{N}\right) (\log_2 N)^{1/2} \left(\frac{E_b}{N_0}\right)^{1/2}$$

Equation C.41 is plotted in Figure 6.16 for PSK systems with various numbers of phases.* To determine error rates with respect to signal-power to noise-power ratios use the following:

$$\text{SNR} = \log_2(N)\left(\frac{E_b}{N_0}\right) \qquad (N > 2) \tag{C.42}$$

*For 2-PSK systems, the error rates as specified in Equations C.38 or C.41 should be divided by 2 because only one phase detector is needed and it produces errors for one polarity of noise only.

TRAFFIC TABLES

Table D.1 is a table of maximum offered loads A for various blocking probabilities B and number of servers N. The blocking probabilities are for infinite sources with lost calls cleared (erlang B, Equation 9.8)

Table D.2 is a table of maximum offered loads A for various blocking probabilities B, number of servers N, and finite number of sources M. The offered load A is determined as $M\rho$ where ρ is the average source activity assuming no calls are cleared. The blocking probability for finite sources is determined from Equation 9.13.

Note: Entries in the following tables were obtained from reference 15 of Chapter 9: *Telephone Traffic Theory, Tables, and Charts*, Siemens Aktiengesellschaft, Munich, 1970.

TABLE D.1 MAXIMUM OFFERED LOAD VERSUS B AND M[a]

N/B	0.01	0.05	0.1	0.5	1.0	2	5	10	15	20	30	40
1	.0001	.0005	.001	.005	.010	.020	.053	.111	.176	.250	.429	.667
2	.014	.032	.046	.105	.153	.223	.381	.595	.796	1.00	1.45	2.00
3	.087	.152	.194	.340	.455	.602	.899	1.27	1.60	1.93	2.63	3.48
4	.235	.362	.439	.701	.869	1.09	1.62	2.05	2.50	2.95	3.89	5.02
5	.452	.649	.762	1.13	1.36	1.66	2.22	2.88	3.45	4.01	5.10	6.60
6	.728	.996	1.15	1.62	1.91	2.28	2.96	3.76	4.44	5.11	6.51	8.19
7	1.05	1.39	1.58	2.16	2.50	2.94	3.74	4.67	5.46	6.23	7.86	9.80
8	1.42	1.83	2.05	2.73	3.13	3.63	4.54	5.60	6.50	7.37	9.21	11.4
9	1.83	2.30	2.56	3.33	3.78	4.34	5.37	6.55	7.55	8.52	10.6	13.0
10	2.26	2.80	3.09	3.96	4.46	5.08	6.22	7.51	8.62	9.68	12.0	14.7
11	2.72	3.33	3.65	4.61	5.16	5.84	7.08	8.49	9.69	10.9	13.3	16.3
12	3.21	3.88	4.23	5.28	5.88	6.61	7.95	9.47	10.8	12.0	14.7	18.0
13	3.71	4.45	4.83	5.96	6.61	7.40	8.83	10.5	11.9	13.2	16.1	19.6
14	4.24	5.03	5.45	6.66	7.35	8.20	9.73	11.5	13.0	14.4	17.5	21.2
15	4.78	5.63	6.08	7.38	8.11	9.01	10.6	12.5	14.1	15.6	18.9	22.9
16	5.34	6.25	6.72	8.10	8.88	9.83	11.5	13.5	15.2	16.8	20.3	24.5
17	5.91	6.88	7.38	8.83	9.65	10.7	12.5	14.5	16.3	18.0	21.7	26.2
18	6.50	7.52	8.05	9.58	10.4	11.5	13.4	15.5	17.4	19.2	23.1	27.8
19	7.09	8.17	8.72	10.3	11.2	12.3	14.3	16.6	18.5	20.4	24.5	29.5
20	7.70	8.83	9.41	11.1	12.0	13.2	15.2	17.6	19.6	21.6	25.9	31.2
21	8.32	9.50	10.1	11.9	12.8	14.0	16.2	18.7	20.8	22.8	27.3	32.8
22	8.95	10.2	10.8	12.6	13.7	14.9	17.1	19.7	21.9	24.1	28.7	34.5
23	9.58	10.9	11.5	13.4	14.5	15.8	18.1	20.7	23.0	25.3	30.1	36.1
24	10.2	11.6	12.2	14.2	15.3	16.6	19.0	21.8	24.2	26.5	31.6	37.8
25	10.9	12.3	13.0	16.0	16.1	17.5	20.0	22.8	25.3	27.7	33.0	39.4
26	11.5	13.0	13.7	15.8	17.0	18.4	20.9	23.9	26.4	28.9	34.4	41.1
27	12.2	13.7	14.4	16.6	17.8	19.3	21.9	24.9	27.6	30.2	35.8	42.8
28	12.9	14.4	15.2	17.4	18.6	20.2	22.9	26.0	28.7	31.4	37.2	44.4
29	13.6	15.1	15.9	18.2	19.5	21.0	23.8	27.1	29.9	32.6	38.6	46.1
30	14.2	15.9	16.7	19.0	20.3	21.9	24.8	28.1	31.0	33.8	40.0	47.7
31	14.9	16.6	17.4	19.9	21.2	22.8	25.8	29.2	32.1	35.1	41.5	49.4
32	15.6	17.3	18.2	20.7	22.0	23.7	26.7	30.2	33.3	36.3	42.9	51.1
33	16.3	18.1	19.0	21.5	22.9	24.6	27.7	31.3	34.4	37.5	44.3	52.7
34	17.0	18.8	19.7	22.3	23.8	25.5	28.7	32.4	35.6	38.8	45.7	54.4
35	17.8	19.6	20.5	23.2	24.6	26.4	29.7	33.4	36.7	40.0	47.1	56.0
36	18.5	20.3	21.3	24.0	25.5	27.3	30.7	34.5	37.9	41.2	48.6	57.7
37	19.2	21.1	22.1	24.8	26.4	28.3	31.6	35.6	39.0	42.4	50.0	59.4
38	19.9	21.9	22.9	25.7	27.3	29.2	32.6	36.6	40.2	43.7	51.4	61.0
39	20.6	22.6	23.7	26.5	28.1	30.1	33.6	37.7	41.3	44.9	52.8	62.7
40	21.4	23.4	24.4	27.4	29.0	31.0	34.6	38.8	42.5	46.1	54.2	64.4
41	22.1	24.2	25.2	28.2	29.9	31.9	35.6	39.9	43.6	47.4	55.7	66.0
42	22.8	25.0	26.0	29.1	30.8	32.8	36.6	40.9	44.8	48.6	57.1	67.7
43	23.6	25.7	26.8	29.9	31.7	33.8	37.6	42.0	45.9	49.9	58.5	69.3
44	24.3	26.5	27.6	30.8	32.5	34.7	38.6	43.1	47.1	51.1	59.9	71.0
45	25.1	27.3	28.4	31.7	33.4	35.6	39.6	44.2	48.2	52.3	61.3	72.7
46	25.8	28.1	29.3	32.5	34.3	36.5	40.5	45.2	49.4	53.6	62.8	74.3
47	26.6	28.9	30.1	33.4	35.2	37.5	41.5	46.3	50.6	54.8	64.2	76.0
48	27.3	29.7	30.9	34.2	36.1	38.4	42.5	47.4	51.7	56.0	65.6	77.7
49	28.1	30.5	31.7	35.1	37.0	39.3	43.5	48.5	52.9	57.3	67.0	79.3
50	28.9	31.3	32.5	36.0	37.9	40.3	44.5	49.6	54.0	58.3	68.5	81.0

[a] N is the number of servers. The numerical column headings indicate blocking probability (%).

N/B	0.01	0.05	0.1	0.5	1.0	2	5	10	15	20	30	40
51	29.6	32.1	33.3	36.9	38.8	41.2	45.5	50.6	55.2	59.7	69.9	82.7
52	30.4	32.9	34.2	37.7	39.7	42.1	46.5	51.7	56.3	61.0	71.3	84.3
53	31.2	33.7	35.0	38.6	40.6	43.1	47.5	52.8	57.5	62.2	72.7	86.0
54	31.9	34.5	35.8	39.5	41.5	44.0	48.5	53.9	58.7	63.5	74.2	87.6
55	32.7	35.3	36.6	40.4	42.4	44.9	49.5	55.0	59.8	64.7	75.6	89.3
56	33.5	36.1	37.5	41.2	43.3	45.9	50.5	56.1	61.0	65.9	77.0	91.0
57	34.3	36.9	38.3	42.1	44.2	46.8	61.5	67.1	62.1	67.7	78.4	92.6
58	35.1	37.8	39.1	43.0	45.1	47.8	62.6	68.2	63.3	68.4	79.8	94.3
59	35.8	38.6	40.0	43.9	46.0	48.7	53.6	59.3	64.5	69.7	81.3	96.0
60	36.6	39.4	40.8	44.8	46.9	49.6	54.6	60.4	65.6	70.9	82.7	97.6
61	37.4	40.2	41.6	45.6	47.9	50.6	55.6	61.5	66.8	72.1	84.1	99.3
62	38.2	41.0	42.5	46.5	48.8	51.5	56.6	62.6	68.0	73.4	85.5	101.
63	39.0	41.9	43.3	47.4	49.7	52.5	57.6	63.7	69.1	74.6	87.0	103.
64	39.8	42.7	44.2	48.3	50.6	53.4	58.6	64.8	70.3	75.9	88.4	104.
65	40.6	43.5	45.0	49.2	51.5	54.4	59.6	65.8	71.4	77.1	89.8	106.
66	41.4	44.4	45.8	50.1	52.4	55.3	60.6	66.9	72.6	78.3	91.2	108.
67	42.2	45.2	46.7	51.0	53.4	56.3	61.6	68.0	73.8	79.6	92.7	109.
68	43.0	46.0	47.5	51.9	54.3	57.2	62.6	69.1	74.9	80.8	94.1	111.
69	43.8	46.8	48.4	52.8	55.2	58.2	63.7	70.2	76.1	82.1	95.5	113.
70	44.6	47.7	49.2	53.7	56.1	59.1	64.7	71.3	77.3	83.3	96.9	114.
71	45.4	48.5	50.1	54.6	57.0	60.1	65.7	72.4	78.4	84.6	98.4	116.
72	46.2	49.4	50.9	55.5	58.0	61.0	66.7	73.5	79.6	85.8	99.8	118.
73	47.0	50.2	51.8	56.4	58.9	62.0	67.7	74.6	80.8	87.0	101.	119.
74	47.8	51.0	52.7	57.3	59.8	62.9	68.7	75.6	81.9	88.3	103.	121.
75	48.6	51.9	53.6	58.2	60.7	63.9	69.7	76.7	83.1	89.5	104.	123.
76	49.4	52.7	54.4	59.1	61.7	64.9	70.8	77.8	84.2	90.8	105.	124.
77	50.2	53.6	55.2	60.0	62.6	65.8	71.8	78.9	85.4	92.0	107.	126.
78	51.1	54.4	56.1	60.9	63.5	66.8	72.8	80.0	86.6	93.3	108.	128.
79	51.9	55.3	56.9	61.8	64.4	67.7	73.8	81.1	87.7	94.5	110.	129.
80	52.7	56.1	57.8	62.7	65.4	68.7	74.8	82.2	88.9	95.7	111.	131.
81	53.5	56.9	58.7	63.6	66.3	69.6	75.8	83.3	90.1	97.0	113.	133.
82	54.3	57.8	59.5	64.5	67.2	70.6	76.9	84.4	91.2	98.2	114.	134.
83	55.1	58.6	60.4	65.4	68.2	71.6	77.9	85.5	92.4	99.5	115.	136.
84	56.0	59.5	61.3	66.3	69.1	72.5	78.9	86.6	93.6	101.	117.	138.
85	56.8	60.4	62.1	67.2	70.0	73.5	79.9	87.7	94.7	102.	118.	139.
86	57.6	61.2	63.0	68.1	70.9	74.5	80.9	88.8	95.9	103.	120.	141.
87	58.4	62.1	63.9	69.0	71.9	75.4	82.0	89.9	97.1	104.	121.	143.
88	59.3	62.9	64.7	69.9	72.8	76.4	83.0	91.0	98.2	106.	123.	144.
89	60.1	63.8	65.6	70.8	73.7	77.3	84.0	92.1	99.4	107.	124.	146.
90	60.9	64.6	66.5	71.8	74.7	78.3	85.0	93.1	101.	108.	126.	148.
91	61.8	65.5	67.4	72.7	75.6	79.3	86.0	94.2	102.	109.	127.	149.
92	62.6	66.3	68.2	73.6	76.6	80.2	87.1	95.3	103.	111.	128.	151.
93	63.4	67.2	69.1	74.5	77.5	81.2	88.1	96.4	104.	112.	130.	153.
94	64.2	68.1	70.0	75.4	78.4	82.2	89.1	97.5	105.	113.	131.	154.
95	65.1	68.9	70.9	76.3	79.4	83.1	90.1	98.6	106.	114.	133.	156.
96	65.9	69.8	71.7	77.2	80.3	84.1	91.1	99.7	108.	116.	134.	158.
97	66.8	70.7	72.6	78.2	81.2	85.1	92.2	101.	109.	117.	135.	159.
98	67.6	71.5	73.5	79.1	82.2	86.0	93.2	102.	110.	118.	137.	161.
99	68.4	72.4	74.4	80.0	83.1	87.0	94.2	103.	111.	119.	138.	163.
100	69.3	73.2	75.2	80.9	84.1	88.0	95.2	104.	112.	121.	140.	164.

TABLE D.2 MAXIMUM OFFERED LOAD VERSUS B, N, AND FINITE SOURCES M^a

N	M	.01	.05	0.1	0.5	1.0	2	5	10	15	20	30	4
	2	.0002	.0010	.0020	.0100	.0200	.0400	.100	.202	.307	.417	.659	.9
1	3	.0002	.0008	.0015	.0075	.0151	.0303	.077	.159	.246	.341	.559	.8
	4	.0001	.0007	.0013	.0067	.0134	.0270	.069	.143	.224	.312	.519	.7
	5	.0001	.0006	.0013	.0063	.0126	.0254	.065	.136	.213	.298	.498	.7
	3	.030	.067	.095	.212	.300	.425	.678	.980	1.23	1.47	1.97	2.
	4	.023	.052	.074	.167	.238	.342	.560	.832	1.07	1.30	1.78	2.
2	5	.021	.046	.065	.149	.213	.308	.510	.767	.997	1.22	1.70	2.
	6	.019	.043	.061	.139	.200	.289	.482	.731	.955	1.18	1.65	2.
	7	.018	.041	.058	.133	.191	.277	.464	.707	.928	1.15	1.62	2.
	4	.186	.317	.400	.685	.864	1.09	1.50	1.95	2.31	2.65	3.35	4.
	5	.148	.254	.322	.561	.715	.918	1.30	1.72	2.08	2.42	3.13	3.
	6	.131	.227	.288	.505	.648	.837	1.20	1.62	1.97	2.31	3.02	3.
3	7	.122	.211	.268	.473	.609	.790	1.14	1.55	1.90	2.24	2.95	3.
	8	.116	.201	.255	.452	.583	.759	1.10	1.51	1.85	2.19	2.90	3.
	9	.111	.194	.246	.437	.565	.737	1.07	1.47	1.82	2.16	2.86	3.
	10	.108	.188	.240	.426	.551	.720	1.05	1.45	1.79	2.13	2.84	3.
	15	.100	.174	.222	.396	.514	.675	.994	1.38	1.72	2.06	2.76	3.
	5	.500	.748	.889	1.33	1.59	1.89	2.24	2.98	3.43	3.86	4.76	5.
	6	.408	.617	.737	1.12	1.36	1.64	2.15	2.71	3.16	3.60	4.51	5.
	7	.365	.554	.665	1.02	1.24	1.52	2.01	2.56	3.02	3.46	4.38	5.
4	8	.340	.517	.621	.963	1.17	1.44	1.92	2.47	2.93	3.37	4.30	5.
	9	.323	.492	.592	.922	1.13	1.39	1.86	2.41	2.87	3.31	4.24	5.
	10	.310	.474	.571	.892	1.09	1.35	1.82	2.36	2.82	3.27	4.20	5.
	15	.280	.429	.518	.816	1.00	1.25	1.71	2.24	2.70	3.14	4.08	5.
	6	.951	1.31	1.51	2.08	2.40	2.77	3.39	4.04	4.58	5.09	6.17	7.
	7	.794	1.11	1.28	1.80	2.10	2.45	3.07	3.73	4.28	4.80	5.91	7.
	8	.716	1.00	1.16	1.66	1.94	2.29	2.90	3.56	4.12	4.65	5.77	7.
5	9	.668	.940	1.09	1.56	1.84	2.18	2.78	3.45	4.01	4.55	5.68	7.
	10	.635	.896	1.04	1.50	1.77	2.11	2.71	3.37	3.94	4.48	5.61	6.
	12	.592	.839	.979	1.42	1.68	2.01	2.60	3.27	3.84	4.38	5.52	6.
	15	.556	.791	.924	1.35	1.60	1.92	2.51	3.18	3.74	4.29	5.44	6.
	20	.525	.748	.876	1.28	1.53	1.84	2.42	3.09	3.66	4.21	5.37	6.
	7	1.51	1.97	2.21	2.90	3.26	3.69	4.38	5.12	5.73	6.32	7.59	9.
	8	1.28	1.70	1.91	2.55	2.90	3.32	4.02	4.78	5.41	6.02	7.32	8.
	9	1.17	1.55	1.76	2.37	2.71	3.12	3.82	4.59	5.24	5.85	7.17	8.
6	10	1.09	1.46	1.65	2.25	2.58	2.98	3.69	4.47	5.12	5.74	7.07	8.
	15	.926	1.25	1.43	1.97	2.29	2.68	3.38	4.17	4.84	5.48	6.84	8.
	20	.865	1.17	1.34	1.87	2.17	2.56	3.26	4.05	4.72	5.37	6.74	8.
	30	.813	1.10	1.27	1.77	2.07	2.45	3.15	3.94	4.62	5.28	6.66	8.
	8	2.15	2.70	2.98	3.76	4.17	4.63	5.39	6.20	6.89	7.56	9.01	10.
	9	1.85	2.36	2.62	3.36	3.75	4.22	5.00	5.85	6.56	7.25	8.74	10.
	10	1.70	2.17	2.42	3.13	3.52	3.99	4.78	5.64	6.37	7.07	8.58	10.
7	15	1.39	1.80	2.02	2.68	3.05	3.51	4.31	5.21	5.97	6.70	8.25	10.
	20	1.28	1.67	1.88	2.52	2.88	3.33	4.14	5.05	5.81	6.56	8.13	10.
	30	1.19	1.56	1.76	2.38	2.74	3.18	3.99	4.90	5.68	6.44	8.03	9.

^{a}The numerical column headings indicate blocking probability (%).

N	M	.01	.05	0.1	0.5	1.0	2	5	10	15	20	30	40
8	9	2.85	3.48	3.80	4.65	5.09	5.59	6.41	7.30	8.05	8.80	10.4	12.
	10	2.49	3.08	3.37	4.20	4.64	5.14	6.00	6.92	7.71	8.48	10.2	12.
	11	2.29	2.85	3.14	3.94	4.38	4.89	5.76	6.71	7.51	8.29	9.99	12.
	12	2.16	2.70	2.97	3.77	4.20	4.17	5.59	6.56	7.37	8.17	9.88	12.
	15	1.93	2.43	2.70	3.46	3.89	4.40	5.29	6.29	7.13	7.94	9.69	11.
	20	1.76	2.24	2.49	3.23	3.65	4.16	5.06	6.07	6.93	7.76	9.54	11.
	30	1.63	2.08	2.32	3.03	3.45	3.95	4.86	5.89	6.77	7.61	9.41	11.
9	10	3.59	4.30	4.64	5.57	6.03	6.56	7.44	8.39	9.22	10.0	11.9	14.
	11	3.18	3.84	4.17	5.07	5.54	6.09	7.00	8.01	8.87	9.72	11.6	13.
	12	2.94	3.57	3.89	4.79	5.26	5.81	6.75	7.78	8.66	9.52	11.4	13.
	13	2.78	3.39	3.71	4.59	5.06	5.61	6.57	7.62	8.51	9.39	11.3	13.
	14	2.66	3.26	3.57	4.44	4.91	5.47	6.43	7.50	8.41	9.29	11.2	13.
	16	2.50	3.08	3.38	4.23	4.70	5.27	6.24	7.33	8.25	9.15	11.1	13.
	18	2.39	2.95	3.25	4.09	4.56	5.13	6.11	7.21	8.15	9.06	11.0	13.
	20	2.31	2.87	3.15	3.99	4.46	5.02	6.02	7.13	8.07	8.99	11.0	13.
	30	2.12	2.64	2.91	3.73	4.19	4.76	5.77	6.90	7.87	8.81	10.8	13.
10	11	4.38	5.15	5.52	6.49	6.98	7.54	8.47	9.49	10.4	11.3	13.3	15.
	12	3.91	4.64	5.00	5.97	6.47	7.04	8.02	9.09	10.0	11.0	13.0	15.
	13	3.63	4.34	4.69	5.65	6.16	6.74	7.75	8.86	9.81	10.8	12.8	15.
	14	3.45	4.13	4.47	5.43	5.94	6.54	7.56	8.69	9.66	10.6	12.7	15.
	15	3.31	3.98	4.32	5.27	5.78	6.38	7.41	8.56	9.55	10.5	12.6	15.
	16	3.20	3.86	4.19	5.14	5.65	6.25	7.30	8.46	9.46	10.4	12.6	15.
	18	3.04	3.68	4.01	4.95	5.46	6.07	7.13	8.31	9.33	10.3	12.5	15.
	20	2.93	3.56	3.88	4.81	5.32	5.93	7.01	8.21	9.23	10.2	12.4	15.
	25	2.75	3.36	3.68	4.59	5.10	5.72	6.81	8.04	9.08	10.1	12.3	14.
	30	2.65	3.25	3.56	4.47	4.98	5.59	6.69	7.93	8.99	10.0	12.2	14.
11	12	5.19	6.01	6.41	7.44	7.95	8.53	9.50	10.6	11.6	12.5	14.7	17.
	13	4.68	5.46	5.85	6.88	7.40	8.01	9.04	10.2	11.2	12.2	14.4	17.
	14	4.37	5.13	5.51	6.54	7.07	7.69	8.76	9.94	11.0	12.0	14.2	17.
	15	4.15	4.90	5.27	6.30	6.84	7.47	8.56	9.77	10.8	11.9	14.1	17.
	16	3.99	4.72	5.09	6.12	6.66	7.30	8.40	9.63	10.7	11.7	14.0	16.
	17	3.86	4.59	4.95	5.98	6.52	7.17	8.28	9.53	10.6	11.7	14.0	16.
	18	3.76	4.48	4.84	5.86	6.41	7.06	8.18	9.44	10.5	11.6	13.9	16.
	20	3.60	4.31	4.66	5.68	6.23	6.88	8.03	9.31	10.4	11.5	13.8	16.
	25	3.36	4.04	4.39	5.40	5.95	6.62	7.79	9.10	10.2	11.3	13.7	16.
	30	3.32	3.90	4.24	5.24	5.79	6.46	7.64	8.98	10.1	11.2	13.6	16.
12	13	6.03	6.90	7.31	8.39	8.92	9.52	10.5	11.7	12.7	13.8	16.1	19.
	14	5.47	6.31	6.72	7.80	8.35	8.98	10.1	11.3	12.4	13.4	15.8	18.
	15	5.13	5.95	6.35	7.44	8.00	8.65	9.77	11.0	12.1	13.2	15.7	18.
	16	4.88	5.69	6.09	7.18	7.75	8.42	9.56	10.8	12.0	13.1	15.5	18.
	17	4.70	5.50	5.90	6.99	7.56	8.24	9.40	10.7	11.8	13.0	15.5	18.
	18	4.56	5.35	5.74	6.83	7.41	8.09	9.27	10.6	11.8	12.9	15.4	18.
	20	4.34	5.12	5.51	6.60	7.18	7.87	9.08	10.4	11.6	12.8	15.3	18.
	25	4.03	4.77	5.16	6.24	6.83	7.54	8.78	10.2	11.4	12.6	15.1	18.
	30	3.85	4.58	4.96	6.04	6.63	7.35	8.61	10.0	11.3	12.5	15.0	18.

GLOSSARY

Many of the following definitions have been obtained from CCITT recommendations and are so indicated.

Abbreviated dialling. A subscriber can dial a short code which is converted in the originating exchange into a form which will establish connection with the local, national, or international number required. (CCITT)

Added channel framing. *See* Bunched frame alignment signal.

Added digit framing. *See* Distributed frame alignment signal.

Alternate mark inversion signal (AMI) (bipolar signal). A pseudo-ternary signal, conveying binary digits, in which successive "marks" are normally of alternative, positive, and negative, polarity but equal in amplitude, and in which "space" is of zero amplitude. (CCITT)

Alternate mark inversion violation (AMI) (bipolar violation). A "mark" which has the same polarity as the previous "mark" in the transmission of AMI signals. (CCITT)

Amplitude distortion. Distortion of a transmission signal caused by nonuniform passband attenuation as a function of frequency.

Amplitude modulation (AM). Modulation in which the amplitude of an alternating current is the characteristic varied. (CCITT)

Antipodal signalling. A technique of encoding binary signals so that the symbol for a 1 is the exact negative of the symbol for a 0. Antipodal signalling provides optimum error performance in terms of the signal-to-noise ratio.

APSK. Amplitude and phase shift keying, a form of digital modulation using a combination of both amplitude and phase modulation.

ARQ. *See* Request repeat system.

Asynchronous transmission. A mode of communication characterized by start/stop transmissions with undefined time intervals between transmissions.

Automatic call distributor (ACD). A switching system used to distribute automatically and evenly incoming calls to a number of stations on a first-come, first-served basis. Applications include operator assistance and airline reservations.

Automatic number identification (ANI). The process of automatically identifying a calling number.

Availability. (1) With respect to switching systems: the number of outlets available from a particular inlet. (2) With respect to equipment in general: the percentage of time the equipment is operational.

Balanced code. A line code whose digital sum variation is finite (balanced codes have no dc component in their frequency spectrum). (CCITT)

Baseband. Literally, the frequency band of an unmodulated signal. A baseband signal is an information bearing signal that is either transmitted as is or used to modulate a carrier.

Baud. Unit of signaling speed (symbols per second).

Bipolar coding. *See* Alternate mark inversion.

Blocked calls cleared (BCC). A service discipline in which unserviceable requests are rejected by the system without service. Also called lost calls cleared (LCC).

Blocked calls held (BCH). A service discipline in which unserviceable requests stay "in the system" without being serviced but having a portion of their desired service time elapse until service begins. Also called lost calls held (LCH).

Bridged tap. An extra pair of wires connected in shunt to a main cable pair. The extra pair is normally open circuited but may be used at a future time to connect the main pair to a new customer. Short bridged taps do not affect voice frequency signals but can be extremely detrimental to higher frequency digital signals.

Bunched frame alignment signal (added channel framing). A frame alignment signal in which the signal elements occupy consecutive digit time slots. (CCITT)

Burst isochronous. A transmission process that may be used where the information bearer channel rate is higher than the input data signalling rate. The binary digits being transferred are signaled at the digit rate of the information bearer channel and the transfer is interrupted at intervals in order to produce the required mean data signalling rate. The interruption is always for an integral number of digit periods. (CCITT)

Busy hour (of a group of circuits, a group of switches, or an exchange, etc.). The busy hour is the uninterrupted period of 60 minutes of which the traffic is at the maximum. (CCITT)

C-Message weighting. Selective attenuation of voiceband noise in accordance with the subjective effects as a function of frequency.

Call congestion. Blocking probability of a trunk group.

Call forwarding. A customized feature available on computer controlled switches that allows customers to direct all incoming calls to another number.

Call waiting. Subscriber A whose termination is in the active state associated with a call to subscriber B is given a call waiting indication (CWI) that a caller subscriber C, is attempting to obtain connection. He may either: (*a*) ignore or reject the indication and continue with the existing call; (*b*) terminate the existing call and answer subscriber C; or (*c*) hold the existing call and answer subscriber C. Thus the three party service functions are available to the extent that they are provided in the system. (CCITT)

Calling number indication. A service whereby the calling subscriber's number can be identified by means of a visual or verbal indication at the called terminal. (CCITT)

Carterfone decision. A 1968 decision by the FCC that allows telephone company customers to connect their own equipment to the network as long as it does not adversely affect the network.

CBX. Computer controlled PBX.

CCIS. Common channel interoffice signaling.

CCS. A measure of traffic intensity expressed as so many hundred call seconds per hour. 36 CCS per hour = 1 erlang.

Central office. A switching system in the public network.

Centrex service. The provision to subscribers, by means of a specially equipped public telephone exchange, of services normally available only in PABXs (e.g., internal dialing of PABX type, operators' desk, direct access to the network, direct dialing-in, transfer of calls). (CCITT)

Channel associated signalling. A signalling method in which the signals necessary for the traffic carried by a single channel are transmitted in the channel itself or in a signalling channel permanently associated with it. (CCITT)

Channel bank. Terminal equipment for a transmission system used to multiplex individual channels using FDM or TDM techniques.

Circuit switching. The principle of establishing an end-to-end connection between users of a network. The associated facilities are dedicated to the particular connection and held for the duration of the call.

Clock. Equipment providing a time base used in a transmission system to control the timing of certain functions such as the control of the duration of signal elements, and the sampling. (CCITT)

Codec. An assembly comprising an encoder and a decoder in the same equipment. (CCITT)

Coded mark inversion (CMI). A 2-level nonreturn-to-zero code in which binary zero is coded so that both amplitude levels A_1 and A_2 are attained consecutively, each for half a unit time interval $(T/2)$. Binary one is coded so that either of the amplitude levels, A_1 and A_2, are attained alternately for one full unit time interval (T). (CCITT)

Coherent demodulation. Demodulation using a carrier reference that is synchronized in frequency and phase to the carrier used in the modulation process.

Common channel signalling. A signalling method using a link common to a number of channels for the transmission of signals necessary for the traffic by way of these channels. (CCITT)

Common control. A form of automatic control for a switching system that concentrates all control functions into one equipment shared by all connections.

Community dial office (CDO). A small, normally unattended switching system that is used in small communities and is controlled from a larger central office.

Companding. The process of compressing a signal at the source and expanding it at the destination to maintain a given end-to-end dynamic range while reducing the dynamic range between the compressor-expandor.

Concentration. The process of switching some number of lightly used channels or sources onto a smaller number of more heavily used channels.

Conference call. A call between more than two participants. (CCITT)

CPFSK. Continuous phase frequency shift keying.

Crosstalk. Unwanted signal transfer from one circuit into another.

Datagram. A single packet message.

Data set. *See* Modem.

Dataphone digital service (DDS). A service offering of the Bell System providing digital channels at 2.4, 4.8, 9.6, 56, or 1544 kbps.

dBm. Power level in decibels relative to 1 mW.

dBrnC. Power level of noise with C-message weighting expressed in decibels relative to reference noise. Reference noise power is -90 dBm $= 10^{-12}$ W.

dBrnC0. Noise power measured in dBrnC but referenced to the zero-level transmission level point.

DDD. Direct distance dialing.

Delta modulation. Technique for digitally encoding an analog signal by continuously measuring and transmitting only the polarity of the signal slope. The analog signal is reconstructed as a staircase approximation of the input.

Demand traffic. First attempt traffic. (Offered traffic not including re-tries.)

Despotic (synchronized) network (master clock). A network synchronizing arrangement is despotic when a unique master clock exists with full power of control of all other clocks. (CCITT)

DID. Direct inward dialing.

Digital multiplex equipment (muldem). Equipment for combining, by time division multiplexing (multiplexer), a defined integral number of digital input signals into a single digital signal at a defined digit rate and also for carrying out the inverse function (demultiplexer). *Note:* When both functions are combined in one equipment at the same location the abbreviation "MULDEX" may be used to describe this equipment. (CCITT)

Digital signal. A signal constrained to have a discontinuous characteristic in time and a set of permitted discrete values. (CCITT)

Digital switching. A process in which connections are established by operations on digital signals without converting them to analogue signals. (CCITT)

Direct inward dialling. Calls can be dialed from a telephone connected to the public network directly to extensions on a PABX. (CCITT)

Distributed frame alignment signal (added digit framing). A frame alignment signal in which the signal elements occupy nonconsecutive digit time slots. (CCITT)

Double-sideband modulation. A modulation technique in which a baseband signal with no dc energy directly modulates a carrier to produce both upper and lower sidebands but no carrier energy.

DS-1: Digital signal-1. Primary multiplex level in North American TDM hierarchy. Ascending levels are DS-2, DS-3, and DS-4.

DTE. Data terminal equipment.

Dual tone multifrequency (DTMF) signalling. Generic name for push-button telephone signalling equivalent to the Bell System's TOUCH-TONE.®

Dynamic range. The range of power levels (minimum to maximum) achievable by a signal or specified for equipment operation.

Echo cancellor. A device that removes talker echo in the return branch of a four-wire circuit by subtracting out a delayed version of the signal transmitted in the forward path.

Echo suppressor. A voice-operated device placed in the four-wire portion of a circuit and used for inserting loss in the echo path to suppress echo. (CCITT)

Elastic store. A storage buffer designed to accept data under one clock but deliver it under another, so short term instabilities (jitter) in either clock can be accommodated.

Electronic automatic exchange (EAX). Designation of stored program control switching machines manufactured by General Telephone and Electronics.

Electronic switching system (ESS). Designation of stored program control switching machines in the Bell System.

End office. Class 5 switching office. Also referred to as a central office.

Envelope delay. Derivative of channel phase response with respect to frequency. Ideally, the phase response should be linear, indicating that all frequencies are delayed equally.

Equalization. The practice of compensating for transmission distortions with fixed or adaptive circuitry.

Erlang. A measure of traffic intensity. Basically, a measure of the utilization of a resource (e.g., the average number of busy circuits in a trunk group, or the ratio of time an individual circuit is busy).

Exchange area. A contiguous area of service defined for administrative purposes that typically comprises an entire town or city and includes the immediate countryside and suburbs. An exchange area may have one end office or many end offices interconnected by trunks and tandem offices.

Expansion. The switching of a number of input channels onto a larger number of output channels.

Facsimile. The black and white reproduction of a document or a picture transmitted over a telephone connection. (CCITT)

Far-end crosstalk (FEXT). Unwanted energy coupled from one channel or circuit into another circuit and appearing at the far end of the transmission link.

FIFO. First-in first-out service discipline for a queue.

Flow control (in data network). The procedure for controlling the rate of transfer of packets between two nominated points in a data network, for example between a DTE and a data-switching exchange. (CCITT)

Foreign exchange circuit (FX). An extension of service from one switching office to a subscriber normally serviced by a different switching office.

Four-wire circuit. A circuit using two separate channels for each direction of transmission. When wire-line transmission is involved, each direction of transmission is provided by a separate pair of wires.

Fractional speech loss. The fraction of speech that gets clipped in a TASI system because all channels are in use when talk spurts begin.

Frame. A set of consecutive digit time slots in which the position of each digit time slot can be identified by reference to a frame alignment signal. The frame alignment signal does not necessarily occur, in whole or in part, in each frame. (CCITT)

Frame alignment. The state in which the frame of the receiving equipment is correctly phased with respect to that of the received signal. (CCITT)

Frame alignment recovery time. The time that elapses between a valid frame alignment signal being available at the receive terminal equipment and frame alignment being established. *Note:* The frame alignment recovery time includes the time required for replicated verification of the validity of the frame alignment signal. (CCITT)

Frame alignment signal. The distinctive signal used to enable frame alignment to be secured. (CCITT)

Frequency diversity. In radio systems the use of one or more back up transmitters, channels, and receivers to protect against atmospheric (multipath) fading.

Frequency shift keying (FSK). Digital modulation using discrete frequencies (tones) to represent discrete symbols.

Full-duplex. Transmission in two directions simultaneously.

Full-echo suppressor. An echo suppressor in which the speech signals on either path control the suppression loss in the other path. (CCITT)

Glare. Simultaneous seizure of both ends of a two-way trunk.

Half-duplex transmission. Transmission in both directions but only in one direction at a time.

Half-echo suppressor. An echo suppressor in which the speech signals of one path control the suppression loss in the other path but in which this action is not reciprocal. (CCITT) Half-echo suppressors are normally used in pairs—one at each end of the long-distance circuit.

HDB3. A modified AMI (bipolar) line code in which strings of four zeros are encoded with an AMI violation in the last bit. (CCITT)

Highway. A common path or a set of parallel paths over which signals from a plurality of channels pass with separation achieved by time division. (CCITT)

Hook flash. A momentary depression of the switchhook to alert an operator or equipment, but not so long as to be interpreted as a disconnect.

Hot standby. Redundant equipment kept in an operational mode as backup for primary equipment. Usually, automatic switching to the standby equipment occurs when the primary equipment fails.

Hybrid. A device used to connect a two-wire, bidirectional circuit to a four-wire circuit.

Idle character. A control character that is sent when there is no information to be sent. (CCITT)

Inband signalling. Signalling transmitted within the same channel and band of frequencies used for message traffic.

Integrated digital network. A network in which connections established by digital switching are used for the transmission of digital signals. (CCITT)

Integrated services digital network. An integrated digital network in which the same digital switches and digital paths are used to establish connections for different services, for example, telephony, data, etc. (CCITT)

Intercept. Calls which, for reasons such as those listed below, cannot reach the wanted number may be intercepted and diverted to an operator, an answering machine, or a tone to give the caller appropriate information. (CCITT) (*a*) Change of a particular number including advice of new number. (*b*) Renumbering of a group of numbers or a change of dialling code. (*c*) Wrong information in telephone directory. (*d*) Dialling of an unallocated code. (*e*) Dialling of a number or numbers allowed by the numbering plan but not yet allocated or no longer in service. (*f*) Route(s) out of order. (*g*) Route(s) congested. (*h*) Subscriber's line temporarily out of order. (*i*) Suspension of service owing to nonpayment.

Intersymbol interference. Interference in a digital (or any time division) transmission system caused by a symbol in one signalling interval being spread out and overlapping the sample time of a symbol in another signal interval.

Isochronous. A signal is isochronous if the time interval separating any two significant instants is theoretically equal to the unit interval or to a multiple of the unit interval. *Note:* In practice, variations in the time intervals are constrained within specified limits. (CCITT).

Jitter. Short-term variations of the significant instants of a digital signal from their ideal positions in time. (CCITT)

Junctor. A connecting circuit between switching stages or frames of a single switching office. A junctor may also provide signalling supervision and talking battery for connected subscriber instruments.

Justifiable digit time slot (stuffable digit time slot). A digit time slot that may contain either an information digit or a justifying digit. (CCITT)

Justification (pulse stuffing). A process of changing the rate of a digital signal in a controlled manner so that it can accord with a rate different from its own inherent rate, usually without loss of information. (CCITT)

Justification ratio (stuffing ratio). The ratio of the actual justification rate to the maximum justification rate. (CCITT)

Justification service digits (stuffing service digits). Digits that transmit information concerning the status of the justifiable digit time slots. (CCITT)

Justifying digit (stuffing digit). A digit inserted in a justifiable digit time slot when that time slot does not contain an information digit. (CCITT)

Key telephone set. A telephone set with special buttons (keys) to provide such capabilities as switching between lines, call holding, or alerting of other telephones.

Line code. A code chosen to suit the transmission medium and giving the equivalence between a set of digits generated in a terminal or other processing equipment and the pulses chosen to represent that set of digits for line transmission. (CCITT)

Load balancing. Adjusting the assignment of very active customer lines so that all groups of customer lines in a multistage switch receive approximately the same amount of traffic.

Loading coils. Lumped element inductors inserted at periodic points in cable pairs to flatten out their voice frequency response. Although loading coils improve voice frequency transmission they severely attenuate higher frequencies as required for digital transmission.

Lost calls cleared (also loss system). A mode of operation in which blocked calls are rejected by the network and may or may not return.

Lost calls held. A mode of operation that holds blocked call requests until a channel becomes available. The portion of call that gets blocked is lost.

Master clock. A clock which generates accurate timing signals for the control of other clocks and possibly other equipments. (CCITT)

Mesochronous. Two signals are mesochronous if their corresponding significant instants occur at the same average rate. *Note:* The phase relationship between corresponding significant instants usually varies between specified limits. (CCITT)

Message switching. The practice of transporting complete messages from a source to a destination in non real time and without interaction between source and destination, usually in a store-and-forward fashion.

Modem. A contraction of the terms modulation and demodulation. A device used to generate "voicelike" data signals for transmission over telephone lines. A modem is referred to as a "data set" in Bell System terminology.

Modified alternate mark inversion. An AMI signal that does not strictly conform with alternate mark inversion but includes violations in accordance with a defined set of rules. Examples of such signals are HDB, B6ZS, B3ZS. (CCITT)

Muldem. (*See* Digital multiplex equipment).

Multiframe (Masterframe). A set of consecutive frames in which the position of each frame can be identified by reference to a multiframe alignment signal. The multiframe alignment signal does not necessarily occur, in whole or in part, in each multiframe. (CCITT)

Multifrequency signaling (MF). A signaling method used for interoffice applications. MF signaling uses two of six possible tones to encode 10 digits and five special auxiliary signals.

Multiplexing. The process of combining multiple signals into a single channel for transmission over common facilities.

Mutually synchronized network. A network synchronizing arrangement is mutually synchronized when each clock in the network exerts a degree of control on all others. A mutually synchronized system is democratic when all clocks in the network are of equal status and exert equal amounts of control on the others, the network operating frequency (digit rate) being the mean of the natural (uncontrolled) frequencies of the population of clocks. (CCITT)

Near-end crosstalk (NEXT). Unwanted energy coupled from a transmitter in one circuit into a receiver of another circuit at the same location (near end).

Network management. Network management is the function of supervising a communications network to ensure maximum utilization of the network under all conditions. Supervision requires monitoring, measuring and, when necessary, action to control the flow of traffic. The objective of network management is to provide service protection and to maximize the number of paid conversations by fully utilizing available equipment and facilities during normal and abnormal periods. (CCITT)

Nonblocking. A switching network that always has a free path from any idle incoming trunk or line to any idle outgoing trunk or line.

Nonreturn to zero (NRZ). A line code that switches directly from one level to another. Each level is held for the duration of a signal interval.

Nonsynchronous network (asynchronous network). A network in which the clocks need not be synchronous or mesochronous. (CCITT)

Off hook. The state or condition that a telephone receiver is requesting service-or in use. Also a supervisory signal to indicate active status of a telephone or line.

Offered traffic. It is necessary to distinguish between traffic offered and traffic carried. The traffic carried is only equal to the traffic offered if all calls are immediately handled (by the group of circuits or group of switches being measured) without any call being lost or delayed on account of congestion. The flow of traffic offered, and of traffic carried, is expressed in erlangs. The amount of traffic offered and of traffic carried is expressed in erlang-hours, (CCITT)

On hook. The inactive status of a telephone or line.

One-way trunk. A trunk circuit that can be seized at only one end.

Other common carrier (OCC). Common carriers offering services in competition with established carriers, principally AT&T long lines. (**See** Specialized common carrier).

Out-of-band signalling. A signalling technique that uses the same path as message traffic but a portion of the channel bandwidth above or below that used for voice.

Out-of-frame alignment time. The time during which frame alignment is effectively lost. That time will include the time to detect loss of frame alignment and the alignment recovery time. (CCITT)

PABX. Private automatic branch exchange (also a PBX).

Packet-mode operation. The transmission of data by means of addressed packets whereby a transmission channel is occupied for the duration of transmission of the packet only. The channel is then available for use by packets being transferred between different data terminal equipments. (CCITT)

Pair-gain system. A subscriber transmission system that serves a number of subscribers with a smaller number of wire pairs—using concentration, multiplexing, or both.

Parity. The process of adding a redundant bit to a group of information bits to maintain either odd or even numbers of 1's in the composite group of bits. A parity error results if an odd number of 1's is detected when even parity is transmitted or vice versa.

Partial response signalling (PRS). The use of controlled intersymbol interference to increase the signalling rate in a given bandwidth.

Permanent virtual circuit. A user facility in which a permanent association exists between the DTEs that is identical to the data transfer phase of a virtual call. No call setup or clearing procedure is possible or necessary. (CCITT)

Phase distortion. Signal distortion resulting from nonuniform delay of frequencies within the passband.

Phase reversal keying (PRK). A special case of phase shift keying involving only two phases 180° apart.

Phase shift keying (PSK). A form of digital modulation that uses 2^n distinct phases to represent n bits of information in each signal interval.

Plesiochronous. Two signals are plesiochronous if their corresponding significant instants occur at nominally the same rate, any variation in rate being constrained within specified limits. *Notes:* (1) Two signals having the same nominal digit rate, but not stemming from the same clock are usually plesiochronous. (2) There is no limit to the phase relationship between corresponding significant instants. (CCITT)

Polar signalling. Two-level line-coding for binary data using balanced (symmetric) positive and negative levels. Also called NRZ coding.

Power spectral density. The distribution of signal power as a function of frequency.

Private Branch Exchange (PBX). Switching equipment used by a company or organization to provide in-house switching and access to the public network.

Protection switching. That category of restoration in which one transmission path is substituted for another to permit maintenance operations for the protection against component failure, or remedy temporary conditions such as fading. This is intended to reflect a configuration in which M paths protect N paths on the same route. (CCITT)

Psophometric weighting. Noise weighting filter recommended by CCITT.

Pulse amplitude modulation (PAM). The process of representing a continuous analog waveform with a series of discrete time samples. The amplitudes of the samples are continuous and therefore analog in nature.

Pulse cose modulation (PCM). A process in which a signal is sampled, and the magnitude of each sample with respect to a fixed reference is quantized and converted by coding to a digital signal. (CCITT)

Pulse stuffing. *See* Justification.

QPRS. Quadrature channel modulation using partial response signalling on each channel.

QPSK. Quaternary phase shift keying (4-PSK).

Quadrature amplitude modulation (QAM). Independent amplitude modulation of two orthogonal channels using the same carrier frequency.

Quantization noise. The difference between the discrete sample value represented by a digital code and the original analog sample value.

Quantizing. A process in which samples are classified into a number of adjacent intervals, each interval being represented by a single value called the quantized value. (CCITT)

Raised cosine channel. A digital transmission channel with a particular type of pulse response that produces no intersymbol interference at the sample times of adjacent signalling intervals. The designation "raised cosine" is derived from the form of the analytical frequency spectrum (1 + cosine or cosine squared).

Reference clock. A clock of high stability and accuracy which is used to govern the frequency of a network of mutually synchronizing clocks of lower stability. The failure of such a clock does not cause loss of synchronism. (CCITT)

Reframing time. The time that elapses between a valid frame alignment signal being available at the receive terminal equipment and frame alignment being established. (CCITT)

Regeneration. The process of recognizing and reconstructing a digital signal so that the amplitude, waveform, and timing are constrained within stated limits. (CCITT)

Regenerative repeater. A device used to detect, amplify, reshape, and retransmit a digital bit stream.

Relative (power) level. The expression in transmission units of the ratio P/P_0, where P represents the power at the point concerned and P_0 the power at the transmission reference point. (CCITT)

Request repeat system (automatic repeat request: ARQ). A system employing an error-detecting code and so arranged that a signal detected as being in error automatically initiates a request for retransmission of the signal detected as being in error. (CCITT)

Return loss. The difference in dB between reflected and incident energy at a signal reflection point.

Ringback. The signaling tone returned by switching equipment to a caller indicating that a called party telephone is being alerted (ringing).

Robbed digit signaling. *See* Speech digit signaling.

Sidetone. The portion of a talker's signal that is purposely fed back to the earpiece so that the talker hears his or her own speech.

Signalling. The exchange of electrical information (other than by speech) specifically concerned with the establishment and control of connections and management in a communication network.

Significant instants. The instants at which a digital line code changes state. (The boundaries of a signal interval.)

Simplex transmission. A mode of operation involving transmission in one direction only.

Singing. An audible oscillation of a telephone circuit caused by a net amount of gain in a four-wire segment of the circuit.

SLIC. Subscriber loop interface circuit.

Slip. At the interface between two digital systems, the insertion or deletion of data into or from the data stream caused by an offset in clock frequencies.

Space diversity. In radio systems, the use of two receiving antennas and possibly two separate receivers to provide protection against atmosphere induced signal attenuation (fading).

Span line. A repeatered T1 line from end to end but not including channel banks.

Speech digit signalling (robbed digit signalling). Signalling in which digit time slots primarily used for the transmission of encoded speech are periodically used for signalling. (CCITT)

Speed dialling. The use of a short address code to represent an often-called telephone number. The common control computer in a PBX or end office providing such a service translates the short code into the desired number.

Specialized common carrier (SCC). A common carrier company originally offering specialized leased line services. More recently, the SCCs have been competing in established markets, including dial-up services. Also referred to as "other common carriers."

Star network. A network with a single node to which all other nodes are connected.

State store. A memory map of the connection status of a switching matrix.

STS. Space-time-space digital switching structure.

Stuffing character. A character used on isochronous transmission links to take account of differences in clock frequencies (CCITT)

Supervisory signal. A signal used to indicate the status of a line or to control equipment on the line.

Switchhook. The hook or buttons upon which a telephone handset rests when it is not being used.

Symbol rate. The reciprocal of the unit interval in seconds. (This rate is expressed in bauds.) *Note:* This definition is the same as "modulation rate." The term "symbol rate" is preferred in the case of line transmission of digital signals. (CCITT)

Synchronous. Two signals are synchronous if their corresponding significant instants have a desired phase relationship. (CCITT)

Synchronous network. A network in which the clocks are controlled so as to run, ideally, at identical rates, or at the same mean rate with limited relative phase displacement. *Note:* Ideally the clocks are synchronous, but they may be mesochronous in practice. By common usage such mesochronous networks are frequently described as synchronous. (CCITT)

Synchronous transmission. A mode of digital transmission in which discrete signal elements (symbols) are transmitted at a fixed and continuous rate.

Talkoff. An inadvertent disconnect caused by speech sounds being interpreted as in-channel control signals (disconnects).

Tandem office. In general, any intermediate switch used to establish a connection. In specific terminology, a tandem office is a switch used to interconnect end offices in an exchange area.

TASI. Time Assignment Speech Interpolation, the practice of concentrating a group of voice signals onto a smaller group of channels by dynamically switching active voice signals to idle channels.

Ternary coding. The use of all states of a three-level code to send more than one bit of information in a single symbol. This is in contrast to bipolar coding, which uses three levels, but only one of two in any particular interval. One method of interfacing binary data to a ternary line code is to encode 4 bits with three ternary symbols (4B-3T).

Tie line. A dedicated circuit connecting two private branch exchanges.

Time congestion. The ratio of time that all facilities of a system are busy (congested). Time congestion refers to the status of the system and does not necessarily imply that blocking occurs.

Time division multiplexing (TDM). Sharing a transmission link among multiple users by assigning time intervals to individual users during which they have the entire bandwidth of system.

Time expansion. The use of more time slots on internal links of a switch than exist on external links.

Time out. A network parameter related to an enforced event designed to occur at the conclusion of a predetermined elapsed time. (CCITT)

Traffic carried. The amount of traffic carried (by a group of circuits or a group of switches) during any period is the sum of the holding times expressed in hours. (CCITT)

Traffic flow. The traffic flow (on a group of circuits or a group of switches) equals the amount of traffic divided by the duration of the observation, provided that the period of observation and the holding times are expressed in the same time units. Traffic flow calculated in this way is expressed in erlangs. (CCITT)

Transhybrid loss. The amount of isolation (in decibels) between go and return paths on the four-wire side of a four-wire to two-wire hybrid.

Transmission level point (TLP). A specification, in decibels, of the signal power at a point in a transmission system relative to the power of the same signal at a zero transmission level point (0-TLP).

Transmission reference point (TLP). A hypothetical point used as the zero relative level point in the computation of nominal relative levels. (CCITT)

Transmultiplexer (transmux). An equipment that transforms signals derived from frequency-division-multiplex equipment (such as group or supergroup) to time-division-division-multiplexed signals having the same structure as those derived from PCM multiplex equipment (such as primary or secondary PCM multiplex signals) and vice versa. (CCITT)

Transversal equalizer. A time domain equalizer utilizing a tapped delay line and weighting coefficients at each of the taps to remove intersymbol interference.

Traveling class mark. A code that accompanies connection setup messages indicating the nature of the service request and any special provisions that may be desired.

Trunk. A circuit or channel between two switching systems.

TST. Time-space-time. Digital switching structure.

Two-wire circuit. A circuit consisting of a single pair of wires and capable of simultaneously carrying two signals in opposite directions.

Two-way trunk. A trunk circuit that can be seized at either end of the circuit.

Unipolar. A binary line code using single polarity pulses and zero voltage for the two coding levels.

Vestigial sideband transmission. A form of single-sideband transmission that includes a vestige of the deleted sideband and a small amount of carrier energy.

Virtual call (circuit). A user facility in which a call setup procedure and a call clearing procedure will determine a period of communication between two DTEs in which the user's data will be transferred in the network in the packet mode of operation. All the user's data are delivered from the network in the same order in which they are received by the network (CCITT)

Wait on "busy". A subscriber making a call to a busy number holds the call and is connected when the number is free. (CCITT)

Wide area telephone service (WATS). For a flat-rate a subscribed area from a particular telephone termination is charged without the registration of call charges. (CCITT)

Waiting time jitter. Arrival time jitter in a digital signal produced by pulse stuffing operations and occurring because timing adjustments must wait for predefined time slots to occur.

BIBLIOGRAPHY

BOOKS

Digital Communications Theory

1 W. R. Bennett and J. R. Davey, *Data Transmission*, McGraw-Hill, New York, 1965.
2 R. W. Lucky, J. Salz, and E. J. Weldon, Jr., *Principles of Data Communications*, McGraw-Hill, New York, 1968.
3 R. E. Ziemer and W. H. Tranter, *Principles of Communications*, Houghton Mifflin Company, Boston, 1976.
4 L. R. Rabiner and B. Gold, *Theory and Application of Digital Signal Processing*, Prentice-Hall Inc., 1975.

Communications Theory and Practice

5 Members of Technical Staff, Bell Telephone Laboratories, *Transmission Systems For Communications*, Bell Telephone Laboratories, 1971.
6 Technical Personnel of AT&T and Bell Laboratories, *Telecommunications Transmission Engineering*, Western Electric, 1977, 3 volumes.
7 *Engineering and Operations in the Bell System*, Bell Telephone Laboratories, 1977.
8 R. L. Freeman, *Telecommunication System Engineering, Analog and Digital Network Design*, John Wiley, New York, 1980.
9 M. T. Hills, *Telecommunications Switching Principles*, MIT Press, Cambridge Mass., 1979.
10 John McNamara, *Technical Aspects of Data Communications*, Digital Equipment Corporation, Maynard, Massachusetts, 1977.
11 James Spilker, *Digital Communications by Satellite*, Prentice-Hall, Englewood Cliffs, New Jersey, 1977.

Speech Digitization

12 N. S. Jayant, Ed, *Waveform Quantization and Coding*, IEEE Press, New York, 1976.
13 L. R. Rabiner and R. W. Schafer, *Digital Processing of Speech Signals*, Prentice-Hall, Englewood Cliffs, New Jersey, 1978.

14 Raymond Steele, *Delta Modulation Systems*, J. W. Arrowsmith, Bristol, England, 1975. *Also:* Halsted Press, a division of John Wiley & Sons, in the United States.

15 J. L. Flanagan, *Speech Analysis, Synthesis, and Perception*, 2nd ed., Springer-Verlag, New York 1972.

Traffic Theory

16 D. Bear, *Principles of Telecommunication Traffic Engineering*, Peter Peregrinus LTD, Southgate House, Stevenage, Herts. SG1 1HQ, England, 1976.

17 R. Syski, *Introduction to Congestion Theory in Telephone Systems*, Oliver and Boyd, London, 1959.

18 Leonard Kleinrock, *Queueing Systems Volume 1: Theory*, John Wiley & Sons, New York, 1975.

19 Leonard Kleinrock, *Queueing Systems Volume 2: Computer Applications*, John Wiley & Sons, New York, 1976.

20 Ramses R. Mina, *Introduction to Teletraffic Engineering*, Telephony Publishing Corp., Chicago, 1974.

Traffic Tables

21 *Telephone Traffic Theory Tables and Charts*, Siemens Aktiengesellschaft, Munich, 1970.

22 *CCITT Recommendations, Orange Book*, Vol. 2, International Telecommunications Union, Geneva, 1977.

TECHNICAL PERIODICALS

1 Special Issue: "PCM Transmission in the Exchange Plant," *Bell System Technical Journal*, February, 1962.

2 Special Issue: "The T1 Carrier System," *Bell System Technical Journal*, September, 1965.

3 Special Issue: "D2 Channel Bank," *Bell System Technical Journal*, October, 1972.

4 Special Issue: "LD-4 High Capacity Digital Cable System," *Telesis*, Vol. 3, No. 10, November/December 1974.

5 Special Issue: DMS-1 First of a Digital Switching Family," *Telesis*, Vol. 5, No. 4, August, 1977.

6 Special Issue: "DRS-8 Digital Radio System," *Telesis*, Vol. 5, No. 6, December, 1977.

7 Special Issue: "Digital Transmission," *Siemens Telecom*, Report 2, 1979.

8 Special Issue: "Fibre Optics," *IEEE Transactions on Communications*, July, 1978.

9 Special Issue: "Digital Radio," *IEEE Transactions on Communications*, December, 1979.

10 Special Issue: "SL-1," *Telesis*, Vol. 4, No. 3, Fall 1975.

11 Special Issue: "No. 4ESS," *Bell System Technical Journal*, September, 1977.

12 Special Issue: "Telecommunications Switching," *Proceedings of IEEE*, September, 1977.

13 Special Issue: "DMS-10 Digital Community Dial Office," *Telesis*, Vol. 5, No. 10, August, 1978.

14 Special Issue: "Digital Switching," *IEEE Transactions on Communications*, July, 1979.

15 Special Issue: "Synchronization," *IEEE Transactions on Communications*, August, 1980.

16 Special Issue: "Common Channel Interoffice Signaling," *Bell System Technical Journal*, February, 1978.

17 Special Issue: "Digital Data System," *Bell System Technical Journal*, May–June, 1975.

18 Special Issue: "Packet Communication Networks," *Proceedings of IEEE*, November, 1978.

19 Special Issue: "Solid-State Circuits for Telecommunications," *IEEE Transactions on Communications*, February, 1979.

20 Special Issue: "Electronic Systems in Subscriber Loop," *IEEE Transactions on Communications*, July, 1980.

21 Special Issue: "Digital Signal Processing," *IEEE Transactions on Communications*, May, 1978.

22 Special Issue: "Digital Encoding of Graphics," *Proceedings of IEEE*, July, 1980.

INDEX